EEG/ERP ANALYSIS

T0179953

METHODS AND APPLICATIONS

EEG/ERP ANALYSIS

METHODS AND APPLICATIONS

EDITED BY

Nidal Kamel

Centre for Intelligent Signal & Imaging Research (CISIR)
Department of Electrical & Electronic Engineering
Universiti Teknologi PETRONAS, Malaysia

Aamir Saeed Malik

Centre for Intelligent Signal & Imaging Research (CISIR)
Department of Electrical & Electronic Engineering
Universiti Teknologi PETRONAS, Malaysia

CRC Press
Taylor & Francis Group
Boca Raton   London   New York

CRC Press is an imprint of the
Taylor & Francis Group, an **informa** business

CRC Press
Taylor & Francis Group
6000 Broken Sound Parkway NW, Suite 300
Boca Raton, FL 33487-2742

First issued in paperback 2017

© 2015 by Taylor & Francis Group, LLC
CRC Press is an imprint of Taylor & Francis Group, an Informa business

No claim to original U.S. Government works

Version Date: 20140915

ISBN 13: 978-1-138-07708-9 (pbk)
ISBN 13: 978-1-4822-2469-6 (hbk)

Library of Congress Cataloging-in-Publication Data

EEG/ERP analysis : methods and applications / edited by Nidal Kamel, Aamir Saeed Malik.
  p. ; cm.
Includes bibliographical references and index.
ISBN 978-1-4822-2469-6 (hardcover : alk. paper)
I. Kamel, Nidal, editor. II. Malik, Aamir Saeed, 1969- , editor.
[DNLM: 1. Electroencephalography--methods. 2. Evoked Potentials. 3. Functional Neuroimaging--methods. 4. Image Interpretation, Computer-Assisted--methods. 5. Nervous System Diseases--diagnosis. WL 150]

RC386.6.E43
616.8'047547--dc23
                                                                    2014027312

Visit the Taylor & Francis Web site at
http://www.taylorandfrancis.com

and the CRC Press Web site at
http://www.crcpress.com

Contents

List of Tables vii

Preface ix

Acknowledgments xiii

1. **Introduction** 1
   Humaira Nisar and Kim Ho Yeap

2. **The Fundamentals of EEG Signal Processing** 21
   Nidal Kamel

3. **Event-Related Potentials** 73
   Zhongqing Jiang

4. **Brain Source Localization Using EEG Signals** 91
   Munsif Ali Jatoi, Tahamina Begum, and Arslan Shahid

5. **Epilepsy Detection and Monitoring** 123
   Arslan Shahid and John K.J. Tharakan

6. **Neurological Injury Monitoring Using qEEG** 143
   Waqas Rasheed and Tong-Boon Tang

7. **Quantitative EEG for Brain–Computer Interfaces** 157
   *Anton Albajes-Eizagirre, Laura Dubreuil-Vall, David Ibáñez,
   Alejandro Riera, Aureli Soria-Frisch, Stephen Dunne, and Giulio Ruffini*

8. **EEG and qEEG in Psychiatry** 175
   Likun Xia and Ahmad Rauf Subhani

9. **Perspectives of M-EEG and fMRI Data Fusion** 195
   Jose M. Sanchez-Bornot and Alwani Liyana Ahmad

10. **Memory Retention and Recall Process** 219
    Hafeez Ullah Amin and Aamir Saeed Malik

11. **Neurofeedback** 239
    Mark Llewellyn Smith, Thomas F. Collura, and Jeffrey Tarrant

12. **Future Integration of EEG-fMRI in Psychiatry, Psychology, and Cultural Neuroscience** 269
Mohd Nasir Che Mohd Yusoff, Mohd Ali Md Salim, Mohamed Faiz Mohamed Mustafar, Jafri Malin Abdullah, and Wan Nor Azlen Wan Mohamad

13. **Future Use of EEG, ERP, EEG/MEG, and EEG-fMRI in Treatment, Prognostication, and Rehabilitation of Neurological Ailments** 287
Jafri Malin Abullah, Zamzuri Idris, Nor Safira Elaina Mohd Noor, Tahamina Begum, Faruque Reza, and Wan Ilma Dewiputri

Index 311

List of Tables

Table 1.1 Characteristic Properties of Bioelectric Signals............................ 14

Table 2.1 Some Common Windows and Their Properties 40

Table 4.1 Localization Ability Comparison for Three Algorithms 111

Table 4.2 Summary of Different Techniques for Solution of EEG
Inverse Problem 112

Table 4.3 Comparisons between Various Localization Algorithms 114

Table 4.4 Feature-Based Comparison of Methods.................................... 115

Table 5.1 Summary of Seizure Event Detectors 131

Table 5.2 Summary of Seizure Onset Detectors 134

Table 5.3 EEG Features for Epileptic Seizure Detection 137

Table 6.1 Clinical Evaluation Tests... 146

Table 6.2 Clinical Schemes of Severity Assessment of Brain's
Neuronal Dysfunction 146

Table 6.3 Regional Deficits 147

Table 6.4 Alpha Rhythm and Sleep-Spindle Frequencies Observed
in Preteen Children 151

Table 10.1 Summary of Traditional Models.. 224

Table 10.2 Summary of Factors Affecting Memory Retention
and Recall... 232

Table 11.1 Band-Pass Filters Used for Neurofeedback............................ 247

Table 11.2 Neurofeedback Learning Mechanisms 248

Table 11.3 Goals of Neurofeedback 248

Table 11.4 Standard Inhibit/Enhance Protocols 252

Preface

The first recording of the electric field of the human brain was made by Hans Berger, the German psychiatrist, in 1924. These recordings came to be known as electroencephalogram (EEG). Since then, the EEG has been a useful tool in understanding and diagnosing neurophysiological and psychological disorders.

Recent advances in digital recording and signal processing, together with the leaps in computational power, are expected to spawn a revolution in the processing of measurements of brain activities, primarily EEGs and event-related potentials (ERPs). This will enable the implementation of more complicated denoising techniques of ERP than ensemble averaging and the implementation of more complicated EEG quantification analysis methods (qEEG) than the amplitudes and frequencies, including nonlinear dynamics and higher-order statistics. Furthermore, this will help in the implementation of various techniques describing the interactions between different regions of the brain, which offer more insights into the functional neural networks in the brain.

This book provides an introduction to both the basic and advanced techniques used in EEG/ERP analysis and presents some of their most successful applications. Before we present EEG/ERP methods and applications, in Chapter 1 we introduce the physiological foundations of the generation of EEG/ERP signals and their characteristics. This chapter first explains the fundamentals of brain potential sources and then describes the various rhythms of the brain, the different electrode cap systems, EEG recording and artifacts, and finally outlines the application of EEG in neurological disorders.

In Chapter 2, the basic digital signal processing topics are first explained, including time-domain, frequency-domain, and signal filtering; stochastic processes; power spectrum density, autoregressive modeling of EEG; and the joint time-frequency methods for EEG representation including the short-time Fourier transform and the wavelet. Next, the nonlinear descriptors of the EEG signal are explained, including the higher-order statistics; chaos theory and dynamical analysis using correlation dimension and Lyapunov exponents; and dynamical analysis using entropy. In addition, motivated by research in the field over the past two decades, techniques specifically used in the assessment of the interaction between the various regions of the brain, such as the normalized cross-correlation function, the coherency function, and phase synchronization, are explained. MATLAB® code is provided in order to enable the reader to run the presented techniques with his or her EEG/ERP data.

Chapter 3 explains the extraction of event-related signals using the multi-trials-based ensemble averaging technique and the recently proposed single-trial subspace-based technique, outlining their implementation. In addition, this chapter explains brain activity assessments using ERP signals, including the applications of P100 in the assessment of the functional integrity of the visual system and some psychophysiological disorders, the application of N200 in the assessment of attention-deficit/hyperactivity disorder (AD/HD), and the application of P300 as a biomarker of the severity of alcoholism, Alzheimer's disease, and Parkinson's disease.

Chapter 4 gives a brief description of the forward and the inverse problem, and then outlines the different EEG source localization methods, including the minimum norm, the weighted minimum norm (WMN), low resolution brain electromagnetic tomography (LORETA), standardized LORETA, recursive multiple signal classifier (MUSIC), recursively applied and projected MUSIC, shrinking LORETA-FOCUSS, the hybrid weighted minimum norm method, recursive sLORETA-FOCUSS, and standardized shrinking LORETA-FOCUSS (SSLOFO).

Chapter 5 presents the successful clinical applications of qEEG in the detection and monitoring of epileptic seizures. This chapter describes the various proposed techniques for epileptic seizure onset detection and epileptic seizure event detection and compares them. In addition, it describes a recently proposed method that relies on the singular value decomposition of the EEG data matrix for feature extraction and the support vector machine (SVM) for classification.

The brain undergoes several types of dysfunctions caused by internal or external stimuli. The cause may be an injury or a neurodegenerative disease. Chapter 6 discusses the issue of monitoring a neurological injury with the qEEG. It describes the assessment criteria of a neuronal injury, the post-concussion syndrome (PCS), and common evaluations of functional connectivity, including functional homogeneity, differentiation, or topographic reciprocities. The monitoring process of a brain injury using the qEEG is also outlined, and a review of the various entropy measures, such as information entropy, mutual entropy, and approximate entropy, is given.

The brain–computer interface (BCI) is an emerging field of study where qEEG techniques are used as a direct nonmuscular communication channel between the brain and the external world. In Chapter 7, the different neurotechnologies for the brain–computer interface based on the detection of ERPs such as the P300 or the steady-state visual evoked potentials (SSVEPs) are described. Moreover, the different parameters of the P300 waveform that can be used to quantify attentional processes are outlined in addition to the different technologies for the measurement of arousal, valence, and stress. Experimental work conducted in the field is also provided.

Chapter 8 reviews the use of qEEG in psychiatry and presents its applications as a biomarker for stress, unipolar depression, and alcohol addiction. A case study on stress assessment is included and discussed.

EEG is known to have a high temporal resolution but a low spatial resolution. On the contrary, functional magnetic resonance imaging (fMRI) is known to have a high spatial resolution but a low temporal resolution. Combining EEG with fMRI may provide high spatiotemporal functional mapping of brain activity. Chapter 9 reviews EEG and fMRI fusion techniques, addresses the theoretical and practical considerations for recording and analyzing simultaneous EEG-fMRI, and describes some of the current and emerging applications. Furthermore, it introduces some integrative models for solving the multimodal fusion.

Chapter 10 provides an introduction to memory-related processes. Two important processes are memory retention and memory recall, and these two processes are discussed in this chapter. Various memory models are presented that describe the cognitive loads and memory. The brain anatomical regions associated with memory retention and recall processes are discussed. Both short-term and long-term memory processes are described. Memory experimental design issues are discussed while keeping in view the various factors affecting memory.

Chapter 11 is intended as a general survey of neurofeedback and is written to help the reader understand neurofeeback in a general way. It provides a starting point for deeper investigation by giving a short history of the field and offering an introductory description of the existing clinical modalities that are developed as a result. It also describes basic clinical processes and neurofeedback interventions, including biological, technical, and scientific considerations.

With the advancements in technology, researchers are attempting to combine various modalities to reap the benefits of all those modalities. One example is the combination of EEG with fMRI, which can provide both spatial and temporal resolution. Chapter 12 discusses the EEG-fMRI data fusion and analysis with respect to psychiatry, psychology, and cultural neuroscience. The chapter begins by giving an introduction to cultural neuroscience. One section in the chapter is dedicated to psychopathy and criminal behavior. The final section describes the role of EEG-fMRI in the rehabilitation of developmental and psychological disorders such as autism and obsessive compulsive disorder.

Chapter 13 discusses the future of EEG, ERP, and EEG-fMRI. These modalities can result in better treatment, prognostication, and rehabilitation for various neurological ailments. This chapter provides an insight into how EEG technology can be used in neurosurgery. In addition, clinical applications of EEG are also discussed. Future clinical applications of EEG-fMRI and EEG-MEG are presented with a specific emphasis on epilepsy, stroke, and traumatic brain injury.

Dr. Nidal S. Kamel and Dr. Aamir S. Malik
Editors
Universiti Teknologi PETRONAS
Tronoh, Malaysia

The MathWorks, Inc.
3 Apple Hill Drive
Natick, MA, 01760-2098 USA
Tel: 508-647-7000
Fax: 508-647-7001
E-mail: info@mathworks.com
Web: www.mathworks.com

Acknowledgments

The editors thank all the contributors for their professional support in developing this book. The editors thank especially Prof. Jafri Abdullah from Universiti Sains Malaysia; Dr. Mark Smith from Neurofeedback Services of New York; Dr. Aureli Soria-Frisch from Starlab Barcelona; Prof. Zhongqing Jiang from Liaoning Normal University, China; and Dr. Humaira Nisar from Universiti Tunku Abdul Rahman, Malaysia.

The editors also acknowledge the support of Mission Oriented Research Biomedical Technology, Centre for Intelligent Signal and Imaging Research, and Department of Electrical and Electronic Engineering at Universiti Teknologi PETRONAS, Malaysia.

Finally, the editors express their appreciation for CRC for the opportunity to edit this book and also thank the entire staff of CRC for their professional support during all the phases of development of this book.

Nidal Kamel and Aamir Saeed Malik

1

Introduction

Humaira Nisar and Kim Ho Yeap
Universiti Tunku Abdul Rahman

CONTENTS

1.1 Introduction to EEG and ERP Signals 2
    1.1.1 History.. 2
    1.1.2 Source of Neural Activities 3
        1.1.2.1 Human Brain 3
        1.1.2.2 Structure and Function of the Neuron 4
1.2 Brain Rhythms and Wave Patterns 6
1.3 Techniques of EEG and ERP Recording 8
    1.3.1 Electrodes... 8
    1.3.2 Montages.. 10
        1.3.2.1 Bipolar Montage 10
        1.3.2.2 Referential Montage.. 10
        1.3.2.3 Average Reference Montage 10
        1.3.2.4 Laplacian Montage.. 10
1.4 Recording and Artifacts.. 11
    1.4.1 Biological Artifacts 11
        1.4.1.1 Electrooculographic Artifacts or Ocular Artifacts....... 11
        1.4.1.2 Electromyogram Artifacts 11
        1.4.1.3 Electrocardiogram Artifacts...................................... 13
        1.4.1.4 Other Bio-Artifacts... 13
    1.4.2 Technical Artifacts.. 14
        1.4.2.1 AC or Power Line Artifacts.. 14
        1.4.2.2 DC Noise 14
        1.4.2.3 Artifacts due to Improper Placement of Electrodes..... 14
        1.4.2.4 Artifacts due to High Impedance 14
1.5 Properties of EEG and ERP Signals 15
1.6 Application of EEG in Neurological Disorders 16
References.. 18

1.1 Introduction to EEG and ERP Signals

1.1.1 History

Electroencephalography (EEG) is the recording of electrical activity along the scalp. The flow of current due to firing of neurons in the brain results in a voltage fluctuation that is measured as EEG. The measurement of the brain's response to a stimulus is called event-related potential (ERP). The stimulus can be sensory, motor, or cognitive in nature.

Richard Caton (1842–1926), a physician, deserves credit for the discovery in 1877 of the fluctuation potentials in rabbits, cats, and monkeys that constitute the EEG. However, the first measurement and pictorial demonstration of EEG was performed by Vladimir Pravdich-Neminsky. He measured the electrical activity in the brains of dogs in 1912 and named it *electrocerebrogram* [1,2].

Hans Berger (1873–1941), a neuropsychiatrist, started his study of human EEG in 1920. The first human EEG tracings were shown in his first report in 1929. Berger used a bipolar technique and photographic paper to record electrical activities of the brain and named it *electroenkephalogram* [3]. This term later evolved into electroencephalogram or EEG.

In the United States, the first EEG work was reported in Harvard at Boston in 1934 by Hallowell Davis [1]. During the years 1935 and 1936, Pauline and Hallowell Davis recorded the first known ERPs on conscious humans. They published their findings in 1939. In 1964, Grey Walter and his colleagues reported the first cognitive ERP component, contingent negative variation (CNV) [4].

The development of clinical and experimental EEG work reached a high point around 1960. The interest of electroencephalographers in academic institutions shifted from tracing to automatic data analysis because of computerization. By 1967, people thought that the traditional EEG reading would soon become obsolete, being eventually replaced by fully automatic interpretation. In the 1970s, studies on the evoked potential progressed greatly. The introduction of the pattern changer in the visually evoked potential (VEP) technique by Speckrejse and Spehlmann made this method very reliable [1]. The 1970s and 1980s saw the emergence of structural neuroimaging techniques, computed tomography, and magnetic resonance imaging.

For 30 years after its discovery, EEG was primarily used for assistance in clinical diagnostics, the study of neurological disorders, and brain function assessments. Different areas of the brain involved in specific neurological activities were studied. In 1973, Jacques J. Vidal from the University of California introduced a new concept of brain–computer communication [5,6]. In his study, he proposed the concept of VEP and the route for the future of the brain–computer interface (BCI).

Currently, ERP is the most widely used method in cognitive neuroscience research. It is used to study the physiological correlation associated with the processing of information, such as sensory, perceptual, and cognitive activity [7].

1.1.2 Source of Neural Activities

To understand the origins of the EEG signal, a brief introduction to the human brain and the neuron, the most fundamental cell in neuropsychology, has been presented in the following sections.

1.1.2.1 Human Brain

The human brain may be divided into three major parts: cerebrum, cerebellum, and brain stem. Here, we will only consider the actions of the cerebrum and cerebellum. The cerebellum mainly controls complex body movements, involving coordination and muscle tone modulation. The cerebrum may be subdivided into six parts: frontal lobe, parietal lobe, temporal lobe, occipital lobe, insular lobe, and limbic lobe. The parietal lobe perceives pain and taste sensations, and is involved in problem-solving activities. The temporal lobe is concerned with hearing and memory. The occipital lobe mainly contains the regions used for vision-related tasks. The frontal lobe is mainly associated with emotions, problem solving, speech, and movement. It contains the primary motor cortex located anterior to central gyrus as shown in Figure 1.1 [8,9].

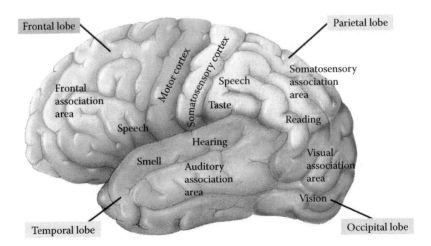

FIGURE 1.1
Functional diagram of brain lobes. (From Chen, P. 2011. Principles of biological science. Accessed September 28, 2013. http://bio1152.nicerweb.com/Locked/src/chap48_g.html.)

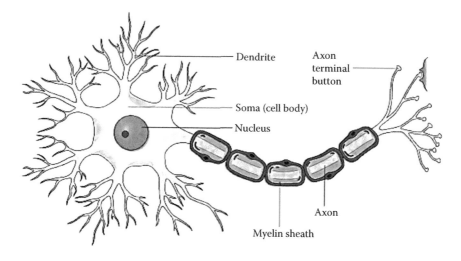

FIGURE 1.2
Structure of neuron or nerve cell. (From Sanei, S. and Chambers, J. *EEG Signal Processing*. Chichester, England: John Wiley; 2007.)

The primary motor cortex is located between the somatosensory cortex and premotor cortex. Different areas of the primary motor cortex control various movements of the body. Upper extremity movements also have a different representation for the shoulder, elbow, wrist, fingers, and thumb. Facial movements, including the neck, eye, face, lips, jaw, tongue, and swallowing, occupy a large area of the primary motor cortex. Any body movement or sensory activity of the body is accompanied by specific signals originated from the primary motor cortex. Thus, to assess motor-related brain activity, the EEG recording should originate from the appropriate location in the spatial domain to ensure the correct reading of each electrode. The auditory cortex is located below the primary motor cortex. Therefore, audio stimulation may affect studies of motor imaginary tasks. Motor imaginary tasks, such as hand movements, are controlled by the contralateral parts of the brain. Left-hand and right-hand movements are controlled by the right and left hemispheres of the brain, respectively [10]. The central nervous system (CNS) mostly consists of nerve cells and glia cells. Every nerve cell (neuron) consists of dendrites, cell bodies, and axons as shown in Figure 1.2. Nerve cells respond to stimuli and information transmission over distances. An axon is a cylindrical tube that sends and transmits an electrical signal in vertebrates. The transport system of an axon is responsible for delivering proteins to the end of the cell.

1.1.2.2 Structure and Function of the Neuron

Neurons (or nerve cells) are functional units of the nervous system. An adult human brain contains, on an average, 100 billion neurons [11]. Neurons process and transmit information through electrical and chemical signals.

Neurons have a resting membrane potential of about –70 to 60 mV [12]. A neuron consists of a cell body (also known as soma), dendrites, and an axon (see Figure 1.2) [13]. Depending on the information the dendrites receive from other neurons, the neuron makes a decision that is then sent to other neurons' dendrites over the axon.

1.1.2.2.1 Action Potentials

The neuron is surrounded by the cell membrane, which controls the in and out movement of charged sodium (Na+) and potassium (K+) ions. The cell body is negatively charged on the outside and has a resting potential of –70 mV [8]. The membrane potential becomes less negative because of the incoming electrical current from the dendrites (see A in Figure 1.3) [8,14]. The cell membrane completely opens up for Na+ ions if this depolarization reaches –55 mV, when the ions now enter the cell, resulting in a momentarily positive action potential (see B in Figure 1.3) [8]. The cell membrane also opens up again, and the K+ ions present in the cell leave (see C in Figure 1.3), which causes the repolarization of the membrane's potential [9]. Owing to the loss of permeability of the cell membrane for the K+ ions, the potential temporarily falls below 70 mV (this is known as hyperpolarization; see D in Figure 1.3). Finally, the action potential stabilizes at the resting potential. This action potential is carried over by the axon to other neurons. Hence, the neuron becomes activated if the total electrical current from all the incoming axons exceeds a certain threshold. This results in the transfer of information (the action potential) to subsequent neurons.

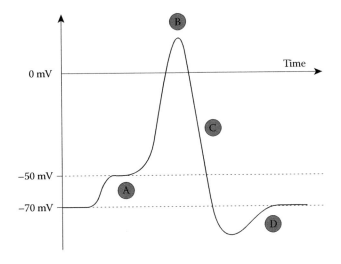

FIGURE 1.3
Action potential. (From Ward, J. *The Student's Guide to Cognitive Neuroscience.* London: Taylor and Francis; 2010, p. 453.)

The EEG activity cannot pick the electrical potential generated by a single neuron [15]. In fact, the EEG activity is a reflection of the summation of the synchronous activity of a big group of neurons, probably thousands or millions, having a similar spatial orientation. A similar spatial orientation is necessary for the ions to line up and create waves that will be strong enough to pass the detection threshold. It is known that the voltage fields fall off with the square of the distance, and hence the pyramidal neurons of the cortex that are well aligned and fire together are thought to produce most EEG signals. The EEG activity from deep sources is difficult to detect as compared to the activity that happens near the skull [16].

1.2 Brain Rhythms and Wave Patterns

Neural oscillations are observed throughout the central nervous system. These are generated by large groups of neurons and can be characterized by the frequency, amplitude, and phase of the oscillations. Cognitive functions such as information transfer, perception, motor control, and memory are in one way or another related to neural oscillations and synchronization [17–19].

EEG recordings are commonly used to investigate neural oscillations. Neurons can generate action potentials or spikes in a rhythmic pattern. Some neurons have the tendency to fire at particular frequencies and are called resonators. Spiking patterns that are the result of bursting are considered fundamental for information coding. In many neurological disorders, the cause is excessive neural oscillation. For example, in seizures, excessive synchronization has been observed. Similar phenomena have been observed in tremor patients in Parkinson's disease. Oscillatory activity can also be used in BCIs to control external devices.

The human brain can produce five major brain waves, classified by their frequency ranges. These five major waves can range from low frequency to high frequency. These are known as alpha (α), theta (θ), beta (β), delta (δ), gamma (γ), and mu (μ). Berger, in 1929, discovered alpha and beta waves; gamma waves were discovered by Jasper and Andrews in 1938, while delta and theta waves were discovered by Walter in 1936, both of which represent waves with a frequency below the alpha range [11].

Delta waves fall within the range of 0.5–4.0 Hz. These waves have the highest amplitude among the other waves but have the lowest frequency. These occur frontally in adults and posteriorly in children. These waves are primarily associated with deep sleep. These delta rhythms may also be associated with subcortical lesions, deep midline lesions, or metabolic encephalopathy hydrocephalus. Figure 1.4 shows delta waves in the interval 0 to 1 s.

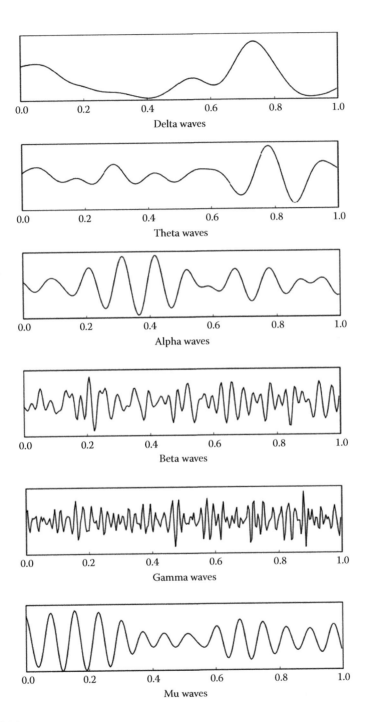

FIGURE 1.4
Different brain rhythms.

Theta waves fall in the range of 4–8 Hz. Theta waves are normally observed in young children. However, in older children and adults, these waves are observed in a state of drowsiness or arousal, as well as in meditation. These waves may occur anywhere and are not related to the tasks at hand. An excess of theta waves may represent abnormal activity.

Alpha waves lie in the frequency range of 8–13 Hz. These waves were discovered by Dr. Hans Berger in 1908. Because these waves were the first to be discovered, they are called alpha waves (first waves). Alpha waves are associated with wakefulness, closing the eye, effortless alertness, and creativity. These waves normally appear in the posterior half of the head and have higher amplitude over the occipital areas.

Beta waves lie in the frequency range 14–26 Hz. Found only in normal adults, these waves are correlated with active attention, active thinking, solving critical problems, or focusing on the outside world and, therefore, are also known as sensory motor rhythm. Rhythmical beta waves are experienced mainly in the frontal and central regions. Beta waves are low in amplitude and are normally under 30 μV.

Gamma waves lie in the ranges above 30 Hz (up to 100 Hz). These waves help to determine the binding of different populations of neurons together. They occur rarely in the human brain. They occur only during crossmodal sensory processing, that is, the process of combining different senses such as sound and sight.

Mu wave ranges from 8 to 13 Hz. These waves are mixed with other waves and sometimes partly overlap other rhythms. It shows the synchronous firing of motor neurons over the sensorimotor cortex. The different brain rhythms are shown in Figure 1.4.

1.3 Techniques of EEG and ERP Recording

An EEG recording could be noninvasive or invasive. The noninvasive procedures use surface electrodes and hence are safe and painless. Human EEG can be measured using special electrodes with a typical diameter of 0.4 cm to 1.0 cm. The electrodes are held on the scalp with a paste (wet or dry), depending on how these electrodes are designed.

1.3.1 Electrodes

Electrodes are placed on the scalp in special positions (following the International 10/20 system as shown in Figure 1.5) [11] to acquire the EEG data. These are small metallic disks, usually made of silver, gold, tin, or stainless steel; and covered with a silver chloride coating. The positions are measured on the scalp relative to the known skull landmarks, from the front

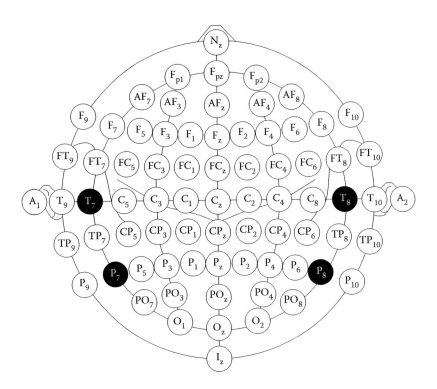

FIGURE 1.5
10–20 electrodes setting for 75 electrodes. (From Patel, N. D. An EEG-based dual-channel imaginary motion classification for brain computer interface. Master of Engineering Science, Thesis, Lamar University, 2011.)

at the nasion to the inion at the back of the head and side to side from the two ear canals. This allows researchers and clinicians in different labs to standardize their measurements and reporting. The location of each electrode is labeled with a letter and a number. The letter refers to the area of brain in which the electrode is located, for example, P for the parietal lobe and T for the temporal lobe. Even numbers denote the right side of the head, whereas odd numbers denote the left side of the head.

In some applications such as BCI and ERP studies, a minimal number of electrodes (usually 14 recorded electrodes) is often used and placed at the movement-related area or selection area in which the signals are strong using the conventional electrode positioning system. In advanced research or neural laboratories, more than 16 electrodes (often 64 to 131 recording electrodes) are used to get more detailed data. However, adding more electrodes may generate less useful data unless computer algorithms are used to manage the raw EEG data. Often, the potential at each location can be measured by taking the average of all the potential differences when a large number of electrodes are used [20].

1.3.2 Montages

The placement of the electrodes is referred to as a montage. Either a bipolar montage or a referential montage is used to monitor the EEG. In a bipolar montage, there are two electrodes per channel, which implies that there is a reference electrode for each channel. In a referential montage, there is a common reference electrode for all channels.

1.3.2.1 Bipolar Montage

In a bipolar montage, each waveform represents the difference between two adjacent electrodes. For example, the channel Fp1–F3 represents the difference in voltage between the Fp1 and the F3 electrodes. Similarly, F3–C3 represents the difference in voltage between the F3 and C3 electrodes.

1.3.2.2 Referential Montage

In this montage, the difference between a certain electrode and a designated reference electrode is measured. This difference is represented by the channel. The reference electrode has no standard position. However, the position of the reference electrode is different from the recording electrodes. Midline positions are often used to avoid the amplification of signals in one hemisphere relative to the other. Another popular reference that is used considerably is "linked ears." It is the physical or mathematical average of the electrodes attached to both the earlobes and the mastoids.

1.3.2.3 Average Reference Montage

In this montage, an average signal is obtained by summing and averaging the outputs of all the amplifiers, which is then used as the common reference for each channel.

1.3.2.4 Laplacian Montage

In this montage, the difference between an electrode and a weighted average of the surrounding electrodes is used to represent a channel. When digital EEG is used, all signals are typically digitized and stored in a particular (usually referential) montage. Because the stored information is digital, it is possible to mathematically construct any montage from any other montage. However, for the case of analog EEG, which is stored on paper, the person in charge has to switch between the montages during the recording to focus or highlight special features of the EEG.

1.4 Recording and Artifacts

"Potential fluctuations of non-neural origin are called artifacts" [5]. EEG signals that originate from noncerebral origins are called EEG artifacts. These signals are detected along the scalp by EEG. The amplitude of the signals that constitute an artifact can be quite large relative to the amplitude of the signals of interest.

EEG noise or artifacts may be subdivided into two different classes. Biological artifacts arise from the internal brain functions. The technical artifacts are introduced into the EEG signal by the experimental equipment at the surrounding, or various human errors. The EEG artifacts are briefly discussed below.

1.4.1 Biological Artifacts

These artifacts arise from biological activity in the brain.

1.4.1.1 Electrooculographic Artifacts or Ocular Artifacts

Electrooculographic (EOG) artifacts, or simply ocular artifacts, arise from vision-related stimuli. In simple words, this means that EOG artifacts are induced by eye movements and blinks. Visual stimuli are highly localized in the occipital lobe [8]. Artifacts such as eye blinks, eyeball movements, and extraocular muscle activities can be found in the frontal lobe, whereas VEPs are usually associated with the occipital lobe.

Eye blink artifacts may introduce observable alterations in both the amplitude and the frequency content of the EEG. These artifacts may considerably reduce the classification accuracy. Eye blink artifacts also fall in the alpha band frequency region. For motor classification, the alpha rhythm is the most important frequency band [10]. We may also view eye blink artifacts as muscle artifacts.

A VEP is induced when the subject is exposed to different visual stimuli. A VEP is phase-locked to the event, so it may affect the temporal resolution. We need the precise temporal location of the event in a classification process. A VEP also affects the primary motor cortex in some subjects by producing VEP artifacts at the time of a motor imaginary event. Because of its high amplitude, a VEP may completely obscure the desired evoked potential in the frequency domain.

As shown in Figure 1.6 [21], eyeball movements can cause muscular artifacts in EEG signals. This artifact may be avoided by keeping the viewing angle constant during the recording time.

1.4.1.2 Electromyogram Artifacts

The electromyogram (EMG) artifacts are found when electrical current is generated by the contraction of muscles. This represents neuromuscular activities. An example of EEG signals recorded from the EMG is depicted

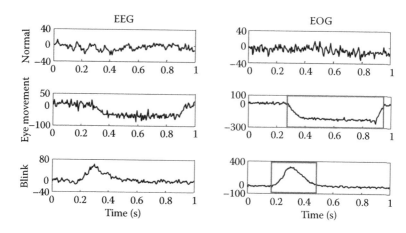

FIGURE 1.6
Effect of eye movements and blinks (right) on the EEG recordings (left). (From Kierkels, J. J. M. Validating and improving the correction of ocular artifacts in electro-encephalography. PhD thesis, Technische Universiteit Eindhoven, 2007. http://alexandria.tue.nl/extra2/200711857.pdf.)

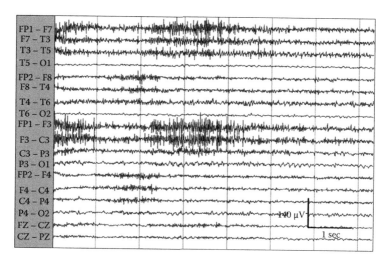

FIGURE 1.7
EMG artifacts in an EEG recording. (From Rielo, D. and Benbadis, S. R. EEG artifacts, March 2010. http://emedicine.medscape.com/article/1140247-overview.)

in Figure 1.7 [22]. The subjects are instructed to avoid movements as much as possible during signal recordings. Muscle artifacts may include the following.

Hand/leg muscle artifacts may cause a slow drift in the EEG amplitude. This drift is usually high compared to other brain activities. To minimize this artifact, we have avoided hand or leg movements during EEG recordings.

Facial muscle artifacts are also undesirable unless the facial expression is used as a signal. Different swallowing motions produce different artifact shapes.

1.4.1.3 Electrocardiogram Artifacts

Electrocardiogram (ECG) artifacts occur when there is a potential difference in the cardiac muscle cell. In each cycle of the heartbeat, such a potential difference occurs during depolarization and repolarization. The ECG influence depends on the length and width of a person's neck. An example of ECG artifacts is shown in Figure 1.8 [23].

These artifacts arise from heart activity. Cardio artifacts are present in every subject, while their magnitude and location might vary depending on the subject. We can remove cardio artifacts using an ECG signal. Specific methods are proposed for ECG artifact mitigation [11]. Using these methods, one can approximately reconstruct the electrical signal produced by the heart function. Depending on the experimental setup, ECG noise may severely damage EEG recordings.

1.4.1.4 Other Bio-Artifacts

Acoustic noise produces an artifact when the subject encounters acoustic stimuli such as noise coming from the outside, people moving inside the lab, telephone calls, and so on. Therefore, any form of distraction capturing the subject's attention should be avoided while capturing EEG signals.

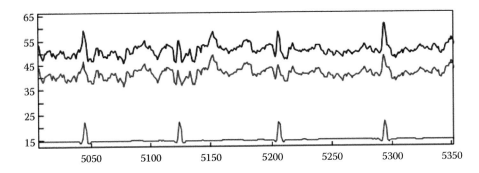

FIGURE 1.8
ECG artifacts in an EEG recording. ECG signal (bottom, 3 mV/sample), EEG signal (top, 1 μV/sample), and the EEG signal after ECG suppression (middle, 1 μV/sample). (From Schlögl, A. and Pfurtscheller, G. EOG and ECG minimization based on regression analysis. Report siesta-task310 artifact detection and denoising. en.scientificcommons.org, Jan 2007. http://en.scientificcommons.org/42786168.)

1.4.2 Technical Artifacts

1.4.2.1 AC or Power Line Artifacts

Power line artifacts have a significant peak of about 50/60 Hz in the power spectrum. High-impedance electrodes can cause the wire running from them to function as antennae that receive electrostatic noise. Shielding the power source can reduce the electrostatic noise. A notch filter can be used to remove this artifact [7].

1.4.2.2 DC Noise

A DC offset is frequently observed in recorded signals. Only hardware is responsible for these offsets. For each trial, the DC noise might differ in amplitude.

1.4.2.3 Artifacts due to Improper Placement of Electrodes

In bipolar recordings, placing electrodes at slightly different locations during each recording may produce additional artifacts in EEG signals.

1.4.2.4 Artifacts due to High Impedance

The high impedance of EEG electrodes (>5 kΩ) may result in artifacts. The EEG is a very weak signal, having an amplitude up to approximately 100 µV. The contact resistance of electrodes is generally kept below 5 kΩ to avoid high-impedance artifacts [24]. Table 1.1 gives a brief overview of the different types of artifacts [21].

TABLE 1.1

Characteristic Properties of Bioelectric Signals

Signal	Amplitude Range	Frequency Range (Hz)
EEG	2 µV–100 µV	0.5–100
EEG (EP)	0.1 µV–20 µV	1–3000
EOG	10 µV–5 mV	0–100
EMG	50 µV–5 mV	2–500
ECG	1 mV–10 mV	0.05–100

Source: Kierkels, J. J. M. Validating and improving the correction of ocular artifacts in electro-encephalography. PhD thesis, Technische Universiteit Eindhoven, 2007. http://alexandria.tue.nl/extra2/200711857.pdf.

1.5 Properties of EEG and ERP Signals

The EEG has the ability to monitor human brain activity noninvasively, with a precision of milliseconds. This is necessary for understanding the foundations of cognitive functions. The EEG reflects thousands of ongoing brain processes.

Continuous EEG recordings consist largely of oscillations at different frequencies that fluctuate over time and provide valuable information about a subject's brain state. Brain oscillations such as EEG alpha activity (8–13 Hz) clearly respond to sensory stimulation (e.g., alpha suppression). To what extent these oscillations contribute to event-related EEG signals such as ERPs is a matter of ongoing research. It has become evident that an ERP does not necessarily capture all event-related information present in the EEG. For instance, the oscillations induced by, but not perfectly phase-locked to, an event of interest, zero out in the process of ERP calculation. It is therefore helpful to distinguish between evoked, phase-locked oscillations and induced, non-phase-locked signals.

In pure EEG data, it is difficult to focus or isolate individual neurocognitive processes; hence, such processes are considered difficult to assess and analyze. To extract more specific information related to sensory, cognitive, and motor events, ERPs are used.

The common way of analyzing event-related EEG signals is the calculation of ERPs. This can be done by repeating an event of interest such as a visual stimulus on a computer screen and analyzing a small fraction of the EEG activity that is evoked by this event.

An ERP in response to sensory and cognitive events usually consists of a number of peaks and deflections, which, if characterized by latency, morphology, topography, and experimental manipulation [25], are called ERP components. Early components typically reflect sensory processing and can be associated with the respective sensory cortical areas, whereas later ERP components convey information about cognitive effects of brain function. ERP components are usually small in magnitude (i.e., 1 to 20 μV). They show substantial interindividual variation and are susceptible to various artifacts. It is therefore necessary to carefully evaluate ERP properties before any conclusions can be drawn.

ERP components are referred to with different acronyms. Most components are represented by a letter N, representing negative polarity, and P, representing positive polarity. This is followed by a number that indicates either the latency (measured in milliseconds) or the component's ordinal position in the waveform.

For example, the first substantial negative peak in the waveform that often occurs about 100 milliseconds after a stimulus is often referred to as the N100 or N1. N100 indicates that it is a negative peak, and its latency is 100 ms.

N1 indicates that it is the first negative peak. This is often followed by a positive peak, usually called the P200 or P2. However, it is to be noted that the stated latencies for ERP components are often quite variable. For example, the P300 component may exhibit a peak of anywhere between 250 ms and 700 ms [26].

ERPs are caused by processes that might involve memory, attention, or changes in the mental state, among others, whereas the processing of the physical stimulus is reflected by evoked potentials. ERPs are used extensively in neuroscience, cognitive psychology and science, and psychophysiological research. Dementia, Parkinson's disease, multiple sclerosis, stroke, head injuries, and obsessive compulsive disorder [27–32] all have shown abnormalities in ERP components.

In the recent past, P300 is used to develop BCIs. This is done by arranging many signals in a grid. The rows of the grid are randomly flashed, and the P300 responses of a subject staring at the grid are observed. In this manner, the subject can communicate which stimulus he is looking at [33].

1.6 Application of EEG in Neurological Disorders

The nervous system consists of the brain, nerves, and the spinal cord. A disorder in any of the components of the nervous system of a structural, biochemical, or electrical nature is called a neurological disorder. There are many common and rare neurological disorders. General symptoms of neurological disorders may include seizures, paralysis, muscle weakness, poor coordination, altered levels of consciousness, loss of sensation, pain, and confusion.

The specific causes of neurological problems may vary from genetic disorders to environmental and lifestyle problems, including trauma, spinal cord, brain or nerve injury, and malnutrition. The origin of the problem is not necessarily the nervous system; it may start in any other body system that interacts with the nervous system. For example, problems in cardiovascular systems, that is, blood vessels supplying the brain, may result in cerebrovascular disorders; the immune system of the body may cause autoimmune disorders, and so on. However, in some cases, no neural cause can be identified in spite of the presence of neurological symptoms. Such conditions invite different theories about the causes of the condition.

The brain and spinal cord are physically enclosed in the bones of the skull and spinal vertebrae. Chemically, these are isolated by the blood–brain barrier. They are very susceptible to damage if compromised. Similarly, nerves tend to lie deep under the skin but can still become exposed to damage. The peripheral nervous system may be healed by the process of neuroregeneration and thus overcome the injuries to some extent, but this is thought to be rare in the case of the brain and spinal cord.

Parkinson's disease (PD) is a chronic and progressive degenerative disorder of the central nervous system. The general symptoms are impairment of motor skills and speech, resulting in rigidity of muscles, tremor, slowing of physical movements, and in extreme cases total loss of physical movement. The primary symptoms are the result of decreased stimulation of the motor cortex, and secondary symptoms may include a high level of cognitive dysfunction and speech problems.

Alzhemier's disease is a neurodegenerative disease. It causes progressive cognitive deterioration along with symptoms of behavior changes and a decline in daily activities. Minor forgetfulness or loss of short-term memory is one of the early symptoms. However, the progression of the disease results in the impairment of skilled movements, recognition, decision-making, and planning.

Myasthenia gravis is a disease that causes muscle weakness and fatigue while performing simple daily activities. Demyelination is the loss of myelin sheath insulating the nerves. It results in the impairment or loss of the signals along the nerves. The nerves eventually whither, which leads to disorders such as multiple sclerosis.

Brain rhythms can help to treat different brain disorders. Recent studies have led to the discovery of brain rhythms for PD [34]. By implanting electrodes within the brain, irregular brain rhythms can be detected. The measurement of signals from the cerebral cortex, that is, the outermost layer of brain that helps govern memory, physical movement, and consciousness, can be used for the detection of the disease. In normal circumstances, the cells of the brain work independently, but in the case of PD, the cells display synchronization most of the time, which is not an appropriate behavior. This leads to different characteristic symptoms displayed by PD. Deep brain stimulation is now also used to treat depression and obsessive compulsive disorder. It requires a surgeon to implant electrodes inside tiny parts of the brain to deliver an electrical current. In PD, these electrodes are implanted normally in patients who cannot fully benefit from drugs owing to unknown reasons or complications, and have mid-range disease. Deep brain stimulation can free them from severe immobility and other related symptoms, so that they may have a better quality of life for many years.

In addition to PD, it has been observed that many psychiatric and neurological conditions are caused by an abnormality in the brain rhythms, that is, the firing mechanism of neurons. Hence, various symptoms can be generated depending on the location of the abnormality in the brain. Rhythm abnormalities may happen anytime a group of neurons begin to generate oscillatory activity at a lower frequency, and coherently at a longer wavelength, than in a normal alert brain. A pattern very similar to this happens in the brain in the sleeping state. However, when it occurs in the awakened state, then it results in abnormality; that part of the brain remains fixed at a lower frequency, and it stops responding to stimulants and external inputs. Hence, the brain becomes nonresponsive, as though it is under general anesthesia.

PD, depression, obsessive–compulsive disorder, and many other neuropsychiatric conditions are being considered for treatment by deep brain stimulation (DBS). The principle underlying DBS is that the firing pattern of neurons is changed by the targeted stimulation of the brain.

Epilepsy is one of the most common neurological diseases. Seizures are the characteristic symptom of epilepsy and could be recurrent and unprovoked [35]. Epilepsy is the most common condition among children and ranks third among adults after Alzheimer's and stroke. Currently, there is no cure for epilepsy; however, it can be controlled using medication. Abnormal, excessive, or synchronous neuronal activity in the brain results in transient epochs called seizures. Epilepsy has widely divergent symptoms that are the result of abnormal brain electrical activity. The cause for 30% of epilepsy cases is known, whereas 70% of them have no known causes. Among the known causes are brain tumor, stroke, head trauma, gunshot wounds, accidents, falls, poisoning, infection, and the genetic factor. It has been observed that the EEG of the epileptic brain is a nonlinear signal with deterministic properties [35]. A mathematical modeling of the nonlinear brain can help pave the way for avenues of treatment where the early onset of the seizures can be identified and preventive measures can be taken to control it.

References

1. Niedermeyer, E. and Lopes da Silva, F. H. *Electroencephalography: Basic Principles, Clinical Applications, and Related Fields.* Lippincott Williams and Wilkins, PA; 2005.
2. Swartz, B. E. and Goldensohn, E. S. Timeline of the history of EEG and associated fields. *Electroencephalography and Clinical Neurophysiology,* 106, 173–176, 1998.
3. Millett, D. Hans Berger: From psychic energy to the EEG. In *Perspectives in Biology and Medicine.* Chicago, IL: John Hopkins University Press; 2001, 44, pp. 522–542.
4. Walter, W. G., Cooper, R., Aldridge, V. J., McCallum, W. C., and Winter, A. L. Contingent negative variation: An electric sign of sensorimotor association and expectancy in the human brain. *Nature,* 203(4943), 380–384, 1964.
5. Vidal, J. Real-time detection of brain events in EEG. *IEEE Proceedings Special Issue Biological Signal Processing and Analysis,* 65, 633–664, 1977.
6. Vidal, J. Towards direct brain-computer communications. *Annual Review of Biophysics and Bioengineering,* 2, 157–180, 1973.
7. Handy, T. C. *Event Related Potentials: A Methods Handbook.* Cambridge, MA: Bradford/MIT Press; 2005.
8. Gray, H. *Gray's Anatomy: The Classic Collector's Edition.* New York: Random House; 1988.
9. Chen, P. Principles of biological science. 2011. Accessed September 28, 2013. http://bio1152.nicerweb.com/Locked/src/chap48_g.html

10. Blankertz, B., Tomioka, R., Lemm, S., Kawanabe, M., and Muller, K.-R. Optimizing spatial filters for robust EEG single trial analysis. *IEEE Signal Processing Magazine*, 41(1), 41–56, 2008.

11. Patel, N. D. An EEG-based dual-channel imaginary motion classification for brain computer interface. Master of Engineering Science, Thesis, Lamar University, 2011.

12. Guyton, A. C. and Hall, J. E. *Textbook of Medical Physiology*. Philadelphia, PA: Elsevier Saunders; 2006.

13. Sanei, S. and Chambers, J. *EEG Signal Processing*. Chichester, England: John Wiley; 2007.

14. Ward, J. *The Student's Guide to Cognitive Neuroscience*. London: Taylor and Francis; 2010, p. 453.

15. Nunez, P. L. and Srinivasan, R. *Electric Fields of the Brain: The Neurophysics of EEG*. New York: Oxford University Press; 1981.

16. Klein, S. and Thorne, B. M. *Biological Psychology*. New York: Worth; 2006.

17. Fries, P. A mechanism for cognitive dynamics: Neuronal communication through neuronal coherence. *Trends in Cognitive Sciences*, 9, 474–480, 2001.

18. Fell, J. and Axmacher, N. The role of phase synchronization in memory processes. *Nature Reviews Neuroscience*, 12, 105–118, 2011.

19. Schnitzler, A. and Gross, J. Normal and pathological oscillatory communication in the brain. *Nature Reviews Neuroscience*, 6(4), 285–296, 2005.

20. Gevins, A. S. and Cutillo, B. C. Neuroelectric evidence for distributed processing in human working memory. *Electroencephalography and Clinical Neurophysiology*, 87, 128–143, 1993.

21. Kierkels, J. J. M. Validating and improving the correction of ocular artifacts in electro-encephalography. PhD thesis, Technische Universiteit Eindhoven, 2007. http://alexandria.tue.nl/extra2/200711857.pdf

22. Rielo, D. and Benbadis, S. R. EEG artifacts. 2010. http://emedicine.medscape.com/article/1140247-overview

23. Schlögl, A. and Pfurtscheller, G. EOG and ECG minimization based on regression analysis. Report siesta-task310 artifact detection and denoising. en.scientificcommons.org. 2007. http://en.scientificcommons.org/42786168

24. Freye, E. *Cerebral Monitoring in the OR and ICU*. Dordrecht: Springer; 2005.

25. Gutberlet, I., Debener, S., Jung, T.-P., and Makeig, S. Techniques of EEG recording and preprocessing. In N. T. Thokar, Ed., *Quantitative EEG Analysis Methods and Clinical Applications*. London: Artech House; 2009, pp. 1–439.

26. Luck, S. *An Introduction to the Event-Related Potential Technique*. Cambridge, MA: MIT Press; 2005.

27. Boutros, N., Torello, M. W., Burns, E. M., Wu, S.-S., and Nasrallah, H. A. Evoked potentials in subjects at risk for Alzheimer's disease. *Psychiatry Research*, 57(1), 57–63, 1995.

28. Prabhakar, S., Syal, P., and Srivastava, T. P300 in newly diagnosed nondementing Parkinson's disease: Effect of dopaminergic drugs. *Neurology India*, 48(3), 239–242, 2000.

29. Boose, M. A. and Crandford, J. L. Auditory event-related potentials in multiple sclerosis. *American Journal of Otology*, 17(1), 165–170, 1996.

30. Duncan, C. C., Kosmidis, M. H., and Mirsky, A. F. Event-related potential assessment of information processing after closed head injury. *Psychophysiology*, 40(1), 45–59, 2003.

31. D'Arcy, R. C. N., Marchand, Y., Eskes, G. A., Harrison, E. R., Phillips, S. J., Major, A., and Connolly, J. F. Electrophysiological assessment of language function following stroke. *Clinical Neurophysiology*, 114(4), 662–672, 2003.

32. Hanna, G. L., Carrasco, M., Harbin, S. M., Nienhuis, J. K., LaRosa, C. E., Chen, P., Fitzgerald, K. D., and Gehring, W. J. Error related negativity and tic history in pediatric obsessive-compulsive disorder. *Child Adolescent Psychiatry*, 51(9), 902–910, 2012.

33. Farwell, L. A. and Donchin, E. Talking off the top of your head: Toward a mental prothesis utilizing event-related brain potentials. *Electroencephalography and Clinical Neurophysiology*, 70(6), 510–523, 1988.

34. de Hemptinne, C., Ryapolova-Webb, E. S., Air, E. L., Garcia, P. A., Miller, K. J., Ojemann, J. G., Ostrem, J. L., Galifianakis, N. B., and Starr, P. A. Exaggerated phase-amplitude coupling in the primary cortex in Parkinson's disease. *Proceedings of National Academy of Sciences of the United States of America*, 110(12), 4780–4785, 2013.

35. Fisher, N., Talathi, S., Cadotte, A., Meyers, S., and Carney, P. R. Epilepsy detection and monitoring. In J. Greenfield, J. Geyer, and P. Carney, Eds., *Reading EEGs: A Practical Approach*. Philadelphia, PA: Lippincott Williams and Wilkins; 141–145, 2008.

2

The Fundamentals of EEG Signal Processing

Nidal Kamel

Universiti Teknologi PETRONAS

CONTENTS

2.1 Introduction 22
2.2 Continuous-Time and Discrete-Time Signals 23
    2.2.1 Typical Discrete-Time Sequences and Their Representation 24
    2.2.2 Classification of Discrete-Time Signals 25
2.3 Continuous-Time and Discrete-Time Systems 25
    2.3.1 Classification of Discrete-Time Systems 26
        2.3.1.1 Linear Systems 26
        2.3.1.2 Time-Invariant Systems 26
        2.3.1.3 Linear Time-Invariant System 26
        2.3.1.4 Causal Systems 27
        2.3.1.5 Stable System 27
2.4 Representation of Discrete-Time Signals in Frequency Domain 27
    2.4.1 The Discrete-Time Fourier Transform 27
        2.4.1.1 The Convolution Property 28
2.5 Stochastic Processes 29
    2.5.1 Mean and Autocorrelation Functions of Stochastic
        Processes 30
    2.5.2 Stationarity and Ergodicity of Random Processes 31
        2.5.2.1 Stationary Processes 31
        2.5.2.2 Ergodic Processes 32
2.6 Gaussian Processes 33
2.7 The Power Spectrum of Random Processes 34
2.8 Linear Filtering of Random Processes 35
2.9 The EEG Signals and Their Characteristics 36
    2.9.1 Time and Frequency Domains Methods 37
        2.9.1.1 Periodogram for Power Spectrum Estimation 38
        2.9.1.2 Commonly Used Windows 39
        2.9.1.3 MATLAB Functions for Welch Spectrum
        Estimation 42
        2.9.1.4 Correlogram for Power Spectrum Estimation 42

2.9.2   The Parametric Methods for EEG Modeling 43
        2.9.2.1   Time and Frequency Domains Representation of
                  EEG Signals Using AR...44
        2.9.2.2   Selecting the Order of the AR Model............................45
2.9.3   The Relationship between the AR Process and the Linear
        Prediction...46
        2.9.3.1   Burg Algorithm 47
        2.9.3.2   The Modified Covariance Method 48
2.9.4   MATLAB Functions for AR Parameters and Spectrum
        Estimation 49
2.10 Joint Time–Frequency Representation of EEG Signals........................50
2.10.1 The Short-Time Fourier Transform..51
2.10.2 Wavelet Transform (WT) 51
2.11 Nonlinear Descriptors of the EEG Signals...52
2.11.1 Higher-Order Statistics for EEG Data Analysis 53
        2.11.1.1 Definitions and Properties...53
2.11.2 MATLAB Functions for AR Parameters and Spectrum
        Estimation 58
2.11.3 Chaos Theory and Dynamical Analysis 59
        2.11.3.1 Definitions..61
        2.11.3.2 Reconstruction of the State Space 61
        2.11.3.3 Correlation Dimension...62
        2.11.3.4 Lyapunov Exponents..63
2.11.4 Dynamical Analysis of EEG Signals Using Entropy 64
2.12 Bivariate Analysis of EEG Signals...65
2.12.1 Normalized Cross-Covariance Function 65
2.12.2 The Coherence Function 66
2.12.3 Phase Synchronization...67
References..68

2.1 Introduction

It is common knowledge that a signal contains information about the behavior or nature of some phenomenon. The task of biomedical signal processing in general is to extract useful information contained in the signal. The method used to extract this information depends on the nature of the signal and the kind of information contained in it. In fact, there is no unique way to process signals for information extraction but rather a wide range of approaches. Generally speaking, the first step in extracting information from the signals is selecting the right analytical representation. This representation can be achieved through basis functions in the original domain of occurrence or in a transformed domain. Similarly, the process of information extraction may be performed in either domain.

Before discussing the representation of signals in the time and frequency domains, it is necessary to distinguish between deterministic and nondeterministic signals. A deterministic signal is one whose future values can be exactly predicted if enough information about its past is available. Often, one only requires a small amount of information related to the past. This means that the values of the signal may be calculated by means of closed mathematical expressions as a function of time or extrapolated by the knowledge of a certain number of preceding samples of the signal. On the contrary, the signal is called nondeterministic or stochastic if it is impossible to predict an exact future value even if one knows its entire past history. Generally speaking, with stochastic signals, every sample is considered a random variable with a specific probability distribution function. The entire sequence of the samples of the stochastic signals is called a stochastic process, and its measurement at a certain time is called its realization.

With the EEG signal, despite its predominantly unpredictable (nondeterministic) nature, we could theoretically rescue some deterministic information through direct recording in certain cortical areas. In practice, the EEG signal is generally processed with statistical methods, extending from first-order to second-order and to higher-order statistics.

In this chapter, we outline the time and frequency domain representations of the deterministic signals, the representation of stochastic processes in the time and frequency domains, parametric models of nondeterministic signals, joint time–frequency representation of EEG signals, higher-order statistics for EEG data analysis, nonlinear analysis of EEG signals, and bivariable analysis of EEG signals.

2.2 Continuous-Time and Discrete-Time Signals

Many biomedical measurements such as electroencephalography (EEG) and electrocardiography (ECG) or arterial blood pressure are inherently defined at all instants of time. The signals resulting from these measurements are called continuous-time signals, and it is traditional to represent their functions of time in the form $x(t)$, $p(t)$, and so forth. On the contrary, we could sample the values of the continuous-time signal at integral multiples of a fundamental time increment, called the sampling time, and obtain a signal that consists of a sequence of samples, each corresponding to one instant of time. This signal is called a discrete-time signal, and it is traditional to indicate its function of time as $x[n]$, where n is an integer.

Despite the continuous nature of the majority of biomedical signals, the continuing development of high-speed digital computers, integrated circuits, and sophisticated high-density device fabrication techniques have made it increasingly advantageous to consider processing *continuous-time* signals by converting them first to *discrete-time* signals. This process includes sampling the continuous-time signals at a suitable sampling time related to the maximum frequency in the signal. For the aforementioned reasons, we will consider the *discrete-time* domain for the representation and processing of EEG signals in the rest of this chapter.

2.2.1 Typical Discrete-Time Sequences and Their Representation

We shall now give a brief explanation of several basic discrete-time sequences that play an important role in signal and systems analysis.

The first sequence is the unit impulse (or the unit sample), which is defined as

$$\delta[n] = \begin{cases} 0, & n \neq 0 \\ 1, & n = 0 \end{cases} \tag{2.1}$$

The second sequence is the discrete unit step, denoted by $u[n]$ and defined as

$$u[n] = \begin{cases} 0, & n < 0 \\ 1, & n \geq 0 \end{cases} \tag{2.2}$$

It is possible to represent the unit step sequence in terms of the unit impulse sequence and vice versa. This is given as

$$u[n] = \sum_{k=0}^{\infty} \delta[n-k] \tag{2.3}$$

and

$$\delta[n] = u[n] - u[n-1] \tag{2.4}$$

Another important signal in discrete time is the complex exponential signal or sequence, defined as

$$x[n] = A e^{j\omega_0 n} \tag{2.5}$$

where A is, in general, a complex number, and ω_0 is a real number. Using Euler's relation allows us to relate the complex exponential in Equation 2.5 and the sinusoidal:

$$A e^{j\omega_0 n} = A \cos \omega_0 n + jA \sin \omega_0 n \tag{2.6}$$

and

$$A \cos(\omega_0 n) = \frac{A}{2} e^{j\omega_0 n} + \frac{A}{2} e^{-j\omega_0 n} \tag{2.7}$$

A point to be noted when discussing discrete sinusoidals is that they are periodic if and only if ω_0 is a rational multiple of 2π, or equivalently

$$\frac{\omega_0}{2\pi} = \frac{n}{m} \tag{2.8}$$

where n and m are positive integers. Accordingly, the angular frequency of the discrete-time sinusoidal signal is given as

$$\omega_0' = \frac{2\pi}{N} = \frac{\omega_0}{m} \tag{2.9}$$

2.2.2 Classification of Discrete-Time Signals

In this section, we outline the often-used classes of discrete-time signals.
The energy of a sequence $x[n]$ over an infinite interval, N, is given as

$$E_\infty = \lim_{N \to \infty} \sum_{n=-N}^{N} |x[n]| \tag{2.10}$$

and the average power is defined as

$$P_\infty = \lim_{N \to \infty} \frac{1}{2N+1} \sum_{n=-N}^{N} |x[n]|^2 \tag{2.11}$$

The signal $x[n]$ is classified as an *energy signal* if $E_\infty < \infty$, or equivalently, $P_\infty = 0$.

The signal $x[n]$ is classified as a *power signal* if $P_\infty < \infty$.

The signal $x[n]$ is classified as a periodic signal with fundamental period N if

$$x[n] = x[n \pm kN] \tag{2.12}$$

where k is an integer. The periodic signal has infinite energy but finite average power, and therefore it is an example of a power signal.

2.3 Continuous-Time and Discrete-Time Systems

In contexts ranging from signal processing to electromechanical motors and chemical processing plants, a system can be viewed as a process in which input signals are transformed by the system or cause the system to respond in some way, resulting in other signals as outputs.

Because the discrete-time domain will be used for EEG signal representation and processing in this chapter, we have confined our discussion to
discrete-time systems.

2.3.1 Classification of Discrete-Time Systems

Discrete-time systems are typically classified based on their input–output
relationships into the following classes.

2.3.1.1 *Linear Systems*

A discrete-time system is called linear if it possesses the important property
of superposition: if the input consists of the weighted sum of several signals,
then the output is the weighted sum of the responses of the system to each
of those signals. More precisely, let $y_1[n]$ be the response of the system to an
input $x_1[n]$, and $y_2[n]$ be the output corresponding to the input $x_2[n]$. Then the
system is linear if the response to

$$ax_1[n] + bx_2[n] \text{ is } ay_1[n] + by_2[n] \tag{2.13}$$

2.3.1.2 *Time-Invariant Systems*

The system is time invariant if the behavior and characteristics of the system are fixed over time. Specifically, a system is time invariant if a time
shift in the input signal results in an identical time shift in the output signal. That is, if $y[n]$ is the output of the system when $x[n]$ is the input, then
the system is time invariant if $y[n-n_0]$ is the output when $x[n-n_0]$ is applied
at the input.

2.3.1.3 *Linear Time-Invariant System*

We call a system that possesses the properties of both linearity and time
invariance a linear time-invariant (LTI) system. Such a system is amendable
to simple mathematical analysis and characterization, which makes it widely
used.

The main advantage of an LTI system is that the input–output relationship
is fully described by the system response to the unit impulse function. This
relationship is given as

$$y[n] = \sum_{k=-\infty}^{\infty} x[k]\, h[n-k] \tag{2.14}$$

where $h[n]$ is called the impulse response of the system and is the output of
the system to the unit impulse, $\delta[n]$.

2.3.1.4 Causal Systems

A system is causal if the output at any time depends on values of only the present and past inputs. Specifically, the system is causal if the response at n_0, namely $y[n_0]$, depends only on input samples, $x[n]$, for $n \leq n_0$ and not on input values $n > n_0$. In words, the system is causal if it does not precede the stimulus (input) with its response (output).

In terms of an LTI system, it can be shown that the system is causal if $h[n] = 0$ for $n \leq 0$ [1,2].

2.3.1.5 Stable System

Stability is another important system property. In general, a system is stable if a bounded input produces a bounded output. This type of stability is called bounded-input bounded-output (BIBO) stability. For the case of an LTI system, it can be shown [1,2] that BIBO stability is equivalent to satisfying the condition

$$\sum_{n=-\infty}^{\infty} h[n] < \infty \tag{2.15}$$

where $h[n]$ is the impulse response of the system.

2.4 Representation of Discrete-Time Signals in Frequency Domain

It is well known that different domain descriptions of the same phenomenon reveal different aspects of the phenomenon and enable us to gain more insight into it. This is equally true of the mathematical representation of brain signals, where the time domain alone is inadequate. More insight can be obtained if the frequency domain is used to represent the signal in terms of its spectrum.

2.4.1 The Discrete-Time Fourier Transform

The discrete-time Fourier transform (DTFT) of a discrete-time signal, $x[n]$, of N samples is defined as

$$X\left(e^{j\omega}\right) = \sum_{n=0}^{N-1} x(n) e^{-j\omega n} \tag{2.16}$$

In general, $X(e^{j\omega})$ is a complex-valued function of the real variable ω and has the following two properties:

It is a continuous function of ω.

It is periodic with period 2π.

Consequently, $x(n)$ can be determined from $X(e^{j\omega})$ through the inverse discrete-time Fourier transform (IDTFT), given as

$$x[n] = \frac{1}{2\pi} \int_{2\pi} X\left(e^{j\omega}\right) e^{j\omega n} d\omega \tag{2.17}$$

The DTFT equation in (2.16) is called the *synthesis equation*, and the IDTFT equation in (2.17) is called the *analysis equation*.

2.4.1.1 The Convolution Property

One of the important aspects of the DTFT is its effect on the operation of convolution and its use in representing and analyzing discrete-time LTI systems. Specifically, if $x[n]$, $h[n]$, and $y[n]$ are the input, impulse response, and output, respectively, of an LTI system, so that

$$y[n] = x[n] * h[n] \tag{2.18}$$

then

$$Y(e^{j\omega}) = X(e^{j\omega}) H(e^{j\omega}) \tag{2.19}$$

where $X(e^{j\omega})$, $H(e^{j\omega})$, and $Y(e^{j\omega})$ are the Fourier transforms of $x[n]$, $h[n]$, and $y[n]$, respectively.

Clearly, Equation 2.19 maps the convolution of two signals to the simple algebraic operation of multiplying their Fourier transforms, a fact that both facilitates the analysis of signals and systems and adds significantly to our understanding of the way in which an LTI system responds to the input signals that are applied to it. In particular, from Equation 2.19, we see that the frequency response $H(e^{j\omega})$ captures the change in the complex amplitude of the Fourier transform of the input at each frequency ω. Thus, in frequency-selective filtering, we require $\left|H(e^{j\omega})\right| \approx 1$ over the desired range of frequencies and $\left|H(e^{j\omega})\right| \approx 0$ over the range of undesired frequencies, or frequencies to be eliminated. Figure 2.1 plots the frequency responses of four types of ideal filters: (1) low pass, (2) high pass, (3) band pass, and (4) band reject. The frequency range for which the gain is 1 is the passband of the ideal filter.

As we previously mentioned, the EEG signal has a predominantly non-deterministic nature, so it is generally processed with statistical methods,

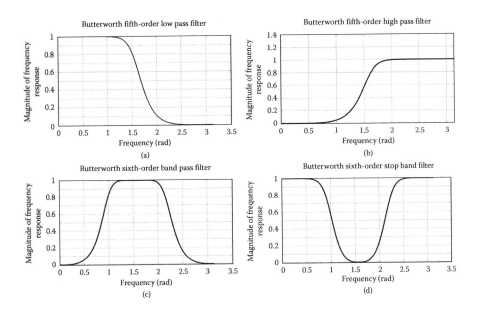

FIGURE 2.1
The frequency responses of the four ideal types of filters.

extending from first-order to second-order and to higher-order statistics. In the following section, we outline the definition of the stochastic process and the main descriptors in the time and frequency domains.

2.5 Stochastic Processes

In probability, we usually conduct experiments consisting of procedures and observations. When we study random variables, each of the observations corresponds to one or more values, whereas when we study stochastic processes, each observation corresponds to a function of time. The word *stochastic* means random, and the word *process* in this context means a function of time.

Definition: A *stochastic process* $x(n)$ consists of an experiment defined on the sample space Ω and a function that assigns a time function $x_i(n)$ to each outcome ω_i in the sample space of the experiment.

Definition: A *sample function* $x_i(n)$ is the time function associated with the outcome ω_i of an experiment.

Definition: The *ensemble* of a stochastic process is the set of all possible time functions that can result from an experiment.

Figure 2.2 shows the correspondence between the sample space of an experiment and the ensemble of sample function $x_i(n)$ of a stochastic process.

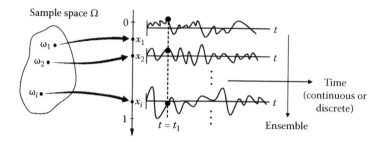

FIGURE 2.2
Conceptual representation of a random process.

2.5.1 Mean and Autocorrelation Functions of Stochastic Processes

In studying random variables, we frequently refer to properties of the probability model, such as expected value, variance, covariance, and correlation. In the case of stochastic processes, deterministic functions of time provide corresponding summaries of the properties of the model. For a stochastic process $x(n)$, $x(n_1)$ is a random variable with *pdf* $f_{x(n_1)}(x)$ and expected value $E\{x(n_1)\}$. Because $E\{x(n_1)\}$ is simply a number, the function $E\{x(n)\}$ is a deterministic function of n.

In general, we define the *kth*-order joint *cdf* function of $x(n)$ by

$$F_{x(n_1),\ldots,x(n_k)}(\alpha_1,\ldots,\alpha_k) = \Pr\{x(n_1) \le \alpha_1, \ldots, x(n_k) \le \alpha_k\} \quad (2.20)$$

where Pr{.} indicates the probability. On the contrary, the *kth*-order joint *pdf* function of $x(n)$ is given *as*

$$f_{x(n_1),\ldots,x(n_k)}(\alpha_1,\ldots,\alpha_k) = \frac{\partial F_{x(n_1),\ldots,x(n_k)}(\alpha_1,\ldots,\alpha_k)}{\partial x(n_1),\ldots,\partial x(n_k)} \quad (2.21)$$

A complete characterization of $x(n)$ requires knowledge of all distributions as $k \to \infty$. Fortunately, often much less information is sufficient [3].

Definition: The *expected value*, $E\{.\}$, of a stochastic process $x(n)$ is the deterministic function

$$\mu_x(n) = E\{x(n)\} \quad (2.22)$$

Definition: The *autocorrelation function* of a random process $x(n)$ is given as

$$r_x(k,l) = E\{x(k)x(l)\} \quad (2.23)$$

Definition: The *autocovariance function* of a stochastic process $x(n)$ is

$$c_x(k,l) = E\{(x(k) - \mu_x(k))(x(l) - \mu_x(l))\}$$

$$= r_x(k,l) - \mu_x(k)\mu_x(l) \quad (2.24)$$

The prefix *auto* of "autocorrelation" emphasizes the correlation between two samples of the same process $x(n)$.

The *correlation coefficient* is given as

$$\rho_x(k,l) = \frac{c_x(k,l)}{\sigma_x(k)\sigma_x(l)} \qquad (2.25)$$

where σ_x is the standard deviation of x. Owing to normalization by the standard deviation of the random process x at instants k and l, the correlation coefficient is bounded by one in magnitude.

In applications involving more than one stochastic process, it is often of interest to determine the cross-correlation and cross-covariance functions. These two functions are given for the two random variables $x(n)$ and $y(n)$, respectively, by

$$r_{xy}(k,l) = E\{x(k)y(l)\} \qquad (2.26)$$

$$c_{xy}(k,l) = E\left\{\left(x(k)-\mu_x(k)\right)\left(y(l)-\mu_y(l)\right)\right\}$$
$$= r_{xy}(k,l) - \mu_x(k)\mu_y(l) \qquad (2.27)$$

The two random processes $x(n)$ and $y(n)$ are said to be uncorrelated if $c_{xy}(k,l) = 0$ and *orthogonal* if their *cross-correlation* is zero.

2.5.2 Stationarity and Ergodicity of Random Processes

If a random process $x(n)$ possesses some special probabilistic structure, we can specify it less fully to characterize it. Some random processes are completely characterized by their first- and second-order distributions.

2.5.2.1 Stationary Processes

A random process $x(n)$ is said to be *stationary* of order L if for every set of L time instants, we have

$$f_{x(n_1),\dots,x(n_L)}(\alpha_1,\dots,\alpha_L) = f_{x(n_1+k),\dots,x(n_L+k)}(\alpha_1,\dots,\alpha_L) \qquad (2.28)$$

for any k. Hence, the distribution of a stationary process will not be affected by a shift in time, and $x(n)$ and $x(n+k)$ will have the same distribution for any k. If the process is stationary for all $L > 0$, then it is said to be *stationary in the strict sense* [4,5].

If the *first-order distribution*

$$f_{x(n)}(\alpha) = f_{x(n+k)}(\alpha) \qquad (2.29)$$

is independent of time for all k, then the process is said to be *first-order stationary*. For a first-order stationary process, the first-order statistics will be independent of time. For example, the mean will be constant:

$$\mu_x(n) = E\{x(n)\} = \mu_x \tag{2.30}$$

Similarly, if the *second-order distribution*

$$f_{x(n_1),x(n_2)}(\alpha_1,\alpha_2) = f_{x(n_1+k),x(n_2+k)}(\alpha_1,\alpha_2) \tag{2.31}$$

is invariant to a time shift, then the process is called *second-order stationary*.
    If the random process $x(t)$ is stationary of *first order* and *second order*, then it is said to be *wide-sense stationary* (WSS).
    Definition: A stochastic process $x(n)$ is *wide-sense stationary* (WSS) if its mean function $\mu_x(n)$ is constant and its *autocorrelation* $r_x(k,l)$ and *autocovariance* $c_x(k,l)$ depend only on the difference $k-l$:

$$
\begin{aligned}
\mu_x(n) &= \mu_x \\
r_x(k,l) &= r_x(k-l) \\
c_x(k,l) &= c_x(k-l) = r_x(k-l) - \mu_x^2
\end{aligned}
\tag{2.32}
$$

The autocorrelation sequence (ACS) of a WSS process has a number of useful and important properties:

Property 1. The autocorrelation sequence of a WSS process is a symmetric function of k,

$$r_x(k) = r_x(-k) \tag{2.33}$$

Property 2. The autocorrelation sequence of a WSS process at lag $k = 0$ is equal to the average power of the process,

$$r_x(0) = E\left\{|x(n)|^2\right\} \tag{2.34}$$

Property 3. The value of the autocorrelation sequence of a WSS process at lag k is upper bounded by its value at lag $k = 0$,

$$r_x(0) \geq |r_x(k)| \tag{2.35}$$

2.5.2.2 *Ergodic Processes*

It is clear that the mean and autocorrelation sequences of a random process are an ensemble-averaging-based function, because an ensemble of all possible discrete-time signals is required for their calculation. Although these ensembles are required in finding the first- and second-order statistics of a random process, they are not generally available. In this section, we present some conditions for which it is possible to estimate the mean and autocorrelation using a time average.

Let us begin with the estimation of the mean value of the random process $x(n)$. Consider a random process $x(n)$ with a typical *sample function* $x_i(n)$. If M realizations of the process are available, then the mean value can be estimated as

$$\hat{\mu}_x(n) = \frac{1}{M} \sum_{i=1}^{L} x_i(n) \tag{2.36}$$

However, in most situations, such a collection of M *sample functions* is not available, and the mean value needs to be estimated from single realization. Given only a single realization $x(n)$ of a WSS process, we may consider estimating the ensemble average with a sample mean that is taken over time of $x(n)$ as follows:

$$\hat{\mu}_x = \frac{1}{N} \sum_{n=0}^{N-1} x(n) \tag{2.37}$$

If the sample mean in Equation 2.37 of the process converges to μ_x in the mean-square sense as
$N \to \infty$, then the process is said to be *ergodic in the mean.*

Similarly, the autocorrelation function of a WSS process estimated from a single realization $x(n)$ can be defined as

$$\hat{r}_x(k) = E\{x(n)x(n-k)\} = \frac{1}{N} \sum_{n=0}^{N-1} x(n)x(n-k) \tag{2.38}$$

If $\hat{r}_x(k)$ converges to $r_x(k)$ in the mean-square sense as $N \to \infty$, then the process is said to be *autocorrelation ergodic* [6].

2.6 Gaussian Processes

An important class of random processes that arises in a wide spectrum of applications is the Gaussian processes.

Definition: $x(n)$ is a Gaussian stochastic process if and only if $\mathbf{X} = [x(n_1)$ $x(n_2) \ldots x(n_k)]$ is a Gaussian vector for any integer $k > 0$ and any set of time instants $n_1, n_2, \ldots, n_k$.

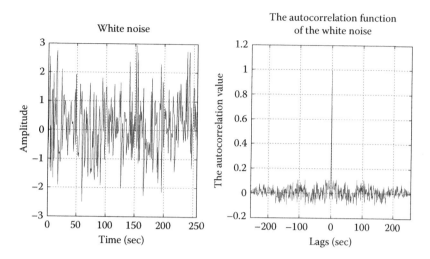

FIGURE 2.3
The autocorrelation function of the white noise random process.

Definition: A discrete-time *white noise process*, $d(n)$, is a WSS *zero-mean* random process whose autocorrelation function is zero for all $k \neq 0$

$$r_d(k) = \sigma^2 \delta(k) \tag{2.39}$$

where $\delta(k)$ is the unit impulse function and σ^2 is the noise variance. This means that the samples $d(n)$ and $d(n + k)$ *remain uncorrelated for any k* no matter how small it is [7]. Figure 2.3 shows the autocorrelation function of the white noise random process.

A random process that consists of a sequence of uncorrelated Gaussian random variables is called white Gaussian noise (WGN).

2.7 The Power Spectrum of Random Processes

As with deterministic discrete-time signals, Fourier analysis plays an important role in the study of random processes. However, because a random process is an ensemble of discrete-time sample functions, the Fourier transform of a random process cannot be computed. Nevertheless, it is possible to develop a Fourier representation if the process is WSS. Because the autocorrelation sequence of a WSS process is a deterministic sequence, its Fourier transform is given as

$$S\left(e^{j\omega}\right) = \sum_{k=-\infty}^{\infty} r_x\left(k\right) e^{-jk\omega} \tag{2.40}$$

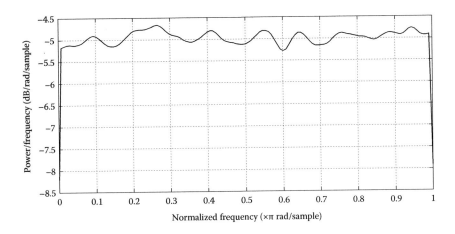

FIGURE 2.4
The power spectral density of white noise.

which is called the *power spectrum density* (PSD) or *power spectral density* of the process. In the case of *real-valued* random process, the autocorrelation sequence is real and even, which means a *real* and *even* power spectrum. In addition, the power spectrum is nonnegative, $P(e^{j\omega}) \geq 0$, and the total power in the zero mean WSS process is proportional to the area under the power spectral density curve.

$$E\left\{ \left| x(n) \right|^2 \right\} = \frac{1}{2\pi} \int_{-\pi}^{\pi} P_x\left(e^{j\omega}\right) d\omega \qquad (2.41)$$

The second definition of PSD is

$$S\left(e^{j\omega}\right) = \lim_{N \to \infty} E\left\{ \frac{1}{N} \left| \sum_{n=0}^{N-1} x(n) e^{-jn\omega} \right|^2 \right\} \qquad (2.42)$$

Figure 2.4 shows the power spectral density of the white noise random process in Figure 2.3.

2.8 Linear Filtering of Random Processes

LTIs are frequently used to process signals in various applications ranging from detection and estimation to signal representation and synthesis. Because the inputs to LTI filters are often random signals, it is important to know how the statistics change as a result of filtering. In this section, we outline the relationships between the mean and autocorrelation of the input random process and the output of an LTI filter.

The output of a stable LTI system having an impulse response $h(n)$ to an input WSS random process with mean μ_x and autocorrelation $r_x(k)$ is given by the convolution sum as

$$y(n) = x(n) * h(n) = \sum_{k=-\infty}^{\infty} h(k) x(n-k) \tag{2.43}$$

The mean value of the output, $y(n)$, is given as

$$E\{y(n)\} = E\left\{ \sum_{k=-\infty}^{\infty} h(k) x(n-k) \right\} = \sum_{k=-\infty}^{\infty} h(k) E\{x(n-k)\}$$

$$\tag{2.44}$$

$$= \mu_x \sum_{k=-\infty}^{\infty} h(k)$$

The relationship between the autocorrelation sequence of the input random process and autocorrelation sequence of the output random process of the LTI system is given as [4,6]

$$r_y(k) = r_x(k) * h(k) * h(-k)$$

$$= r_x(k) * r_h(k) \tag{2.45}$$

where $r_h(k)$ represents the autocorrelation of the filter impulse response, $h(n)$.

2.9 The EEG Signals and Their Characteristics

The electrical activity of the active nerves in the brain produces currents flowing through the head. Some of these currents reach the scalp surface, resulting in voltages on the scalp that can be recorded as the EEG. Because the cerebral cortex lies immediately beneath the scalp, the EEG recorded from the top of the head reflects mostly activities of the cortical neurons. Typically, the EEG signal has amplitude ranging from 10 to 100 mV, and its spectrum extends from 0.4 Hz to 40.0 Hz, because this is the band with most of the power of the EEG signals. Within this frequency band, specific sub-bands are often analyzed, depending on the application. The most commonly used bands are as follows.

Delta: This band has a frequency of 3.5 Hz and below. It tends to be the highest in amplitude and the lowest in frequency. It is the dominant rhythm in infants up to one year and in stages 3 and 4 of sleep [8].

Theta: This band has a frequency of 3.5–7.5 Hz and is classified as "slow" activity. It is normal in children up to 13 years and in sleep, but abnormal in awake adults. Theta arises from emotional stress, especially frustration or disappointment [9].

Alpha: This band has a frequency between 7.5 and 12.5 Hz and is usually best seen in the posterior regions of the head on each side, with the higher amplitude on the dominant side. It appears when closing the eyes and relaxing, and disappears when opening the eyes or when concentrating on any mechanism (thinking, calculating). It is the major rhythm seen in normal relaxed adults. It is present during most of life, especially after the 13th year [10].

Beta: This band has a frequency between 12.5 and 30.0 Hz and is usually seen on both sides in a symmetrical distribution; it is most evident frontally. It is the dominant rhythm in people who are alert or anxious or have their eyes open.

Gamma: This band has a frequency range of 31 Hz and greater. It is thought that it reflects the mechanism of consciousness. Beta and gamma waves together have been associated with attention, perception, and cognition [11].

EEG data are generally considered to be nonstationary because its statistical characteristics change with time owing to the variations in the mental states. In the analysis of the EEG signals for information extraction, two different methods are used. If the EEG is considered stationary over a long time, the classical methods used in the characterization of random processes, such as means, variance, and autocorrelation functions in addition to the power spectrum density estimated using the periodogram and the correlogram, are used. On contrary, if the signal is considered stationary over a short time interval (quasi-stationary), parametric methods, such as autoregressive (AR) and autoregressive and moving average (ARMA) are used to best-fit the EEG data for its analysis and classification [12].

The following two sections explain the classical methods and the parametric models in EEG data analysis.

2.9.1 Time and Frequency Domains Methods

In these methods, the EEG signal is considered stationary over a sufficient time interval for efficient estimation of the different average functions, such as the autocovariance or autocorrelation in time domain and the power spectrum density in the frequency domain.

As mentioned before, estimating the autocovariance or autocorrelation of random processes requires averaging over a number of realizations of an ensemble. However, if the process is WSS and ergodic, then the autocovariance or autocorrelation sequence can be obtained through the time average over a single realization. Because the EEG sequence is limited in time, due to the stationarity issue, the covariance can only be estimated. Practically, two estimators are used to estimate the autocorrelation sequence, $r_x(k)$, of the EEG signal $x(n)$. The first estimator of $r_x(k)$ is called the *biased* estimator and is given as

$$r_x(k) = \frac{1}{N} \sum_{n=0}^{N-|k|-1} x(n)x(n+|m|), \qquad |k| \leq N-1 \tag{2.46}$$

where N is the number of samples. The second estimator of the autocorrelation sequence is called the *unbiased* estimator and is given as

$$r_x(k) = \frac{1}{N-|k|} \sum_{n=0}^{N-|k|-1} x(n)x(n+|m|), \qquad |k| \leq N-1 \tag{2.47}$$

Practically speaking, among the two ACS estimators, the biased one in Equation 2.46 is most commonly used, for the following reasons:

For most stationary signals, the autocorrelation function decays rapidly with large lags k. Comparing the two estimators, it can be seen that the biased estimator tends to produce small values for large k, whereas the unbiased estimator could take large and erratic values for large k. This observation implies that the biased estimator is likely to produce a more accurate estimation for the ACS than the biased one [13].

The ACS obtained by the biased estimator is guaranteed to be positive semi-definite, but this is not the case for the biased estimator [14]. This is specifically important for the PSD estimation, because a non-positive-definite ACS may lead to a negative power estimate, which is undesirable in most applications.

2.9.1.1 Periodogram for Power Spectrum Estimation

The periodogram method relies on the definition of the PSD. Neglecting the expectation and the limit operation in Equation 2.42, which cannot be performed with a limited number of samples, we get

$$S_p(e^{j\omega}) = \frac{1}{N} \left| \sum_{n=0}^{N-1} x(n)e^{-jn\omega} \right|^2 \tag{2.48}$$

In order to reduce the large fluctuations of the periodogram, Bartlett proposed to split up the N available samples into L segments of M samples each and then obtain the periodogram for each segment and find their average [15,16]. To describe the Bartlett method in mathematical form, let $S_i(e^{j\omega})$

represent the periodogram of the *i*th segment, where $i = 1, 2, ..., L$; then the Bartlett spectral estimate is given by

$$S_B\left(e^{j\omega}\right) = \frac{1}{L}\sum_{i=1}^{L} S_i\left(e^{j\omega}\right) \tag{2.49}$$

Because the Bartlett method operates on data segments of M samples, the resolution is reduced by a factor of L in comparison to the original periodogram method. On the contrary, the variance of the estimator is reduced by the same factor L.

In 1967, Welch proposed another method based on the periodogram [17]. This method is similar to that proposed by Bartlett in segmenting the N data samples into L segments of M samples each but differs in two aspects. First, the segments in the Welch method are allowed to overlap. Second, each data segment is windowed prior to computation of the periodogram. To describe the Welch method in mathematical form, let the segments be overlapped with 50%, so that the number of segments $G = 2M/N$.

The windowed periodogram for the *i*th segment is given by

$$S_i\left(e^{j\omega}\right) = \frac{1}{MP_w}\left|\sum_{n=1}^{M} w(n)x_i(n)e^{-j\omega n}\right|^2 \tag{2.50}$$

where P denotes the "power" of the temporal window, $w(n)$, given as

$$P_w = \frac{1}{M}\sum_{n=1}^{M}|w(n)|^2 \tag{2.51}$$

The Welch is found by averaging the S windowed periodograms in (2.50)

$$S_W\left(e^{j\omega}\right) = \frac{1}{G}\sum_{i=1}^{S} S_i\left(e^{j\omega}\right) \tag{2.52}$$

By introducing the window in the periodogram computation, we hope to get more control over the bias or resolution properties of the estimated spectrum. Further, by allowing the segments to overlap and thereby allowing more periodograms to be averaged, we hope to decrease the fluctuations (variance) in the estimated spectrum.

2.9.1.2 Commonly Used Windows

In this section, we list some common window functions and outline their relevant properties. Our intention is not to provide a detailed derivation or listing but rather to provide a quick reference of common windows used in EEG software. More detailed information on the windows can be found in Refs. [13,14]. Table 2.1 lists some common windows along with some useful properties.

In order to show the variation in the power spectral density of the signal with its time content, the EEG recording of an epileptic patient is considered. The data are extracted from the CHB-MIT scalp EEG data set [18]. This data set is also known as Physionet EEG data set and contains EEG recordings of 24 pediatric patients acquired at Children's Hospital, Boston. The patients' age range is between 1.5 years and 19 years. The patients were suffering from intractable epilepsy. The EEG was recorded using the International 10–20 system of electrode placement with bipolar montages.

Figure 2.5 shows the EEG recording of one seizure extracted from the temporal lobe area.

TABLE 2.1

Some Common Windows and Their Properties

Name	Equation	Main Lobe Width (rad)	Side Lobe Level (dB)
Rectangular	$w(n) = 1$	$\approx 2\pi/M$	-13
Bartlett	$w(n) = \dfrac{M-n}{M}$	$\approx 4\pi/M$	-25
Hanning	$w(n) = 0.5 + 0.5\cos\left(\dfrac{n\pi}{M}\right)$	$\approx 4\pi/M$	-31
Hamming	$w(n) = 0.54 + 0.46\cos\left(\dfrac{n\pi}{M-1}\right)$	$\approx 4\pi/M$	-41
Blackman	$w(n) = 0.42 + 0.5\cos\left(\dfrac{n\pi}{M-1}\right)$ $+ 0.08\cos\left(\dfrac{n\pi}{M-1}\right)$	$\approx 6\pi/M$	-57

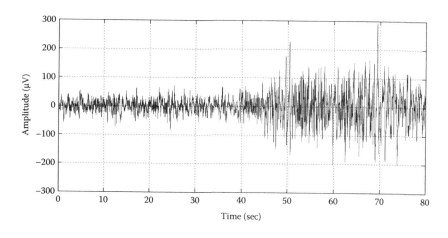

FIGURE 2.5
Epileptic seizure extracted from the temporal lobe area of a patient of the CHB-MIT scalp EEG data set.

Figure 2.6 shows the power spectrum density estimated using the Welch method of the epileptic data in Figure 2.5.

In order to show the variation in the power spectrum density due to the seizure, the data in Figure 2.5 is partitioned into two parts, before and after the start of the seizure. The power spectrum density as estimated by Welch is computed and depicted in Figure 2.7.

The results in Figure 2.7 show clearly the significant increase in the power over all frequencies due to the epileptic seizure.

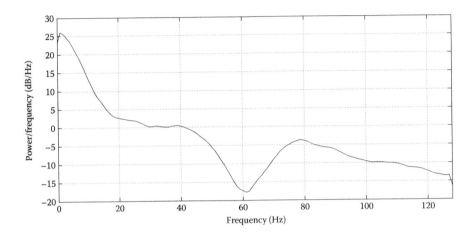

FIGURE 2.6
Welch estimation of the power spectral density of the epileptic data in Figure 2.5.

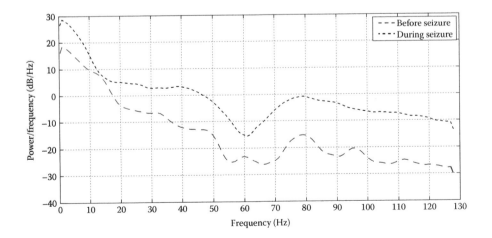

FIGURE 2.7
The power spectral density of the data before the start of the seizure and during the seizure estimated by Welch.

2.9.1.3 MATLAB Functions for Welch Spectrum Estimation

In MATLAB®, the following functions are used to compute Welch's estimation of the power spectral density:

h = spectrum.welch;
Hpsd = psd(h,x,'Fs',256);

where x is the data vector that contains the EEG time series data and 256 is the sampling rate.

2.9.1.4 Correlogram for Power Spectrum Estimation

The correlogram relies on the Fourier transform of the autocorrelation sequence and leads to the correlogram spectral estimation [19]. The formula for the correlogram is given by

$$S_c\left(e^{j\omega}\right) = \sum_{k=-(N-1)}^{N-1} r_x\left(k\right)e^{-jk\omega} \qquad (2.53)$$

where $r_x(k)$ is the ACS estimated using either the biased estimators in Equation 2.46 or the unbiased estimators in Equation 2.47.

As we know, the main problem with the periodogram is the high statistical variability or variance in this spectral estimator. The correlogram has the same problem, which arises from the poor estimation of $r(k)$ values for extreme lags ($k \approx N$). In order to mitigate this effect, Blackman–Tukey proposed to truncate the sum in the correlogram formula in Equation 2.53. This idea implemented with the right window is called the Blackman–Tukey estimator of the PSD, which is given as

$$S_{BT}\left(e^{j\omega}\right) = \sum_{k=-(M-1)}^{M-1} w\left(k\right)r_x\left(k\right)e^{-jk\omega} \qquad (2.54)$$

where $w(k)$ is an even function, with $w(0) = 1$ and $w(k) = 0$ for $|k| \geq M$. $w(k)$ is called a *lag window*. If $w(k)$ is selected to be rectangular, then we simply obtain a truncated version of the basic formula of the correlogram in Equation 2.53. However, $w(k)$ can be chosen in many other ways, giving us the flexibility to improve the accuracy of the Blackman–Tukey spectral estimator. Generally speaking, we may expect smaller values of M to reduce fluctuations in the spectral estimator but at the expense of a reduced resolution. Analysis of the Blackman–Tukey spectral estimator shows that the resolution of the spectrum is on the order of $1/M$, whereas its variance is on the order of M/N [13]. Thus, a compromise between the resolution and the variance should be considered when choosing the window length.

2.9.2 The Parametric Methods for EEG Modeling

This section introduces another powerful approach to signal representation called parametric signal modeling. In this approach, a signal is represented by a mathematical model that has a predefined structure involving a limited number of parameters. Generally speaking, parametric signal modeling has a wide range of applications, including spectrum analysis, signal prediction, system identification, signal detection, and signal classification. The three key elements for successes in all of the applications are an appropriate choice of model, good estimation of its length, and an accurate estimation of its parameters.

The general model for a stationary random process $x(n)$ is given as the input–output relationship of a linear system driven by white noise. This relationship is given as

$$x(n) = -\sum_{k=1}^{p} a_p(k)x(n-k) + \sum_{k=0}^{q} b_q(k)w(n-k) \qquad (2.55)$$

where $w(n)$ is the driving noise of the system of the variance σ_w^2. This form is the general model called the autoregressive and moving average (ARMA). As the name clearly indicates, ARMA includes two models, the autoregressive (AR) of order p and the moving average (MA) of order q. Generally speaking, among the ARMA, AR, and MA models, the AR is the most popular. This is mainly for three reasons:

i. Its p parameters can be obtained as a solution of linear equations.

ii. It tends to exhibit peaks in its spectrum, a feature associated with high resolution.

iii. Its relationship with the linear prediction.

In terms of EEG, it is widely accepted that AR can accurately represent the signal, so its parameters can be used for classification and prediction. For example in Ref. [20], the parameters of the AR model are used as inputs of a neural network for predicting movements during anesthesia. In Ref. [21], an off-line distance measure based on the AR model parameters was proposed to effectively quantify the changes in EEG signals related to brain injury. In Ref. [22], a multivariate AR model was proposed to analyze EEG signals for classification of spontaneous EEG signals during mental tasks. Because multichannel synchronization of EEG signals is one of the important characteristics of brain activities that help in analyzing the brain disorder, its ictal and interictal measures based on the residual covariance matrix of a multichannel AR model are presented in Ref. [23]. In Refs. [24] and [25], the AR parameters are used in brain–computer interfacing for the identification and classification of various mental tasks.

Even though there is a wide range of applications of AR modeling of EEG signals, the AR capability in estimating the power spectrum density from short data is one of its great advantages. Generally speaking, the EEG data are generally considered to be nonstationary because their statistical characteristics change with time owing to the changes in the mental states. In order to handle this feature, one approach is to assume that over short time intervals the signal remains stationary. In this case, the spectrum of the signal can be estimated using the periodogram or correlogram, but it is known that they perform poorly with limited data. This makes the AR a reliable choice in solving the spectrum estimation issue of a quasi-stationary EEG signal.

2.9.2.1 Time and Frequency Domains Representation of EEG Signals Using AR

If we set the order of the MA to zero ($q = 0$) in Equation 2.55, we get the EEG data modeled by the autoregressive model as

$$x(n) = -\sum_{k=1}^{p} a_p(k)x(n-k) + w(n) \tag{2.56}$$

where $b_q(0) = 1$, $w(n)$ is a white noise process with zero mean called the input driving noise, and $a_p(1)$, $a_p(2)$, ..., $a_p(p)$ are called the autoregressive model parameters.

If Equation 2.56 is rearranged as

$$x(n) + \sum_{k=0}^{p} a_p(k)x(n-k) = -w(n) \tag{2.57}$$

and the Fourier transform is applied on both sides of equation, we get

$$X(e^{j\omega}) + \sum_{k=1}^{p} a_p(k)X(e^{j\omega})e^{-jk\omega} = -W(e^{j\omega})$$

$$X(e^{j\omega})\left(1 + \sum_{k=1}^{p} a_p(k)e^{-jk\omega}\right) = -W(e^{j\omega})$$

$$\tag{2.58}$$

$$X(e^{j\omega}) = \frac{-W(e^{j\omega})}{\left(1 + \sum_{k=1}^{p} a_p(k)e^{-jk\omega}\right)}$$

Then the power spectral density of the $x(n)$ process is given as

$$\left|X\left(e^{j\omega}\right)\right|^2 = \frac{\left|-W\left(e^{j\omega}\right)\right|^2}{\left|\left(1+\displaystyle\sum_{k=1}^{p} a_p(k)e^{-jk\omega}\right)\right|^2}$$

$$= \frac{\sigma_w^2}{\left|\left(1+\displaystyle\sum_{k=1}^{p} a_p(k)e^{-jk\omega}\right)\right|^2}$$

(2.59)

where σ_w^2 is the variance of the input driving white noise, $w(n)$.

2.9.2.2 Selecting the Order of the AR Model

One of the critical issues in AR spectrum estimation or data fitting is how to select the order p of the AR process. If the model order is too small, the error in fitting the data will be large and the spectrum will be smoothed, resulting in poor resolution. If, on the contrary, the model order is too large, then the spectrum may have spurious peaks and may lead to spectral line splitting [14]. Therefore, it would be useful to have a criterion that indicates the appropriate model order to be used with the given set of data. Several criteria have been proposed, among them the *Akaike's Information Criterion (AIC)* [26]. The $AIC(p)$ criterion is given as

$$AIC(p) = N \log \varepsilon_p + 2p \tag{2.60}$$

where ε_p is the energy of the modeling error, obtained as

$$e(n) = x(n) - \hat{x}(n)$$

$$\varepsilon_p = \sum_{n=0}^{N-1} \left|e(n)\right|^2$$

(2.61)

where $\hat{x}(n)$ is the estimated value of $x(n)$. The idea, then, is to select the value of p that minimizes $AIC(p)$. It is useful to indicate that the AIC gives too small an estimation of the order if the modeled process is nonautoregressive and tends to overestimate the order as N increases [6].

In summary, using AR to model the EEG data in the time and frequency domains involves three major steps. In step one, the appropriate model order should be obtained. In step two, an estimation of the AR(p) parameters is made. In step three, the estimated parameters are either inserted into Equation 2.56 for time-domain modeling or into Equation 2.58 for frequency-domain modeling.

Among the aforementioned implementation steps of the AR for EEG data modeling, step two is of major importance. A large body of material has been developed on the estimation of AR(p) parameters during the past four decades. The majority of the proposed methods utilize the relationship between the autoregressive model and the *linear predictor* (LP) in estimating AR(p) parameters [27].

2.9.3 The Relationship between the AR Process and the Linear Prediction

Suppose that we are given the EEG samples $x(0)$, $x(1)$, ..., $x(N-1)$. The forward linear prediction (FLP) of the sample $x(n)$ is given as

$$\hat{x}_L^f(n) = -\sum_{k=1}^{L} f_L(k) x(n-k) \qquad (2.62)$$

where $f_L(1), f_L(2), ... f_L(L)$ are the FLP coefficients and L is the predictor order. The error of the FLP is defined as the difference between the accrual and the predicted values, as given by

$$e_L^f(n) = x(n) - \hat{x}_L^f(n)$$
$$= x(n) + \sum_{k=1}^{L} f_L(k) x(n-k) \qquad (2.63)$$

If we rearrange Equation 2.63 as

$$x(n) = -\sum_{k=1}^{L} f_L(k) x(n-k) + e_L^f(n) \qquad (2.64)$$

then the similarity between the autoregressive model given in Equation 2.56 and the linear predictor becomes clear. Even with the similarity between Equation 2.56 and Equation 2.64, there are two major distinctions: (1) the sequence $w(n)$ of Equation 2.55 is a white noise process used as a driving sequence at the input of the autoregressive model, whereas the $e_L^f(n)$ sequence is the output of the forward linear prediction error (FLPE) filter; (2) the FLPE sequence is not, in general, a white process unless $x(n)$ is generated as an AR(p) process and the order of predictor (L) is equal to the order of the AR process [14].

In addition to a linear prediction formed in the forward direction, a *backward linear prediction* (BLP) given as

$$\hat{x}_L^b(n) = -\sum_{k=1}^{L} b_L(k)x(n-L+k) \tag{2.65}$$

may be formed. In similar way to the FLPE, the backward linear prediction error (BLPE) is given as

$$e_L^b(n) = x(n-L) - \hat{x}_L^b(n-L)$$

$$= x(n-L) + \sum_{k=1}^{L} b_L(k)x(n-L+k) \tag{2.66}$$

From now on, we shall consider that the EEG data are generated as an AR(p) process; thus, for a predictor with order $L = p$, the AR(p) parameters are simply given as the predictor coefficients.

Many algorithms addressing the problem of predictor coefficient estimation have been proposed. Among the proposed algorithms, Burg and the forward–backward linear prediction (FBLP) remain among the best. In the next two sections, we briefly describe these two algorithms.

2.9.3.1 Burg Algorithm

As a consequence of the Levinson recursion [28–31], we may relate the coefficients of a linear predictor of order L to the coefficients of a linear predictor of order L–1 as follows:

$$f_L(k) = f_{L-1}(k) + \Gamma_L f_{L-1}(L-k), \quad \text{for } 1 \leq k \leq L-1 \tag{2.67}$$

where $\Gamma_L = f_L(L)$ is called the *reflection coefficient*. By substituting Equation 2.67 into the FLPE and BLPE equations in Equations 2.63 and 2.66, respectively, we obtain the following alternative expressions for the FBLPE and the BLPE:

$$e_L^f(n) = e_{L-1}^f(n) + \Gamma_L e_{L-1}^b(n-1)$$

$$e_L^b(n) = e_{L-1}^b(n) + \Gamma_L e_{L-1}^f(n) \tag{2.68}$$

This special structure is known as a *lattice filter*. The Burg algorithm exploits the lattice structure to minimize an objective function defined as the mean-squared values of the forward and backward predictor errors, given as

$$\theta = E\left\{ \left| e_L^f(n) \right|^2 + \left| e_L^b(n) \right|^2 \right\} \tag{2.69}$$

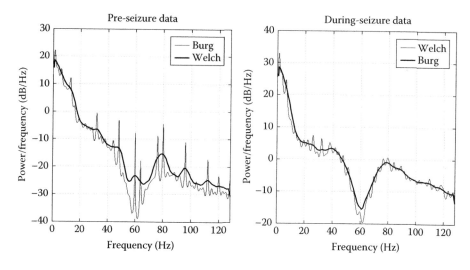

FIGURE 2.8
The power spectrum density of the pre-seizure and during-seizure data sets estimated by Burg and Welch.

By substituting the lattice structure in Equation 2.68 into the objective function and then differentiating θ with respect to Γ_L and setting the result to zero, an optimum value for the reflection coefficients is obtained [14]. Having obtained the optimum value of Γ_L, then by using the lattice structure in Equation 2.68 and the *Levinson recursion*, the coefficients of the forward linear predictor that minimizes the objective function in Equation 2.69 are obtained.

Figure 2.8 shows Burg estimation of the power spectrum density of the epileptic data set in Figure 2.5 before and after the seizure as compared with Welch.

The results in Figure 2.8 show clearly the superior capability of the parametric technique of Burg in showing the frequency components of the epileptic data in Figure 2.5 compared to the averaging technique of Welch.

2.9.3.2 The Modified Covariance Method

The basic idea behind this algorithm is to combine the prediction errors in the forward and backward directions, and then to perform the least squares minimization with respect to all of the linear prediction parameters. This algorithm was apparently proposed independently by Ulrich and Clyton [32] and Nuttal [33].

If $x(0)$, $x(2)$, ..., $x(N-1)$ is the available EEG data set, then the $2(N-L) \times L$ forward–backward LP data matrix is defined as

$$
D = \begin{vmatrix}
x(L-1) & x(L-2) & \cdots & x(0) \\
x(L) & x(L-1) & \cdots & x(1) \\
\vdots & \vdots & \cdots & \vdots \\
x(N-2) & x(M-3) & \cdots & x(M-L) \\
x(1) & x(2) & \cdots & x(L) \\
x(2) & x(3) & \cdots & x(L+1) \\
\vdots & \vdots & \cdots & \vdots \\
x(N-L) & x(N-L+1) & \cdots & x(N-1)
\end{vmatrix}
\tag{2.70}
$$

Let w denote the desired response at the predictor output, defined as

$$
w = [x(L) \quad x(L+1) \quad \cdots \quad x(N-1) \quad x(1) \quad x(2) \quad \cdots \quad x(N-L-1)]^T \tag{2.71}
$$

where $[.]^T$ is the vector transpose. Thus, the predictor equations can be expressed as

$$
Df = w \tag{2.72}
$$

where $f = [f_1 \, f_2 \cdots f_L]^T$. By solving the last equation using the least squares criterion, we obtain the predictor coefficients.

If D is full rank, then $D^T D$ is invertible, and the least squares solution of $\|Df - w\|_2^2$ is given by

$$
f = (D^T D)^{-1} D^T w
$$
$$
= D^+ w
\tag{2.73}
$$

where D^+ is the *pseudoinverse* of D.

If D is rank deficient with rank value $= r$, then the $\text{null}(D^T) \neq 0$, and the LS has an infinite number of solutions. This issue can be resolved by developing a *minimum norm solution* or the *general definition of the pseudoinverse*. This solution guarantees a unique least squares solution in terms of the minimum norm even with $\text{null}(D^T) \neq 0$.

2.9.4 MATLAB Functions for AR Parameters and Spectrum Estimation

In MATLAB, the following functions are used to estimate the AR(p) parameters using the foregoing two methods:

```
a = arburg(x,p)% Burg's method
a = armcov(x,p)% the modified covariance method
```

where x is the data vector that contains the EEG time series data, and p is the order of the AR model.

The following are MATLAB functions for the estimation of the PSD using the Burg and the modified covariance methods:

h = spectrum.burg(p) % Burg's method

h = spectrum.mcov(p)% the modified covariance method.

Hpsd = psd(h,x,'Fs',256);

2.10 Joint Time–Frequency Representation of EEG Signals

The classical spectrum techniques represented by the periodogram and correlogram are efficient when the signal is stationary over a long time. On the contrary, the parametric techniques are efficient when the signal is quasi-stationary or stationary over a short interval. Practically, EEG signals are in general a nonstationary type or process, which means that there will be problems in obtaining the frequency components from the classical spectrum estimators and to some extent the parametric-based estimators. In order to overcome this problem, the nonstationary EEG data can be divided into short durations over which the signal is considered stationary, and then the correlogram or periodogram is computed. The same applies to the parametric methods, where the model parameters are estimated from each division, and the spectrum is obtained. With the parametric methods, the use of a short window may result in poor estimation of the model parameters and a correspondingly inaccurate estimation of the spectrum. On the contrary, with the classical spectrum estimators, the use of a short window leads to poor spectral resolution. Nevertheless, if the length of the window is increased, the frequency resolution will improve, whereas the time resolution will deteriorate. The relationship between the time resolution and frequency resolution is an inverse relationship associated with the Heisenberg's uncertainty principle, which states that simultaneous improvement over time and frequency resolutions cannot be achieved [34]. This relationship is given as

$$\Delta f \Delta t \geq \frac{1}{4\pi} \tag{2.74}$$

where Δf and Δt are the frequency and time resolutions, respectively. An alternative representation of the signal called the *joint time–frequency representation* may provide better EEG information than is possible in either the time or frequency domains. Several *joint time–frequency representation* techniques have been proposed, among them the *short-time Fourier transform* (STFT) and the *wavelet transform* (WT).

2.10.1 The Short-Time Fourier Transform

With the STFT, the EEG data record is first segmented into short segments, and then a Fourier transform is applied on each segment after modifying it with the right window. Accordingly, the STFT can be expressed as

$$X(m,\omega) = \sum_{n=0}^{N-1} x(n)w(n-m)e^{-j\omega n} \qquad (2.75)$$

where the index m refers to the position of the window $w(n)$. In analogy with the periodogram, the spectrogram is defined as

$$S_x(n,\omega) = |X(n,\omega)|^2 \qquad (2.76)$$

Windows are typically selected in order to eliminate discontinuities at the block edges and to improve the spectral resolution.

2.10.2 Wavelet Transform (WT)

The Fourier transform uses exponential functions of the form $e^{j\omega n}$ as the orthogonal basis. With a Fourier transform of a time domain signal $x(n)$, we obtain accurate localization of the basis functions in the frequency domain but with no information about the times at which these frequencies occur. As outlined before, one of the solutions to this problem is to implement the STFT. However, the inverse relationship between time and frequency resolution makes it impossible to improve the resolution simultaneously over the two domains. This issue can be resolved by the wavelet transform.

The Morlet–Grossmann definition of the continuous wavelet transform of signal $x(t)$ is given as [35]

$$WT\{x(t)\} = \frac{1}{\sqrt{|a|}} \int_{-\infty}^{\infty} x(t)\Psi^*\left(\frac{t-b}{a}\right)dt \qquad (2.77)$$

where $\Psi(t)$ is a basic and unique function called the *mother wavelet*, $\{.\}^*$ denotes the complex conjugate, $a > 0$ is the *dilation* parameter, and b is the *translation* parameter. If $a > 1$, then Ψ is stretched along the time axis, and if $0 < a < 1$, then Ψ is contracted. The waveforms generated from the different values of a are called the *analyzing waveforms*. Wavelet coefficients represents the similarity between the wavelets at different dilations and translations and the signal x.

Among the various mother wavelets, or simply wavelets, the Morlet and the Mexican hat are the most popular. Morlet's wavelet is a complex waveform defined as

$$\Psi(t) = \frac{1}{\sqrt{2\pi}} e^{-t^2/2 + j2\pi b_0 t} \qquad (2.78)$$

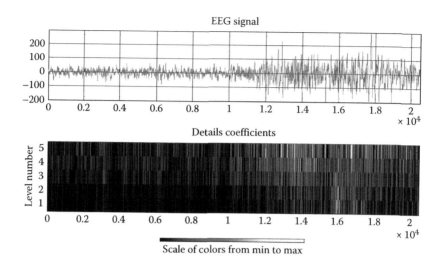

FIGURE 2.9

Wavelet coefficients of the epileptic data in Figure 2.5.

where b_0 is constant. The Mexican hat is defined as [36]

$$\Psi(t) = \left(1 - t^2\right) e^{-0.5t^2} \tag{2.79}$$

which is the second derivative of the Gaussian waveform.

Figure 2.9 shows the wavelet transform of the epileptic EEG signal of Figure 2.5, showing the change in frequency upon the start of the seizure. For a correct view of the time axis, it should be scaled by the sampling time ($T = 1/265$ s).

2.11 Nonlinear Descriptors of the EEG Signals

Generally speaking, the head acts as a mixing medium of a large number of locally generated EEG signals at the different positions of the sensors over the scalp. The changes in brain metabolism as a result of biological and physiological changes may alter the mixing process, making the brain, as a system, act on the signals in a nonlinear way. Analysis of such a complicated nonlinear system through system modeling is extremely difficult. Therefore, it appears that using the existing methods within the framework of nonlinear dynamics and higher-order statistics for the study of nonlinear systems, is a viable alternative. Where the nonlinear dynamics measure complexity (correlation dimension and stability), Lyapunov and Kolmogorov entropy quantifies critical aspects of the dynamics of the brain [34]. On the contrary, with the higher-order statistics (HOS), cumulants and the bispectrum measure the nonlinear interactions between the components of the signal or between signals.

2.11.1 Higher-Order Statistics for EEG Data Analysis

It is clear that power spectrum estimation based on second-order statistics is a powerful tool that provides sufficient information about the frequency content of the undergoing process. However, this is only applicable if the process is a linear Gaussian process and the phase is not of interest. When the process is nonlinear or when accurate detection of the phase differences is required, analysis going beyond the second-order statistics needs to be performed. Generally speaking, there are four major application areas for higher-order statistics: (1) suppress additive colored Gaussian noise with unknown PSD; (2) extract information due to deviations from Gaussianity; (3) reconstruct nonminimum phase information; and (4) characterize nonlinear properties in signals and identify nonlinear systems [37,38].

The first application area stems from the property that the *higher-order spectra*, known as *polyspectra* and defined in terms of higher-order statistics (*cumulants*), are zero for Gaussian signals. If a non-Gaussian signal, such as the event-related potentials (ERPs), is received along with Gaussian-colored noise, such as EEG signals, then the transformation of ERP + EEG into the higher-order cumulants will eliminate the EEG signal from the polyspectra. The second application area is based on the zero higher-order spectra of Gaussian signals, which enables the polyspectra to store information about the non-Gaussian components of the signals. This can be used in signal recognition, where distinct features can be extracted from the cumulants' domain. The third application area is based on the fact that polyspectra preserve the phase information of the signal regardless of whether the phase is minimum phase or not, contrary to second-order moments (autocorrelation), where phase information is preserved only if the phase of the signal is minimum. Finally, the cumulants and higher-order spectra can be used to analyze the nonlinearity of a system operating under random input. Polyspectra can play a major role in the detection and characterization of nonlinearity in a system from its output [38].

2.11.1.1 Definitions and Properties

In this section, the definitions and the properties of the higher-order statistics (moments and cumulants) and their spectra are introduced.

If $x(n)$ is a real stationary and its moments up to order k exist, then the k-order moment function is given as

$$M_k(\tau_1, \tau_2, \ldots, \tau_{k-1}) = E\{x(n)x(n+\tau_1)\ldots x(n+\tau_{k-1})\} \quad (2.80)$$

Clearly, the second-order moment, $M_2(\tau_1)$, is the autocorrelation of $x(n)$, whereas $M_3(\tau_1, \tau_2)$ is the third-order moment and $M_4(\tau_1, \tau_2, \tau_3)$ is the fourth-order moment.

The *kth*-order cumulants function of a non-Gaussian stationary random process $x(n)$ is given as

$$C_k(\tau_1,\tau_2,\ldots,\tau_{k-1}) = M_k(\tau_1,\tau_2,\ldots,\tau_{k-1}) - \tilde{M}_k(\tau_1,\tau_2,\ldots,\tau_{k-1}) \qquad (2.81)$$

where $\tilde{M}_k(\tau_1,\tau_2,\ldots,\tau_{k-1})$ is the *kth*-order moment of an equivalent Gaussian signal that has the same mean value and correlation sequence as $x(n)$. This is the reason why the cumulants have zero value if the process is Gaussian.

The following are the relationships between moments and cumulants for $k = 1, 2, 3$:

First-order cumulants:

$$C_1 = M_1 = E\{x(n)\} \qquad \text{(mean value)} \qquad (2.82)$$

Second-order cumulants:

$$C_2(\tau_1) = M_2(\tau_1) - M_1^2 \qquad \text{(covariance sequence)} \qquad (2.83)$$

Thus, we can see that the second-order cumulant sequence is the covariance, while the second-order moment sequence is the autocorrelation.

Third-order cumulants:

$$C_3(\tau_1,\tau_2) = M_3(\tau_1,\tau_2) - M_1\left[M_2(\tau_1) + M_2(\tau_2) + M_2(\tau_1 - \tau_2)\right] + 2M_1^3 \qquad (2.84)$$

The second- and third-order cumulants at zero lags are given as

$$\gamma_2 = E\{x^2(n)\} = C_2(0) \qquad \text{(variance)}$$
$$\gamma_3 = E\{x^3(n)\} = C_3(0,0) \qquad \text{(skewness)} \qquad (2.85)$$

Cumulants that are higher than third order are also important and have many applications. For example, if the process is symmetrically distributed (e.g., Laplace, uniform, Gaussian), then the third-order cumulants are zero; hence, we must use the fourth-order cumulants. In addition, in some cases the third-order cumulants are extremely small, and much larger values can be obtained through the fourth-order cumulants.

Figure 2.10 shows the third-order cumulants of the epileptic data of Figure 2.5, respectively, before and during the seizure.

Figure 2.10 shows clearly the significant increase in the values of third-order cumulants during epileptic seizure in comparison to their values before the start of the seizure.

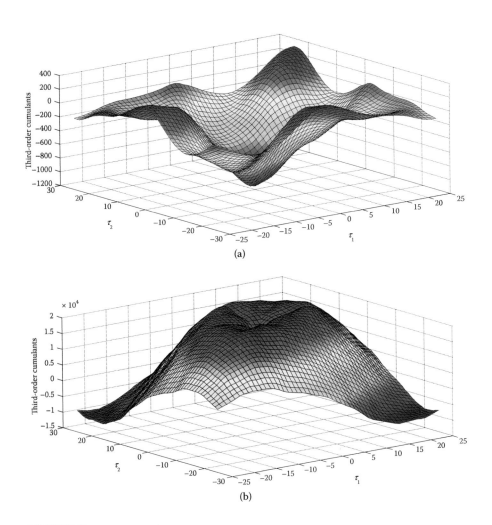

FIGURE 2.10
The third-order cumulants of the epileptic EEG data of Figure 2.5 (a) before the start of the
seizure, (b) during the seizure.

Higher-order spectra are defined either in term of moments (moment
spectra) or in term of cumulants (cumulant spectra). In the case of stochastic signals, cumulant spectra are usually used, and in the case of
deterministic signals, the moment spectra are used. Higher-order spectra
are multidimensional Fourier transform of the higher-order statistics. The
second-order cumulant-related spectrum is called the *power spectrum* and
is given as

$$C_2(\omega) = \sum_{\tau=-\infty}^{\infty} C_2(\tau)e^{-j\omega\tau} \tag{2.86}$$

whereas the third-order cumulant-related spectrum is called the *bispectrum* and is defined as

$$C_3(\omega_1, \omega_1) = \sum_{\tau_1=-\infty}^{\infty} \sum_{\tau_2=-\infty}^{\infty} C_3(\tau_1, \tau_2) e^{-j(\omega_1 \tau_1 + \omega_2 \tau_2)} \qquad (2.87)$$

where $|\omega_1| \leq \pi$, $|\omega_2| \leq \pi$, $|\omega_1 + \omega_2| \leq \pi$.

The bispectra are generally complex functions; that is, they have magnitude and phase.

Cumulant spectra are more efficient in analyzing random processes for the following reasons:

- The bispectrum is zero if the process is Gaussian, so it can be used as a measure of non-Gaussianity.

- In the case of a white noise process, the cumulants are multidimensional impulse functions, and bispectra are multidimensionality flat.

- The bispectrum quantifies the presence of quadratic phase coupling between two frequency components in the signal. Two frequencies' components are quadratically phase-coupled when a third component, whose frequency and phase are the sums of the frequencies and phases of the two components, respectively, exists in the bispectrum. Because the bispectrum represents the product of tuple of three Fourier components in which one frequency equals the sum of the two frequencies, peaks in the bispectrum indicate the presence of quadratic phase coupling.

Figures 2.11 shows the bispectrum of the epileptic EEG data in Figure 2.5 in 2D plots.

Figures 2.12 shows the bispectrum of the epileptic EEG data in Figure 2.5 in 3D plots.

The results in Figures 2.11 and 2.12 show clearly the higher values of the magnitude of the bispectrum and the greater concentration of its energy in certain bands during the seizure than before the start of the seizure.

The normalized bispectrum or third-order coherency index is called biocoherency and is given as

$$BIC(\omega_1, \omega_2) = \frac{C_3(\omega_1, \omega_2)}{\sqrt{C_2(\omega_1) C_2(\omega_2) C_2(\omega_1 + \omega_2)}} \qquad (\text{biocoherency}) \quad (2.88)$$

If all the frequencies are phase-coupled to each other (similar phases), then $C_3(\omega_1, \omega_2) = \sqrt{C_2(\omega_1) C_2(\omega_2) C_2(\omega_1 + \omega_2)}$ and $BIC(\omega_1, \omega_2) = 1$. If there is no coupling at all, the biocoherency is zero.

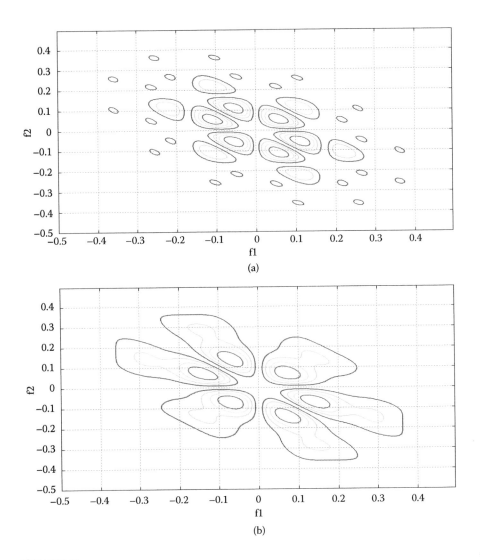

FIGURE 2.11
The 2D bispectrum of the epileptic EEG data of Figure 2.5 (a) before the start of the seizure, (b) during the seizure.

The biocoherency function is also useful in the detection and characterization of nonlinearities in a time series and in discriminating linear processes from nonlinear ones [38]. The signal is a linear non-Gaussian process of order three if the magnitude of the biocoherency function is constant over all frequencies; otherwise, the signal is said to be a nonlinear process [38].

In Figures 2.13 and 2.14, the bispectrum of the data in Figure 2.5 is shown, respectively, in 2D and 3D plots.

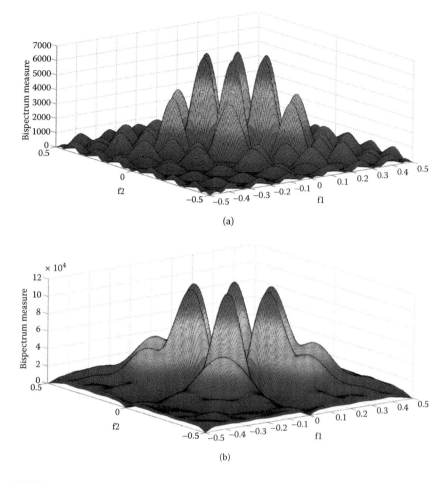

FIGURE 2.12
The 3D bispectrum of the epileptic EEG data of Figure 2.5 (a) before the start of the seizure, (b) during the seizure.

2.11.2 MATLAB Functions for AR Parameters and Spectrum Estimation

The following functions from the HOS toolbox of MATLAB are used to estimate the cumulants of EEG data:

```
n = 25;
for k = -n:n,
cmat(:,k+n+1) = cumest(x,3,n,128,0,'biased',k);
end
surf(-n:n,-n:n, cmat);
```

where x is the data vector that contains the EEG time series data.

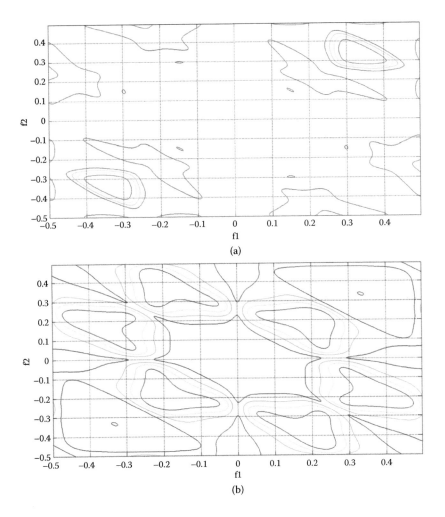

FIGURE 2.13
The 2D bispectrum of the epileptic EEG data of Figure 2.5 (a) before the start of the seizure, (b) during the seizure.

The following are MATLAB functions for the estimation of the bispectrum and biocoherency, respectively:

```
[Bspec,waxis] = bispecd (x, 256, 5, 8, 50);% Bispectrum
surf(waxis,waxis, abs(Bspec));%3D magnitude curve
[Bcoher,waxis] = bicoher (x, 256, 5, 8, 50);% Bicoherence
surf(waxis,waxis, abs(Bcoher));%3D magnitude curve
```

2.11.3 Chaos Theory and Dynamical Analysis

As was previously mentioned, the brain acts on the signals in a nonlinear way, making its analysis through system modeling an extremely difficult task.

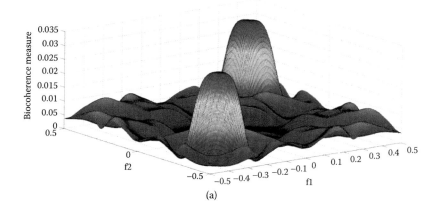

(a)

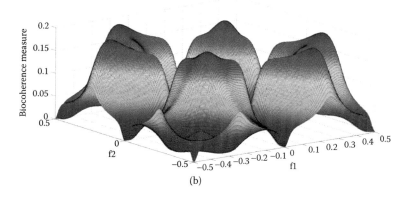

(b)

FIGURE 2.14

The 3D bicoherence of the epileptic EEG data of Figure 2.5 (a) before the start of the seizure, (b) during the seizure.

However, this section will show that a nonlinear dynamical system with chaotic behavior can be studied and understood from the measured EEG data, even if the model is unknown or too complex. This can be achieved through the deterministic chaos theory.

Chaotic systems have an apparently noisy behavior but are in fact ruled by deterministic laws. They are characterized by their sensitivity to initial conditions. This means that similar initial conditions give completely different outcomes after some time. Several methods were developed in order to calculate the degree of determinism (or random nature), complexity, chaoticity, and so on of these signals. Among these, the correlation dimension, Lyapunov exponents, and Kolmogorov entropy have been the most popular. In this section, the mathematical background of these methods and their application to the study of EEG signals have been discussed.

2.11.3.1 Definitions

State space: It is a common practice to generate the state space portrait of the system in order to visualize its multidimensional dynamics. A state space is created by treating each time-dependent variable of the system as one component in the vector space [39]. In classical mechanics, the state space is the space of all possible states of a physical system. By "state," we mean the positions of all the objects in the system and their velocities or momenta.

In general, the state space is identified with a topological manifold, and an n-dimensional state space is spanned by a set of n vectors called *embedding vectors*. Each point in the state space represents the instantaneous state of the system. However, for the most dynamic systems, the state vectors are confined to a subspace of the state space, and their evolution with time creates curves that are commonly called *attractors* [40].

Attractor: A set of states of a dynamic system toward which the system tends to evolve, regardless of the initial conditions of the system. According to their topology, several types of attractors can be distinguished.

Point attractor: The attractor consists of a single state. A typical example is a damped pendulum that has come to rest after some time.

Periodic attractor: The attractor consists of a finite set of states, where the evolution of the system results in cyclic movement through each state. The ideal orbit of a planet around a star is a periodic attractor. A periodic attractor is also called a *limit cycle*.

Strange attractor: It is an attractor that is sensitive to initial conditions and evolves through the set of possible physical states in an aperiodic (chaotic) way. Most real physical systems involve strange attractors, and signals corresponding to strange attractors have a random appearance.

The geometric characteristics of the attractors give valuable information about the nonlinear behavior of the system. Different methods were proposed to study the systems in state space, among them were the correlation dimension and the Lyapunov exponents [41,42].

2.11.3.2 Reconstruction of the State Space

In principle, in order to have a complete description of the nonlinear dynamics of a system, a simultaneous measurement of *all* the state variables is required. Unfortunately, in many real applications, one has no information about these variables and their values at certain times. Instead, we have measurements in terms of a time series of the output of the system, which has a relationship with the state variables of the system, but this relationship is not known. Fortunately, Taken [43] has managed to show that we may reconstruct the dynamics of a deterministic finite-dimensional system from its output measurements.

Let us denote the state space of a dynamic system as M. The state of the system at time n, denoted as $s_n \in M$, evolves according to $s_{n+1} = f(s_n)$. The system is observed using a smooth measurement function $y:M{\to}R$ giving a scalar time series $y_n=g(s_n)$. The purpose of the so-called method of delays is to reconstruct the state space M from the time series. Because M is high dimensional and each component of y_n is only one dimensional, it is evident that in order to obtain a suitable state space, we need to group different elements of the time series. The most straightforward way of doing this is to take successive y_n values to create a vector:

$$y_n = \begin{bmatrix} y_n & y_{n-\tau} & \cdots & y_{n-\tau(m-1)} \end{bmatrix} \qquad (2.89)$$

where τ indicates the interval of the time series that creates the reconstructed state space (lag time). Usually τ is chosen to be the first zero of the autocorrelation function or the first minimum of the mutual information [44]. Commonly, the number of components m is referred to as the *embedding dimension*. Although the data are embedded in m-dimensional space, it is not necessarily to fill in that space. Sometimes the system defines a nonlinear hypersurface in which the state variables exist. The dimension of this hypersurface is often referred to as a topological or local dimension, d. Takens proved that ideal systems (an infinite number of points without noise) converge to the real dimension if $m \geq 2d+1$ [43]. According to Sauer et al., in practical cases $m \geq d+1$ can be enough to reconstruct the original state space [45].

2.11.3.3 Correlation Dimension

The estimation of the local dimension d of an attractor from the corresponding data series is an important measure of its complexity. One of the estimators of d of an attractor is the *correction dimension D*.

The correlation dimension quantifies the complexity of attractor in the state space, where a larger correlation dimension corresponds to a larger degree of complexity and less self-similarity. Generally speaking, the correlation dimension has finite values for deterministic signals and infinite values for stochastic signals, so it is a good parameter for evaluating the deterministic or chaotic nature of a system.

The frequently used method to estimate the correlation dimension was introduced by Grassberger and Procaccia [46,47]. They defined the correlation sum for a collection of points ($i = 1,2, \ldots, N$) in some state space to be the fraction of all possible pairs of points that are closer than a given distance ε in a particular norm:

$$C(m,\varepsilon) = \frac{2}{(N)(N-1)} \sum_{i=1}^{N} \sum_{j=i+1}^{N} \Theta\left(\varepsilon - \left\| y_i - y_j \right\| \right) \qquad (2.90)$$

where Θ is the Heaviside step function, $\Theta(y) = 0$ if $y = 0$ and $\Theta(y) = 1$ if $y > 0$. The summation counts the pairs of points (y_i, y_j) whose distance is smaller the ε. In the limit of an infinite amount of data (i.e., $N \to \infty$) and for small ε, C scales like a power law: $C(\varepsilon) \propto \varepsilon^D$, where D is known as the correlation dimension. Thus, D is defined as

$$D = \lim_{\varepsilon \to 0} \lim_{N \to \infty} \frac{\partial \ln C(m, \varepsilon)}{\partial \ln \varepsilon} \tag{2.91}$$

It is clear that Equation 2.91 cannot be satisfied with real data, and an approximation has to be made.

2.11.3.4 Lyapunov Exponents

Another useful tool for characterizing the attractor is Lyapunov exponents, which provide a quantitative indication of the level of chaos of a system. It measures the exponential divergence (positive exponents: chaotic motion) or convergence (negative exponents: regular motion) of two initially close state space trajectories with time. In other words, they measure the degree of unpredictability of the future. This means that the more rapidly two trajectories diverge, the more chaotic the system and the more sensitive it is to initial conditions.

There are many different Lyapunov exponents for a dynamical system. The most important is known as the *maximal Lyapunov exponent* (L_{max}). Usually, only this exponent is computed.

If it is positive, trajectories will diverge; otherwise, they will get closer, reaching a nonchaotic attractor. Thus, for an attractor to be chaotic, at the very least the maximum Lyapunov exponent, L_{max}, should be positive. Lyapunov exponents also give an indication of the period of time in which predictions are possible, and this is strongly related to the concept of information theory and entropy.

Many techniques were proposed to estimate L_{max}, among them is the Wolf's algorithm [42]. With Wolf's algorithm, a set of m-dimensional embedding vectors that span the state space must be first constructed. These vectors are constructed as follows [43]:

$$s(t) = [\, s(t), s(t-\tau), \ldots, s(t-(m-1)\tau)\,] \tag{2.92}$$

where τ is a fixed time increment and m is the embedding dimension. After reconstruction of the state space through the embedding vectors, the nearest neighbor is searched for one of the first embedding vectors. Once the neighbor and the initial distance (L) is determined, the system is evolved forward for some fixed time (evolution time), and the new distance (L') is calculated. This evolution is repeated, and the distances are calculated until the separation becomes greater than a certain threshold. Then a new vector (replacement vector) with approximately the same orientation as the first neighbor and as close as possible to it is searched. Finally, Lyapunov exponents can be estimated using the following formula [34]:

$$L_{\max} = \frac{1}{N_a \Delta t} \sum_{i=1}^{N_a} \ln \frac{\left| \delta s_{i,j}(\Delta t) \right|}{\left| \delta s_{i,j}(0) \right|} w \tag{2.93}$$

where $\delta s_{i,j}(0) = s(t_i) s(t_j)$, $\left| \delta s_{i,j}(\Delta t) \right| = \left| s(t_i + \Delta t) - s(t_j + \Delta t) \right|$ represents the distance between the reference trajectory (fiducial) and the perturbed trajectory, $s(t_i)$ is a point at the fiducial trajectory, $\delta s_{i,j}(0)$ is the displacement vector at the initial time, Δt is the evolution time,
$t_i = t_0 + (i-1)\,\Delta t$, and $t_j = t_0 + (j-1)\,\Delta t$, where $i \in [1, N_a]$ and $j \in [1, N_a]$ with $j \neq i$. The term N_a represents the number of local $L_{\max}$ estimated every Δt within the duration T data segment.

2.11.4 Dynamical Analysis of EEG Signals Using Entropy

As previously outlined, a system is defined as chaotic if its attractor is sensitive to the initial conditions. This means that it evolves along two completely different trajectories even if the initial conditions are very close to each other. The entropy is another mathematical tool that can be used to assess the level of chaoticity in nonlinear systems, with higher entropy values representing a more chaotic system and lower values representing a close-to-deterministic system. On the contrary, the entropy may also be used to assess the uncertainty in signals, where higher values indicate less predictability (more noisy), and lower values indicate high predictability (less noisy). In EEG applications, entropy has been used to analyze epileptic seizures [48,49], Alzheimer's disease [50], and Parkinson disease [51]. Compared with the correlation dimension and Lyapunov exponents, the calculation of entropy requires less data.

Mathematically, the entropy of a discrete signal $x(n)$ is given as

$$\text{Entropy of } x(n) = -\sum_{n=0}^{N-1} p_{x(n)} \ln p_{x(n)} \tag{2.94}$$

where N is the number of samples, and $p\{.\}$ represents the probability distribution function. One of the simplest methods of obtaining the approximate probability distribution function of the data set $x(n)$ is to use the histogram [34].

So far, we have been discussing tools for single time series analysis (univariate) of the EEG signal. In the rest of this chapter, we extend the analytical tools of the EEG signals to two time series, which is known as bivariate analysis.

2.12 Bivariate Analysis of EEG Signals

Bivariate analysis is one of the simplest forms of the quantitative analysis of two variables in order to determine the relationship between them and accordingly to find to what degree they are related to each other. In this brief discussion of the bivariate analyses of EEG signals, we outline the normalized cross-covariance, coherency, and synchronization because they are the most popular among the bivariate descriptors in EEG applications.

2.12.1 Normalized Cross-Covariance Function

The cross-correlation function is a widely used measure of the interdependence between two signals in neuroscience. Suppose that $x(n)$ and $y(n)$ represent two real-time random variables, then the cross-correlation function is given as

$$r_{xy}(m) = E[x(n+m)\,y(n)] \qquad (2.95)$$

where $E\{.\}$ is the expected value operator. Frequently, the mean values of the two random variable x and y are subtracted, and the resultant formula is called the *cross-covariance*:

$$c_{xy}(m) = E\{[x(n+m)-\mu_x][y(n)-\mu_y]\} \qquad (2.96)$$

Although the covariance function in Equation 2.96 does provide information on the linear relationship between the two random variables x and y, the magnitude of the function does not reflect the strength of the relationship, because it is not scale free. There is a scale-free version of the covariance called the *normalized cross-covariance* that is widely used in statistics and given as

$$\tilde{c}_{xy}(m) = \frac{c_{xy}(m)}{\sigma_x \sigma_y} \qquad (2.97)$$

where σ_x and σ_y are the standard deviations of the random variables x and y, *respectively*. For any two random variables, we have

$$-1 \le \tilde{c}_{xy}(m) \le 1 \qquad (2.98)$$

If the two random variables are closely related to each other, the normalized covariance function will be close to +1, and if they are close but with opposite polarity, the normalized covariance function will take values close to −1. On the contrary, if the two random variables are weakly

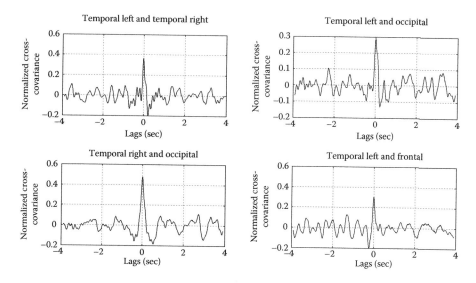

FIGURE 2.15
The normalized cross-covariance functions of EEG signals of the same seizure calculated between the different lobes of the brain.

related to each signal, the normalized covariance function will take values close to zero. However, it is important to indicate that the normalized covariance function can only capture the linear relationship between the random variables, and a value of zero simply means there is no linear relationship.

Different estimators were proposed to estimate the normalized covariance function in Equation 2.98, among them the *unbiased estimator*, which is given by

$$\tilde{c}_{xy}(m) = \frac{1}{N-|m|} \sum_{n=0}^{N-|m|-1} \left(\frac{x(n+|m|)-\mu_x}{\sigma_x} \right) \left(\frac{y(n)-\mu_y}{\sigma_y} \right), \quad |m| \le N-1 \quad (2.99)$$

where N is the number of samples.

In order to show the mutual relationships between the various lobes of the brain during an epileptic seizure, the normalized cross-covariance functions between the temporal, occipital, and frontal lobes for the left-temporal EEG data in Figure 2.5 are calculated and depicted in Figure 2.15.

2.12.2 The Coherence Function

The coherence function provides information about the linear relationships between the different frequencies of the signal. This is particularly interesting in neuroscience because it gives a more detailed picture of the mutual

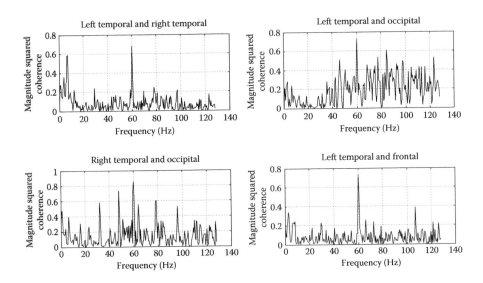

FIGURE 2.16
The coherency functions calculated between the different lobes of the brain.

interaction between the various frequencies in different areas than the normalized covariance function.

The coherence function is given by [52]

$$C_{xy}(\omega) = \frac{S_{xy}(\omega)}{\sqrt{S_{xx}(\omega)S_{yy}(\omega)}} \tag{2.100}$$

where S_{xx} and S_{yy} are the power spectra of signals x and y, and S_{xy} is the cross-power spectrum of the two signals.

The power spectral density of x and y can be estimated using any of the previously described techniques, whereas the cross-power spectrum density can be calculated as

$$S_{xy}(\omega) = S_{xx}(\omega)S_{yy}^*(\omega) \tag{2.101}$$

where $\{.\}^*$ indicate the complex conjugation.

Figure 2.16 shows the coherence functions between the various lobes of the brain during a seizure.

2.12.3 Phase Synchronization

In the classical sense, synchronization means the adjustment of frequencies of periodic self-sustained oscillators due to weak interaction. This effect is also referred to as *phase locking*. During the last 15 years, the notion of synchronization has been generalized to the case of interacting chaotic oscillators.

Where the interaction of two or more chaotic systems means that their states can coincide, while their dynamics in time remains chaotic [53,54]. This interaction between the chaotic oscillators may vary between high and slight synchronization. Generally speaking, the effect of *phase synchronization* of chaotic systems is mostly described in a similar way to synchronization of periodic oscillations, where only the phase locking is important, with no restriction on the amplitudes [55]. This allows one to describe both effects within a common framework.

Classically, the synchronization of two periodic oscillators x and y is understood through their *phase locking*, which reads as follows:

$$|\varphi_{n,m}(t)| < \text{contant}, \quad \text{where} \quad \varphi_{n,m}(t) = n\varphi_x(t) - m\varphi_y(t) \qquad (2.102)$$

where n and m are integers, and φ_x, φ_y are the phases of two oscillators, and $\varphi_{n,m}$ is the phase difference, or relative phase. Note that for the determination of synchronous states, it is irrelevant whether the amplitudes of both oscillators are different or not. We say that x and y are $m{:}n$ synchronized if the (n,m) phase difference of Equation 2.101 remains bounded for all t. And in most cases, only (1:1) phase synchronization is considered. The phase synchronization index is defined as follows [56]:

$$\gamma = \sqrt{\left\langle \cos\varphi_{xy}(t) \right\rangle^2 + \left\langle \sin\varphi_{xy}(t) \right\rangle^2} \qquad (2.103)$$

where the angle brackets denote the average over time. The phase synchronization index is close to zero if the phases are weakly synchronized and close to one for a close-to-constant phase difference.

References

1. A. Oppenheim and R. Schafer, *Discrete-Time Signal Processing*, Prentice Hall, Englewood Cliffs, NJ, 1999.
2. A. Oppenheim, A. Willsky, and H. Nawab, *Signals and Systems*, Prentice Hall, Englewood Cliffs, NJ, 1997.
3. A. Papoulis, *Probability, Random Variables, and Stochastic Processes*, 3rd ed., McGraw-Hill, New York, 1991.
4. C. W. Therrien, *Discrete Random Signals and Statistical Signal Processing*, Prentice Hall, Englewood Cliffs, NJ, 1992.
5. B. Porat, *Digital Processing of Random Signals*, Prentice Hall, Englewood Cliffs, NJ, 1994.
6. M. Hayes, *Statistical Digital Signal Processing and Modeling*, John Wiley, New York, 1996.

7. B. Picinbono, *Random Signals and Noise*, Prentice Hall, Englewood Cliffs, NJ, 1993.

8. D. C. Hammond, What is neurofeedback? *Journal of Neurotherapy*, Vol. 10, No. 4, pp. 11, 2006.

9. L. Zhang, W. He, X. Miao, and J. Yang, Dynamic EEG analysis via the variability of band relative intensity ratio: A time-frequency method, *Proceeding of the IEEE Engineering in Medicine and Biology Society*, Vol. 3, pp. 2664–2667, 2005.

10. E. Niedermeyer, *Electroencephalography: Basic Principles, Clinical Applications, and Related Fields*, 4th ed., Lippincott Williams & Wilkins, Philadelphia, PA, 1999.

11. M. Rangaswamy, et al., Beta power in the EEG of alcoholics, *Society of Biological Psychiatry*, Vol. 51, pp. 831–842, 2002.

12. H. Steinberg, T. Gasser, and J. Franke, Fitting autoregressive models to EEG time-series: An empirical comparison of estimates of the order, *IEEE Transaction on Acoustics, Speech, and Signal Processing*, Vol. 33, No. 1, pp. 143–150, 1985.

13. P. Stoica and R. Moses, *Spectral Analysis of Signals*, Prentice Hall, Englewood Cliff, 2005.

14. S. L. Marple, *Digital Spectral Analysis with Applications*, Prentice Hall, Englewood Cliff, NJ, 1987.

15. M. S. Bartlett, Smoothing periodograms for time series with continuous spectra, *Nature*, Vol. 161, pp. 686–687, 1948.

16. M. S. Bartlett, Periodogram analysis and continuous spectra, *Biometrika*, Vol. 37, pp. 1–16, 1950.

17. P. D. Welch, The use of FFT for the estimation of power spectra: A method based on time averaging over short, modified periodograms, *IEEE Transactions on Audio Electroacoustics*, Vol. 15, pp. 70–76, 1997.

18. A. Goldberger, L. Amaral, L. Glass, J. Hausdorff, and P. Ivanov, PhysioBank, PhysioToolkit, and PhysioNet: Components of a new research resource for complex physiologic signals, *Circulation*, Vol. 101, pp. e215–e220, 2000.

19. R. B. Blackman and J. W. Tukey, *The Measurement of Power Spectra from the Point of View Communication Engineering*, Dover, New York, 1959.

20. A. Sharma and R. Roy, Design of a recognition system to predict movement during anesthesia, *IEEE Transactions on Biomedical Engineering*, Vol. 44, pp. 505–511, 1997.

21. X. Kong, X. Lou, and N. Thakor, Detection of EEG changes via a generalized Itakura distance, *Annual International Conference of the IEEE Engineering in Medicine and Biology*, pp. 1540–1542, 1997.

22. C. W. Anderson, E. A. Stolz, and S. Shamsunder, Multivariate autoregressive models for classification of spontaneous electroencephalographic signals during mental tasks. *IEEE Transactions on Biomedical Engineering*, Vol. 45, pp. 277–286, 1998.

23. P. J. Franaszczuk and G. K. Bergey, An autoregressive method for the measurement of synchronization of interictal and ictal EEG signal, *Biological Cybernetics*, Vol. 81, pp. 3–9, 1999.

24. T. Cassar, K. Camilleri, and S. Fabri, Three-mode classification and study of AR pole variations of imaginary left and right hand movements, *Proceedings of the Biomed 2010 Conference*, Austria, 2010.

25. N. J. Huan and R. Palaniappan, Classification of mental tasks using fixed and adaptive autoregressive models of EEG signals, *Proceedings of the 2nd International IEEE EMBS Conference on Neural Engineering*, Arlington, VA, pp. 633–636, 2005.

26. H. Akaike, Information theory and an extension of the maximum likelihood principle, *Proceedings of the 2nd International Symposium on Information Theory*, 1973.

27. J. Makhoul, Linear prediction: A tutorial review, *Proceedings of the IEEE*, Vol. 63, No. 4, pp. 561–580, 1975.

28. S. Kay, *Modern Spectral Estimation*, Prentice Hall Englewood Cliffs, JN, 1987.

29. S. Marple, A new autoregressive spectrum analysis algorithm, *IEEE Transactions on Acoustics Speech and Signal Processing*, Vol. 28, No. 4, pp. 441–454, 1980.

30. J. Burg, Maximum entropy spectral analysis, *Proceeding of the 37th Meeting of the Society of Exploration Geophysicists*, 1967.

31. J. Burg, Maximum entropy spectral analysis, PhD dissertation, Department of Geophysics, Stanford University, Stanford, CA, 1975.

32. J. Ulrich and W. Clyton, Time series modeling and maximum entropy, *Physics of the Earth and Planetary Interiors*, Vol. 12, No. 2–3, pp. 188–200, 1976.

33. H. Nuttal, *Spectral Analysis of Univariate Process with Bad Data Points via Maximum Entropy and Linear Predictive Technique*, Technical Report, TR-5303, Naval Underwater System Center, New London, CT, 1976.

34. S. Tong and N. Thakor, *Quantitative EEG Analysis Methods and Clinical Applications*, Artech House, Norwood, MA, 2009.

35. A. Grossmann and J. Morlet, Decomposition of Hardy functions into square integrable wavelets of constant shape, *SIAM Journal on Mathematical Analysis*, Vol. 15, pp. 723–736, 1984.

36. M. Stephanie, *A Wavelet Tour of Signal Processing*, 2nd ed., Academic Press, New York, 1999.

37. C. L. Nikias and A. Petropulu, *Higher-Order Spectra Analysis*, Prentice Hall, Englewood Cliff, NJ, 1993.

38. C. L. Nikias and J. Mendel, Signal processing with higher-order spectra, *IEEE Signal Processing Magazine*, pp. 10–37, 1993.

39. A. Holden, *Chaos-Nonlinear Science: Theory and Applications*, Manchester University Press, Manchester, UK, 1986.

40. H. D. Abarbanel, *Analysis of Observed Chaotic Data*, Springer-Verlag, New York, 1996.

41. G. Mayer-Kress, *Dimensions and Entropies in Chaotic Systems*, Springer, New York, 1986.

42. A. Wolf, J. Swift, H. Swinney, and J. Vastano, Determining Lyapunov exponents from time series, *Physica D: Nonlinear Phenomena*, Vol. 16, pp. 285–317, 1985.

43. F. Takens, Detecting strange attractors in dynamic systems and turbulence, *Lecture Notes in Mathematics*, D. A. Rand and L. S. Young, Eds., Springer-Verlag, New York, pp. 366–376, 1980.

44. X. Jiang and H. Adeli, Fuzzy clustering approach for accurate embedding dimension identification in chaotic time series, *Integrated Computer-Aided Engineering*, Vol. 10, pp. 287–302, 2003.

45. T. Sauer, J. A. Yorke, and M. Casdagli, Embedology, *Journal of Statistical Physics*, Vol. 65, No. 3/4, pp. 579–616, 1991.

46. P. Grassberger and I. Procaccia, Characterization of strange attractors, *Physical Review Letters*, Vol. 50, pp. 346–349, 1983.

47. P. Grassberger and I. Procaccia, Measuring the strangeness of strange attractors, *Physica D: Nonlinear Phenomena*, Vol. 9, pp. 189–208, 1983.

48. T. Inouye, K. Shinosaki, A. Yagasaki, and A. Shimizu, Quantification of EEG irregularity by use of the entropy of power spectrum, *Electroencephalography Clinical Physiology*, Vol. 79, pp. 204–210, 1991.
49. X. Li, Wavelet spectral entropy for indication of epileptic seizure in extracranial EEG, *Lecture Notes in Computer Science*, Vol. 4234, p. 66, 2006.
50. D. Abasolo, et al., Entropy analysis of the EEG background activity in Alzheimer's disease patients, *Physiological Measurement*, Vol. 27, pp. 241–253, 2006.
51. B. Jelles, H. A. M. Achtereekte, J. P. J. Slaets, and C. J. Stam, Specific patterns of cortical dysfunction in entropy of the EEG, *Clinical Electroencephalography*, Vol. 26, pp. 188–192, 1995.
52. R. Challis and R. Kitney, The power spectrum and coherence function, *Medical and Biological Engineering Computing*, Vol. 29, pp. 225–241, 1991.
53. H. Fujisaka and T. Yamada, Stability theory of synchronized motion in coupled oscillator systems, *Progress of Theoretical and Experimental Physics*, Vol. 69, pp. 32–47, 1983.
54. A. Pikovsky, On the interaction of strange attractors, *Zeitschrift für Physik Condensed Matter*, Vol. 55, pp. 149–154, 267, 1984.
55. M. Rosenblum, A. Pikovsky, and J. Kurths, Synchronization of chaotic oscillators, *Physical Review Letters*, Vol. 76, pp. 1804–1807, 1996.
56. F. Mormann, K. Lehnertz, P. David, and C. E. Elger, Mean phase coherence as a measure for phase synchronization and its application to the EEG of epilepsy patients, *Physica D*, Vol. 144, No. 3–4, pp. 358–369, 2000.

3

Event-Related Potentials

Zhongqing Jiang

Liaoning Normal University

CONTENTS

3.1 Event-Related Potential Extraction 73
    3.1.1 The Averaging Technique..73
    3.1.2 Single-Trial Subspace-Based Technique 77
        3.1.2.1 Estimation of the Signal Subspace Dimension 82
        3.1.2.2 The Optimum Value of the Lagrange Multiplier 83
3.2 Brain Activity Assessment Using ERP 83
    3.2.1 Applications of P100 84
    3.2.2 Applications of N200 84
    3.2.3 Application of P300...85
3.3 Summary and Conclusions 87
References..88

3.1 Event-Related Potential Extraction

3.1.1 The Averaging Technique

In most event-related potential (ERP) studies, the averaging technique is used twice for ERP extraction. First, the trials are averaged separately for each experimental condition and for each participant. We can call this procedure *average within subject*; through this step, we get the *averaged file*. This step extracts the ERPs from the electroencephalogram (EEG) signal, because the EEG signal is usually larger, and the ERPs are contained in it, which prevents the ERPs from being directly revealed in one or too few trials. The second time, all the individual participants' ERP files are averaged into one file. We can call this step *average between subjects*; the result of this step is usually named as the *grand average file*.

Average within subject. Figure 3.1 illustrates this procedure. Although the signal of each trial (left column in Figure 3.1) has already gone through *filter, artifact detection,* and *bad channel replacement,* which help to eliminate the noise and other artifacts to some extent, these epochs varied from one to another, and it is hard to detect the ERP components. As these epochs

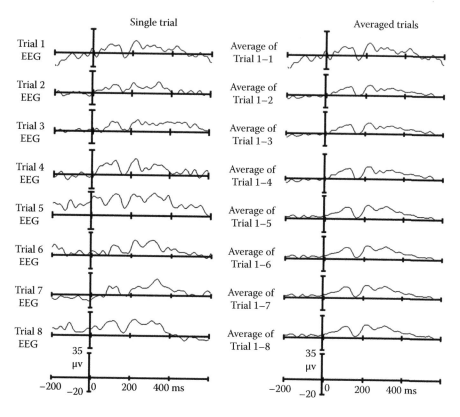

FIGURE 3.1

Example of application of signal averaging within a subject. The left column shows segments
of EEG for a single trial, which has already gone through filter, segment, artifact detection,
and bad channel replacement. The right column shows the effects of the average that includes
from one to eight trials. Data are collected over one sensor (electrode 65, near T5) on the scalp
of one participant watching a picture of an angry face. The EEG acquisition system is an EGI
EEG-ERP system with an HGSN 128 net.

are time-locked and aligned with respect to the stimuli, they can be averaged
together easily in a point-by-point manner. The right column of the figure
shows the ERP waveform gradually presenting when more and more trials
were averaged together.

The logic behind the averaging procedure is as follows. The EEG data of a
single trial consists of an ERP waveform and random noise. The ERP waveform is elicited by the stimulus, and it is assumed to be identical in shape
and phase, whereas the noise is assumed to be completely unrelated to the
time-locking event. Therefore, if several trials are averaged together, the ERP
waveform can be retained, whereas the noise would be canceled out to some
extent. As more and more trials are included for averaging, the noise remaining in the averaged waveform becomes smaller and smaller. Mathematically
speaking, if the amount of noise in a single trial is R and the number of trials

is N, then the size of the noise in an average of the N trials will be decreased to $R/\sqrt{N}$, while the signal elicited by the experimental stimulus is assumed to be unaffected by the averaging processing. That is, the signal-to-noise ratio (SNR) increases as a function of the square root of the number of trials.

However, the relationship between the number of trials and the SNR is rather sobering. Taking Figure 3.1 as an example, the difference between trial 1 alone and the average of trials 1 and 2 is obvious; the ERP waveform looks much better and more stable in the average of trials 1 and 2. Hereafter, the improvement is slow. Even the difference between the average of trials 1 to 8 and the average of trials 1 and 2 is not very obvious. In fact, although the number of trials included into these two averages differ by a factor of 4 (8:2), the SNRs of two averages differ by only a factor of 2 $\left(\sqrt{8}:\sqrt{2}\right)$. On the contrary, considering the participants' psychological variables, increasing the number of trials by repeated presentation of the stimuli is also a constraint, because the meaning of the stimulus might change for the participants if it repeats too many times; the participants may find the dull repetition unbearable. Therefore, improving the SNR of the data by decreasing sources of noise is much easier and more important than by increasing the number of trials.

Furthermore, we should remember that the basic assumptions for the technique of extracting an ERP signal by averaging trials are unrealistic. These assumptions are that (1) the ERP to the time-locking stimulus is the same on every trial and (2) only the EEG noise varies from trial to trial. First, if the EEG noise is due to some constant interference variable, it is not randomly presented and cannot be canceled by averaging. Second, the ERP to the stimulus could vary for many reasons even if the stimuli presented are all the same; the participant cannot react identically to the stimulus for the whole experiment. Therefore, there is also trial-to-trial variability in the amplitude and the latency of the ERP component.

In most cases, such violations of the assumption for the trials averaging are not problematic. For example, even if the amplitude of one ERP component varies from trial to trial, this component wave in the averaged ERP waveform will simply reflect the average amplitude of this component, which is meaningful for most studies. In fact, most studies base their conclusions on averaged data. However, trial-to-trial variability in latency is sometimes a significant problem, especially when the amount of latency variability differs across experimental conditions. If two experimental conditions differ in the amount of latency variability for some component, even if the amplitudes of this component are the same on the single trial, the amplitude of this component in the averaged waveform of these two experimental conditions will be different. If the ERP component has greater latency variability in one experimental condition, the amplitude of the ERP component in the averaged data for this experimental condition would be smaller.

There are several ways to mitigate the effect of latency variability, for example, (1) *by measuring the mean amplitude instead of the peak amplitude for one component in the averaged data.* Usually, the peak amplitude of one component is more sensitive to its latency jitter. It is feasible to cover one component in various trials by measuring the mean amplitude of a specific time window for this component. Sometimes we can use the *adaptive mean amplitude* instead to prevent the mixing of other components in one time window. Two steps are included in *adaptive mean amplitude* measurement. The first is to find the peak of one component; the second step is to measure the mean amplitude in a restricted time window (e.g., 80 ms) centering on the peak. (2) *Adopting response-locked instead of stimuli-locked averages.* In a response-locked average, the response rather than the stimulus is used to align the single-trial EEG epochs during the averaging process. This method is used when the participants' response time varies greatly from trial to trial, and particularly when the response times are correlated with the latency variations of the ERP component of interest. (3) *Time-frequency analysis.* When we perform traditional averaging for trials directly, some neural activities are lost because they are not phase locked to the time-locking event. The non-phase-locked oscillation information could only be obtained by time-frequency analysis; that is, by measuring the power of the neural oscillation activity in a specific frequency band on single trials and then averaging these power measures across trials. *Wavelet* analysis is one of the widely used methods of performing such a time-frequency analysis.

Average between subjects. An ERP experiment usually requires a sufficient number of participants. In most of the current publications, the participants range from 15 to 30 persons for one ERP experiment. Further, the conclusion of the paper is based on the averaged data of all these participants. On the contrary, the figure of this averaged ERP waveform would usually be presented on the publication. We call this averaged ERP waveform the *grand average* ERP waveform, which is used to refer to a waveform created by averaging together the averaged waveforms of the individual subjects. This procedure can be named the *average between subjects.* Similar to averaging trials that help mitigate some random noise in each trial, averaging the ERP of subjects helps to mitigate some random noise in each subject. Figure 3.2 shows the effect of averaging the ERP between subjects.

Figure 3.2 shows that the ERP waveforms become smaller and smoother as more subjects are included for averaging, because averaging masks the variability and spreads the similarity across the subjects. It helps to pick out the ERP component and to set the time window for measuring the mean amplitude of an ERP component. However, it is better to check whether this general time window setting is suitable for each single subject. Furthermore, we should pay attention to the fact that some of the subject's variability might not be attributed as noise. For example, participants' age, mindset, psychopathology, and personality might affect their ERP response. This field of individual difference is worthy of an ERP study.

In summary, the averaging technique is an easy and simple way to extract the ERP component. However, multiple trials are required and many participants as well, as it loses some non-phase-locked neural activity and masks

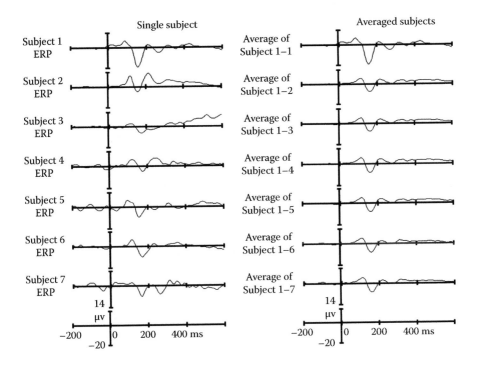

FIGURE 3.2

Example of the application of ERP averaging between subjects. The left column shows the averaged ERP for an individual participant; in addition, these averaged ERPs have also already gone through baseline correction. The right column shows the effects of the average that includes from one to seven subjects. The data collected over one sensor (electrode 65, nearby T5) on the scalp of one participant watching a picture of an angry face. The EEG acquisition system is an EGI EEG-ERP system with an HGSN 128 net.

some valuable variability. How to dig out the ERP information by using the minimum number of trials and participants is a fascinating problem for investigation.

3.1.2 Single-Trial Subspace-Based Technique

The ERPs are usually corrupted by the ongoing background EEG generated in the brain [1]. The EEG is a correlated type of noise (colored) and usually exists at much higher level than the ERP, with a typical SNR of −5 to −10 dB [2]. This makes the extraction of the ERP signal from the EEG noise a challenging issue. The conventional method of ensemble averaging does improve the SNR, but at the expense of a longer recording time causing discomfort and fatigue to the subject under study. In fact, hundreds of response trials need to be gathered to obtain a satisfactorily clean ERP estimate.

In this section, we present a single-trial ERP extraction algorithm based on the signal subspace called the generalized subspace approach (GSA) [1].

The algorithm first uses a linear estimator in order to estimate the clean ERP signal by minimizing signal distortion while maintaining the residual noise energy below some given threshold. Next, the generalized eigendecomposition of the covariance matrices of the ERP signal and brain background EEG noise is used to transform them jointly to diagonal matrices. The generalized subspace is then decomposed into the signal subspace and the noise subspace. Enhancement is performed by removing the noise subspace and estimating the clean signal from the remaining signal subspace.

For clarity, an attempt has been made to adhere to a standard notational convention. Lower case **boldface** letters will generally refer to vectors. Uppercase *BOLDFACE italic* letters will generally refer to matrices. Vector or matrix transposition will be denoted using $(.)^T$ and $(.)^*$ denotes conjugation for complex-valued signals. $\Re^{K \times K}$ denotes the real vector space of $K \times K$ dimensions.

The problem under consideration is to extract the clean ERP signal $x(n)$ from the degraded ERP signal $y(n)$. A K-dimensional vector of ERP samples x is defined as follows:

$$x = [x(0), x(1), \ldots, x(K-1)]^T \tag{3.1}$$

where $(.)^T$ denotes the transpose operation. The ERP samples are usually concealed in the ongoing background EEG generated in the brain. The EEG is a colored and additive form of noise that usually exists at a much higher level than the ERP, making the extraction of the ERP signal from the brain background noise a challenging process.

Let y denote the corresponding K-dimensional vector of the noisy ERP. Because the noise is assumed to be additive, we have

$$\mathbf{y} = \mathbf{x} + \mathbf{n} \tag{3.2}$$

where $y \in \Re^K$, $x \in \Re^K$, and $n \in \Re^K$ are the K-dimensional column vectors of noisy ERPs, clean ERPs, and additive EEG noise vectors, respectively.

Assume a K-dimensional vector of ERP samples x, defined in Equation 3.1, and a noisy ERP vector y, defined according to Equation 3.2. Let H be a $K \times K$ linear estimator of a clean ERP vector as follows:

$$\hat{x} = H \times y \tag{3.3}$$

The error signal obtained in this estimation is given by

$$\varepsilon = \hat{x} - x$$
$$= (H - I)x + H \times n \tag{3.4}$$
$$\overset{\Delta}{=} \varepsilon_x + \varepsilon_n$$

where ε_x represents the signal distortion and ε_n represents the residual noise [3]. The energies of the signal distortion $\overline{\zeta_x^2}$ and the residual noise $\overline{\zeta_n^2}$ are defined as follows:

$$\overline{\zeta_x^2} = tr\left(E\left\{\varepsilon_x \cdot \varepsilon_x^T\right\}\right)$$
$$= tr\left((H-I)R_x(H-I)^T\right)$$

(3.5)

and

$$\overline{\zeta_n^2} = tr\left(E\left\{\varepsilon_n \cdot \varepsilon_n^T\right\}\right)$$
$$= tr\left(HR_nH^T\right)$$

(3.6)

where R_x and R_n are the covariance matrices of the clean signal and noise vector, respectively, and $E(.)$ denotes the expected value. Because the signal distortion and residual noise cannot be simultaneously minimized, the intention is to keep the level of the residual noise below some threshold while minimizing the signal distortion. Accordingly, the optimum linear estimator is obtained from the following constraints:

$$H_{opt} \overset{\Delta}{=} arg\left\{\min_H \overline{\zeta_x^2}\right\},$$
$$\text{Subject to}: \frac{1}{K}\overline{\zeta_n^2} \leq \sigma^2$$

(3.7)

where σ^2 is a positive constant. Reducing the noise in a given noisy ERP signal causes some distortion in the ERP components. In the foregoing optimization, by decreasing the noise threshold level σ^2, we can decrease the amount of residual noise and therefore increase the amount of distortion and vice versa. The optimal distortion in the sense of Equation 3.7 can be found using *Kuhn–Tucker* necessary conditions for constrained minimization. Specifically, H is a stationary feasible point if it satisfies the gradient equation of the Lagrangian

$$L(H,\mu) \overset{\Delta}{=} \overline{\zeta_x^2} + \mu\left(\overline{\zeta_n^2} - K\sigma^2\right)$$

(3.8)

and

$$\left(\overline{\zeta_n^2} - K\sigma^2\right) = 0 \quad \text{for } \mu \geq 0$$

(3.9)

where μ is the Lagrangian multiplier. From $\nabla_H L(H,\mu) = 0$ and Equations 3.15 and 3.16, we obtain

$$H_{\text{opt}} = R_x \, (R_x + \mu R_n)^{-1} \tag{3.10}$$

From Equation 3.9, it is obtained that μ should satisfy

$$\sigma^2 = \frac{1}{K} tr\left(R_x \left(R_x + \mu R_n \right)^{-1} R_n \left(R_x + \mu R_n \right)^{-1} R_x \right) \tag{3.11}$$

Now let the eigendecomposition of R_x be defined as follows:

$$R_x = V_x D_x V_x^T \tag{3.12}$$

where D_x is a diagonal $K \times K$ matrix containing the eigenvalues of Rx; Vx contains its unitary eigenvectors called the inverse Karhunen–Loeve transform (IKLT) in the literature, and the unitary matrix V_x^T is called the Karhunen–Loeve transform (KLT).

In fact, the property of the KLT is that the covariance matrix of $V_x^T x$ (the transformation of the clean ERP signal vector) is diagonal, that is, D_x. Substituting Equation 3.12 in Equation 3.10, we have

$$H_{\text{opt}} = V_x D_x \left(D_x + \mu V_x^T R_n V_x \right)^{-1} V_x^T \tag{3.13}$$

Note that for white noise with variance σ_n^2, the covariance matrix of $V_x^T n$ (the transformation of the noise vector) is a diagonal matrix, equal to $\sigma_n^2 I$, and the aforementioned estimator reduces to the Ephraim and Van Trees estimator [4]. However, in colored noise, each component of $V_x^T n$ has a different variance, and accordingly the covariance matrix ($V_x^T n \, V_x$) is not diagonal. This is not surprising, because V_x is the eigenvector matrix of the symmetric R_x, which diagonalizes R_x and not R_n.

The solution to the colored noise problem is to use a transformation that jointly transforms the matrix pair (R_x, R_n) to diagonal matrices. This can be achieved by using the generalized eigendecomposition transformation.

Theorem 1. (Generalized Eigendecomposition) [5]. If $A \in \Re^{M \times M}$ is a positive-semidefinite matrix and $B \in \Re^{M \times M}$ is a positive-definite matrix, then an eigenvector matrix $V = [v_1 v_2 ... v_M] \in \Re^{M \times M}$ exists such that the v_j column vectors are orthogonal to each other and

$$V^T A V = D$$
$$V^T B V = I \tag{3.14}$$

where the eigenvalue matrix $D = diag(d_1, d_2, ..., d_M) \in \Re^{M \times M}$ with $d_1 > d_2 > ... > d_M$; $I(M \times M)$ is the identity matrix. Equation 3.14 can be manipulated so that

$$AV = BVD \qquad (3.15)$$

Now, according to Theorem 1, the generalized eigendecomposition of the matrix pair (R_x, R_n) is given as

$$R_x V = R_n VD \qquad (3.16)$$

where D is a diagonal $K \times K$ matrix that contains the generalized eigenvalues, and V contains the generalized eigenvectors. The generalized eigenvectors, V, will transform jointly both R_x and R_n to diagonal matrices of the following forms:

$$V^T R_x V = D \qquad (3.17)$$
$$V^T R_n V = I$$

Applying the foregoing generalized eigendecomposition of R_x and R_n, and using Equation 3.10, we can rewrite the optimal (in the sense of Equation 3.7) linear estimator for the colored noise case as

$$H_{opt} = R_n \; VDV^{-1} \left(R_n \; VDV^{-1} + \mu R_n \right)^{-1}$$

$$= \frac{R_n \left(VDV^{-1} \right)}{R_n \left(VDV^{-1} + \mu I \right)} \qquad (3.18)$$

$$= \frac{VDV^{-1}}{VDV^{-1} + \mu I}$$

Before the final form of the optimal estimator, H_{opt}, is considered, it is noteworthy that there is strong empirical evidence indicating that the transformed covariance matrices of most ERP signals transformed by the generalized eigenvectors of the pair (R_x, R_n) have some eigenvalues small enough to be considered as zeros. This can be verified by examining matrix D of the empirical covariance matrices, R_x and R_n. This means that the number of basis vectors for the ERP signal is smaller than the dimension of its vector.

The fact that some of the generalized eigenvalues of matrix D are close to zero indicates that the energy of the clean ERP vector is distributed among a subset of its coordinates, and the signal is confined to a subspace of the noisy Euclidean space. Because all noise-generalized eigenvalues are strictly positive, the noise fills the entire vector space of the noisy ERP. Hence, the vector space of the noisy ERP signal is composed of a signal-plus-noise subspace and a complementary noise subspace. The signal-plus-noise subspace or simply the signal subspace comprises vectors of the clean signal as well as those of the noise process. The noise subspace contains vectors of the noise process only.

Consider the covariance matrix of the clean ERP signal R_x, as a rank-deficient matrix with rank r. Assume matrix V ($K \times K$) as the generalized eigenvectors of the pair (R_x, R_n). Let us partition the generalized eigenvectors matrix V into r vectors that span the signal subspace and $(K - r)$ vectors that span the noise subspace. Mathematically, this can be expressed as $V = [v_1 \ v_2]$, where $V_1 = [v_1 \ v_2 \ \dots \ v_r]$ is a $K \times r$ matrix whose rank is r, and $V_2 = [v_{r+1} \ v_{r+2} \ \dots \ v_K]$ is a $K \times (K - r)$ matrix. Having established the basis of the signal and noise subspaces through the above partition of the eigenvectors, we have to find the oriented energy measured in the direction of both subspaces. This can be achieved by partitioning the matrix of the generalized eigenvalues of (R_x, R_n) as $D = [D_1 \ D_2]$, where $D_1 = diag(d_1, d_2, \dots, d_r)$ and $D_2 = diag(d_{r+1}, d_{r+2}, \dots, d_K)$. The diagonal elements of D_1 represent the oriented energy measured along the column vectors of matrix V_1, and the diagonal elements of D_2 represent the measured energy along the column vectors of V_2.

Having decomposed the vector space of the noisy ERP signal into a signal subspace spanned by the vectors of matrix V_1 with oriented energy D_1, and a noise subspace spanned by V_2 with oriented energy D_2, enhancement in the clean signal estimator H_{opt} in Equation 3.13 is performed by removing the noise subspace and estimating the clean signal from the remaining signal subspace. This results in the following form for the linear estimator in Equation 3.13:

$$H_{opt} = \frac{V_1 D_1 V_1^{-1}}{V_1 D_1 V_1^{-1} + \mu I} \tag{3.19}$$

To implement Equation 3.19 for H_{opt}, the rank of the signal subspace r and the optimum value of the Langrage multiplier μ are considered a priori known. In the next two sections, the optimum values of r and μ are discussed.

3.1.2.1 Estimation of the Signal Subspace Dimension

As stated before, the dimension of the signal subspace is not known a priori and should be estimated. Many criteria have been proposed over the past three decades for the estimation of the signal subspace. The Akaike information criteria (AIC) by Akaike [6], the minimum description length (MDL) by Schwartz and Rissannen [7,8], and the extended AIC by Wax and Kailath [9] are considered to be among the best. Testing the GSA with these criteria, we found no significant difference in its performance, so any one of them can be used to estimate the signal subspace dimension.

3.1.2.2 The Optimum Value of the Lagrange Multiplier

The other parameter that needs to be defined prior to the implementation of the proposed GSA is the Lagrange multiplier μ. Theoretically, finding the optimum value of μ is a complicated topic that is beyond the scope of this chapter. In practice, the GSA reaches its optimum performance, in terms of minimum errors in detecting the peaks P_{100}, P_{200}, and P_{300}, when $\mu = 8$ [1]. Accordingly, this value of μ will be considered for GSA implementation in the area of ERP detection.

3.1.2.2.1 Implementation of GSA Algorithm

Step 1. Compute the covariance matrix of the brain background colored noise R_n using the pre-stimulation EEG sample.

Step 2. Compute the noisy ERP covariance matrix R_y using the post-stimulation EEG sample.

Step 3. Estimate the covariance matrix of the noiseless ERP signal as $\hat{R}_x = R_y - R_n$.

Step 4. Perform the generalized eigendecomposition of the pair $\left(\hat{R}_x, R_n \right)$ as $\hat{R}_x V = R_n VD$.

Step 5. Use the AIC to estimate the dimension of the signal subspace r.

Step 6. Use the estimated r in step 5 and extract the basis vectors of the signal subspace V_1 and their related generalized eigenvalues D_1.

Step 7. Use V_1 and D_1 with Equation 3.41 to find the optimum linear estimator H_{opt}. Consider $\mu = 8$.

Step 8. Estimate the clean ERP signal as $\hat{x} = H_{opt} y$.

3.2 Brain Activity Assessment Using ERP

In the early years of ERP study, the focus was on discovering and understanding ERP components rather than using them to address questions of broad scientific interest. Around the mid-1980s, the ERP technique was widely used to answer various questions in cognitive neuroscience, which has developed explosively since then. The questions have changed from relatively general questions to the more specific step-by-step questions. In other words, ERP is not only used to explore the general process of cognition but is also applied in the field of clinical diagnosis, man–machine interaction, and so on.

3.2.1 Applications of P100

P100 is an early sensory-specific ERP component. For visual stimuli, it is usually identified at the lateral posterior electrodes as the first positive deflection appearing between 50 and 150 ms after stimulus onset [10]. It can provide important diagnostic information regarding the functional integrity of the visual system, and even provide diagnostic information for some psychophysiological disorders.

P100 is used for diagnosing the functionality of the visual system by its stable character in normal people. That is, the peak latency of P100 shows relatively little variation between subjects, minimal within-subject inter-ocular difference, and minimal variation with repeated measurements over time [11]. Therefore, having set up norms including standards of P100 measurement and data based on normal sample people, we can diagnose the dysfunction of the visual system by comparison with the norms.

Because P100 is sensitive to nonpathophysiologic parameters such as pattern size, pattern contrast, pattern mean luminance, refractive error, poor fixation, and so on, it is recommended that each laboratory establish its own normative values using its own stimulus and recording parameters; and the construction of a normal sample for laboratory norms should report the participants' age, gender, and inter-ocular asymmetry. Adult normative data cannot be generalized to pediatric or elderly populations; P100 in children should be recorded when the infant or child is in an attentive behavioral state and compared with appropriate age-related normal values [2].

There are also some comparison studies on P100 between people suffering from psychophysiological illness or disorders and the normal sample. The results might have implications for the diagnosis of these problems. For example, one study revealed that the amplitude of P100 was significantly larger in children with migraine headaches compared with other types of headaches. This result implies that P100 might be helpful in the work-up of a child with headache, particularly a young child, when the signs and symptoms may not be characteristic [2].

Previous study also revealed the P100 is abnormal in schizophrenic patients compared with the normal controls. Campanella et al. compared 14 schizophrenic patients with seven normal controls in an ERP study [4]. Participants were confronted with a visual face-detection task, in which they had to detect deviant faces among a train of standard stimuli (neutral faces) as quickly as possible. Deviant faces changed either with respect to emotion (same identity, happy, fearful, or sad expression) or with respect to identity (different identity, neutral expression). The results revealed that the schizophrenics exhibited a decrease in the amplitude of P100 in response to the deviant faces.

3.2.2 Applications of N200

N200 (or N2) is a negative-going component, which is typically evoked 180 to 325 ms following the presentation of a specific visual or auditory stimulus.

This specifically evoked stimulus is usually a deviation from the prevailing stimulus in the form or context. Several distinct N200 potentials have been characterized: N2a, N2b, and N2c. In repetitive stimulus-presentation, the N2a is a negativity of the anterior cortical distribution; it is evoked by either conscious attention to, or ignoring of, a deviating stimulus; in the auditory oddball detection paradigms, the N2a elicited by the deviating stimulus is more often called the mismatch negativity (MMN). The N2b is a negativity of the central cortical distribution and only presents during conscious stimulus attention; the N2b does not specifically reflect a departure from a collection of standard stimuli, but indexes a mismatch from the template, or a deviation from a mentally stored expectation of the standard stimulus. The N2c arises frontally and centrally during classification tasks. Furthermore, stimuli presented in visual search tasks with specific laterality and that are task relevant may evoke an N2c deflection, as an index of attentional shift, in the occipital–temporal region of the contralateral cortex [12].

Because N200 (N2) is closely related to the attention and inhibitory process, it has been widely used in diagnostic studies of attention-deficit/hyperactivity disorder (AD/HD). AD/HD refers to a variable cluster of hyperactivity, impulsivity, and inattention symptoms, which substantially affects the individual's normal cognitive and behavioral functioning. It can be categorized into three subtypes: the combined type (AD/HDcom), the predominantly inattentive type (AD/HDin), and the predominantly hyperactive/impulsive type (AD/HDhyp).

During a continuous performance test (CPT) in which participants respond to a target stimulus (e.g., when the letter X is preceded by a predefined cue stimulus such as the letter A), and inhibit responses to cued nontargets (e.g., when letters other than X follow the A), with cues requiring orienting and targets requiring selection for action, children with AD/HDcom were reported to show reduced frontal N1 and N2 amplitudes and parietal P2 and P3 amplitudes to target stimuli, indicating diminished evaluative and processing capabilities [13].

3.2.3 Application of P300

The classical P300 deflection, appearing as positivity, typically emerges approximately 300 to 400 ms following stimulus presentation. However, the duration of this component may range widely from 250 to 900 ms, with the amplitude usually varying from 5 to 20 μV for auditory and VEPs. Its typical subcomponents are the *P3a* and the *P3b*.

The *P3a*, typically with shorter latencies and a frontally oriented topography, is usually elicited by the more infrequently appearing stimulus of a two-stimulus oddball task, regardless of attentional (i.e., target or nontarget) status. It is supposed to reflect a passive comparator. The latency of the *P3b* is usually longer than that of the *P3a*, and its distribution is typically over and around the central parietal sites; the amplitude of the *P3a* will be markedly increased in response to the rare stimulus of an oddball experimental paradigm. The *P3a* is proposed to reflect a match/mismatch with a consciously maintained working memory trace by indexing memory storage as well as serving as a link between stimulus characteristics and attention; the *P3b* is the "classical P3"; in most of previous articles, the P300 is used as the synonym of the *P3b*. The P300 might be the ERP component that has been most extensively studied. Beginning in the 1970s, attempts have been made to objectify various aspects of the disease and its pathogenesis via analysis of the P300. Up to now, the P300 has been applied in a wide array of clinical research settings.

Many studies have revealed differences in the P300 between schizophrenics and normal controls. These differences are: the P300 effect size was smaller in amplitude and longer in latency in schizophrenics compared to normal controls. The paranoid subtype demonstrated larger P300 amplitude effect sizes than other disease subtypes, and the P300 latency effect size decreased with disease duration. Psychopathology severity and antipsychotic medications were unrelated to the P300 amplitude effect size [14]. One meta-analysis on 46 studies including 1443 patients and 1251 controls also confirmed that schizophrenics have severe deficits in their P300 amplitude [15].

P300 characteristics have been noted to differ in subjects who engage in addictive behavior. In a longitudinal study, the P300 amplitude was recorded for 1100 adolescent twins at the age of 14 from two midline electrodes during a visual oddball paradigm; semi-structured clinical interviews were used to investigate whether these children met the criteria for any symptoms of a substance use disorder. These data were used to examine the amplitude of the P300 changes in relation to early adolescent (i.e., by age 14) onset of pathological substance use (PSU), late adolescent onset of PSU (i.e., ages 14 to 18), misuse of different classes of substances (PSU–nicotine, PSU–alcohol, PSU–illicit), and the degree of PSU comorbidity. The results revealed that the amplitude of the P300 decreased in relation with the degree of drug class comorbidity, early adolescent onset PSU for all three substance classes, and late adolescent onset PSU for alcohol and illicit PSU [16].

P300 characteristics have even been noted to differ in subjects who are at risk of addictive behavior. For example, one study compared participants with high risk for alcoholism and participants with low risk for alcoholism with respect to the P300 [17]; the high-risk participants were sons of alcoholic fathers; the low-risk participants had no exposure to alcohol or other substances of abuse, and had no history of alcoholism, or other psychiatric disorder in first- or second-degree relatives; the results showed that, compared to a matched low-risk group, the boys with high-risk for alcoholism exhibited attenuated P300 amplitudes [17].

P300 characterization has also shed light on diseases linked etiologically to deep brain structures, including the basal ganglia, as well as clinically evident dysfunction of the superficial cerebral cortex, associated in particular with spreading and advanced disease. For instance, the anterior *P3a* is attenuated in amplitude in patients with Parkinson's disease [18], while the P3b was found to be significantly delayed for the Parkinson's disease group when compared with normal controls [19]. A marked reduction or absence of P300 differentiating target and standard stimuli in visual search tasks is observed in patients with Huntington's disease, which typically demonstrates caudate nuclear atrophy [20]. Patients with Alzheimer's disease, which classically affects temporal and associative cortex regions, exhibit prolonged P300 latency and attenuated P300 amplitude [21].

P300 latency may also be applied clinically as a diagnostic tool and a prognostic marker for recovery after cortical insult. A small study of patients with ischemic stroke has shown that a delay in P300 latency correlated with subclinical damage to the right parietal lobe. Furthermore, the value of the latency in the Cz derivation, nine days after the occurrence of the cerebral infarction, was related to the number of mistakes made by the subjects five months later in the attention test [22].

3.3 Summary and Conclusions

ERPs allow researchers to gain insight into the temporal-spatial characteristics of neural activity related to the component processes of behaviors such as selective attention, information coding, response selection, inhibitory control, and performance monitoring, which are usually respectively reflected by some ERP components with special temporal-spatial distributions. By providing a "microscope" into the sensory, affective, and cognitive processes, ERPs provide a deeper and more objective level of analysis over and above overt behavioral and task performance measures, allowing consideration of typical (e.g., developmental) and atypical processes (e.g., clinical compared to healthy groups). That is the potential of ERP for diagnosis application. Further, the diagnosis provided by ERP is a process diagnosis; that is, it can tell which stage of the participant's affective and cognitive process is abnormal.

When we adopt the ERP technique as a diagnostic tool, some norm data are usually needed for comparison. Therefore, the ERP norm data are critical for its clinical application. As ERP component characteristics (such as timing, amplitude, and topography) are sensitive to various factors such as the physical character of the stimulus, the participants' age, sex, and educational level, and so on, we should compare the participant's ERR results with the norm collected from the strictly matched group and the ERP acquisition with the identical stimulus, procedure, measurement environment, and analysis parameters. It is strongly suggested that each laboratory set up its own norm besides referring to the universal norms.

Furthermore, as many ERP studies are conducted that compare patients and normal subjects, it is time to move beyond describing the ERP deficits associated with the syndrome—rather, we should use ERPs to probe and understand the underlying brain dysfunctions producing the symptom. On the contrary, as dense array nets with more and more electrodes are used for EEG-ERP acquisition, the information on the neural activity will become dauntingly complex and large. Digging out and understanding this information needs good theory guided by advanced calculation techniques, and is a challenging and productive avenue for EEG–ERP study.

References

1. Kamel, N., Yusoff, Z., Fadzil, A., and Hani, B., Single trial subspace-based approach for ERP extraction, *IEEE Transactions on Biomedical Engineering*, Vol. 58, No. 5, pp. 1383–1393, 2011.
2. Lahat, E., Nadir, E., Barr, J., Eshel, G., Aladjem, M., and Biatrilze, T., Visual evoked potentials: A diagnostic test for migraine headache in children, *Developmental Medicine and Child Neurology*, Vol. 39, No. 2, pp. 85–87, 1997.
3. Ephraim, Y. and Van Trees, H. L., A signal subspace approach for speech enhancement, *IEEE Transactions on Speech and Audio Processing*, Vol. 3, No. 4, pp. 251–266, 1995.
4. Campanella, S., Montedoro, C., Streel, E., Verbanck, P., and Rosier, V., Early visual components (P100, N170) are disrupted in chronic schizophrenic patients: An event-related potentials study, *Neurophysiologie Clinique/Clinical Neurophysiology*, Vol. 36, No. 2, pp. 71–78, 2006.
5. Scheick, J. T., *Linear Algebra with Applications*, McGraw-Hill, New York, 1997.
6. Akaike, H., Information theory and an extension of the maximum likelihood principle, *Proceedings of the Second International Symposium on Information Theory, Supplement to Problems of Control and Information Theory*, pp. 267–281, 1973.
7. Schwartz, G., Estimating the dimension of a model, *Annals of Statistics*, Vol. 6, pp. 461–464, 1978.
8. Rissanen, J., Modeling by shortest data description, *Automatica*, Vol. 14, pp. 465–471, 1978.
9. Wax, M. and Kailath, T., Detection of signals by information theoretic criteria, *IEEE Transactions on Acoustics, Speech, and Signal Processing*, Vol. ASSP-33, No. 2, pp. 387–392, 1985.
10. Rutman, A. M., Clapp, W. C., Chadick, J. Z., and Gazzaley, A., Early top–down control of visual processing predicts working memory performance, *Journal of Cognitive Neuroscience*, Vol. 22, No. 6, pp. 1224–1234, 2010.
11. Odom, J. V., Bach, M., Barber, C., Brigell, M., Marmor, M. F., Tormene, A. P., and Holder, G. E., Visual evoked potentials standard, *Documenta Ophthalmologica*, Vol. 108, No. 2, pp. 115–123, 2004.
12. Patel, S. H. and Azzam, P. N., Characterization of N200 and P300: Selected studies of the event-related potential, *International Journal of Medical Sciences*, Vol. 2, No. 4, p. 147, 2005.

13. Lawrence, C. A., Barry, R. J., Clarke, A. R., Johnstone, S. J., McCarthy, R., Selikowitz, M., and Broyd, S. J., Methylphenidate effects in attention deficit/hyperactivity disorder: Electrodermal and ERP measures during a continuous performance task, *Psychopharmacology*, Vol. 183, pp. 81–91, 2005.

14. Jeon, Y. W. and Polich, J., Meta-analysis of P300 and schizophrenia: Patients, paradigms, and practical implications, *Psychophysiology*, Vol. 40, No. 5, pp. 684–701, 2003.

15. Bramon, E., Rabe-Hesketh, S., Sham, P., Murray, R. M., and Frangou, S., Meta-analysis of the P300 and P50 waveforms in schizophrenia, *Schizophrenia Research*, Vol. 70, No. 2, pp. 315–329, 2004.

16. Perlman, G., Markin, A., and Iacono, W. G., P300 amplitude reduction is associated with early-onset and late-onset pathological substance use in a prospectively studied cohort of 14-year-old adolescents, *Psychophysiology*, Vol. 50, pp. 974–982, 2013.

17. Begleiter, H., Porjesz, B., Bihari, B., and Kissin, B., Event-related brain potentials in boys at risk for alcoholism, *Science*, Vol. 225, No. 4669, pp. 1493–1496, 1984.

18. Tsuchiya, H., Yamaguchi, S., and Kobayashi, S., Impaired novelty detection and frontal lobe dysfunction in Parkinson's disease, *Neuropsychologia*, Vol. 38, No. 5, pp. 645–654, 2000.

19. Lagopoulos, J., Gordon, E., Barhamali, H., Lim, C. L., Li, W. M., Clouston, P., and Morris, J. G., Dysfunctions of automatic (P300a) and controlled (P300b) processing in Parkinson's disease, *Neurological Research*, Vol. 20, No. 1, pp. 5–10, 1998.

20. Munte, T. F., Ridao-Alonso, M. E., Preinfalk, J., Jung, A., Wieringa, B. M., Matzke, M., Dengler, R., and Johannes, S., An electrophysiological analysis of altered cognitive functions in Huntington disease, *Archives of Neurology*, Vol. 54, No. 9, p. 1089, 1997.

21. Pokryszko-Dragan, A., Słotwiński, K., and Podemski, R., Modality-specific changes in P300 parameters in patients with dementia of the Alzheimer type, *Medical Science Monitor: International Medical Journal of Experimental and Clinical Research*, Vol. 9, No. 4, p. CR130, 2003.

22. Alonso-Prieto, E., Lvarez-Gonzalez, M. A., Reyes-Verazain, A., Fernández-Concepción, O., Barros-García, E., and Pando-Cabrera, A., Use of event related potentials for the diagnosis and follow up of sub clinical disorders of sustained attention in ischemic cerebrovascular disease, *Revista de Neurologia*, Vol. 34, No. 11, pp. 1017–1020, 2001.

4

Brain Source Localization Using EEG Signals

Munsif Ali Jatoi
Universiti Teknologi PETRONAS

Tahamina Begum
Universiti Sains Malaysia

Arslan Shahid
Universiti Teknologi PETRONAS

CONTENTS

4.1 Introduction 91
    4.1.1 Forward and Inverse Problem 96
    4.1.2 Localization Techniques 99
        4.1.2.1 Minimum Norm 99
        4.1.2.2 Low Resolution Electromagnetic Tomography
                (LORETA) 100
        4.1.2.3 Focal Underdetermined System
                Solution (FOCUSS) 101
        4.1.2.4 Recursive Multiple Signal Classification (MUSIC) 102
        4.1.2.5 Hybrid WMN 103
        4.1.2.6 sLORETA 104
        4.1.2.7 eLORETA 106
        4.1.2.8 WMN-LORETA 107
        4.1.2.9 Recursive sLORETA-FOCUSS 108
        4.1.2.10 Shrinking LORETA-FOCUSS 110
        4.1.2.11 Summary with Comparison Tables 110
4.2 Recommendations for the Future 115
References 119

4.1 Introduction

The human brain generates magnetic signals or electric current when certain parts of the brain are active. Localization of brain signal sources corresponds to the estimation of their location and orientation with a minimum squared

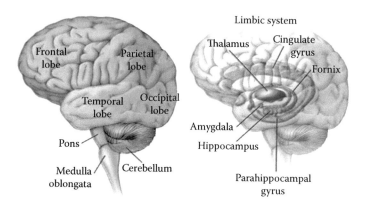

FIGURE 4.1
Brain with the main parts defined.

error between the calculated and measured boundary potential. This process of source estimation, which is also known as the inverse problem, is helpful in understanding the physiological, pathological, mental, and functional abnormalities, and cognitive behavior of the brain. This understanding leads to specifications for diagnoses of various brain disorders such as epilepsy and tumors. The human brain anatomy is shown in Figure 4.1 [1].

Functional imaging techniques are widely used for brain study with high temporal and spatial resolution. These techniques include single photon emission computer tomography (SPECT), positron emission tomography (PET), functional magnetic resonance imaging (fMRI), magneto encephalography (MEG), and electroencephalography (EEG) [2]. The functionalities of these methods are used to discover various diseases related to the central nervous system (CNS) that are responsible for brain dysfunction [3]. Epilepsy is the most important and also the most common neurological disorder related to brain studies, as 1% of the world population is suffering from it. Because epileptic activity propagates very fast, several hyper regions are seen in the images; therefore, a method with high temporal resolution is required to cope with this problem. For this, EEG is regarded as best noninvasive diagnostic tool used at epilepsy surgery centers as it has temporal resolution in milliseconds [4]. The comparison between various neuroimaging techniques in regard to their resolution (temporal and spatial) is shown in Figure 4.2 [5].

EEG is a neuroimaging technique that was developed by Hans Berger, a German physicist, in 1929, which was used to measure the electrical activity of the brain with the help of few electrodes [6–8]. Berger was interested in cerebral localization, particularly the localization of tumors in the brain [9]. However, Kornmuller was the first neuroscientist who discovered the significance of using multichannel recordings for the coverage of a wider brain region.

EEG can be defined as "a noninvasive/invasive neuroimaging technique having a high temporal resolution and a low spatial resolution that records

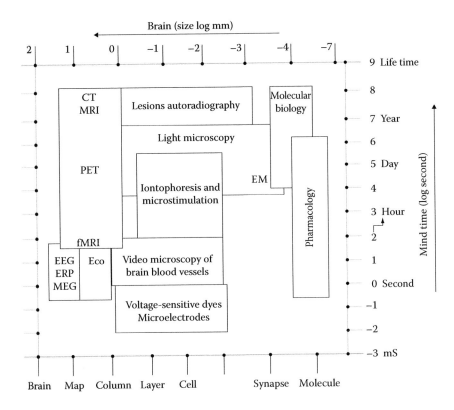

FIGURE 4.2
Overview of neuroimaging techniques with respect to temporal and spatial resolution.

the brain activity by measuring electrical signals generated with the help of electrodes placed on the scalp in order to diagnose/analyze different neural disorders (epilepsy, tumors, locating head injuries, etc.)." In other words, EEG is a functional imaging technique that has millisecond temporal resolution with measured potential differences as linear functions of source strengths and nonlinear functions of dipole locations. The EEG recordings can be used for direct, real-time monitoring of spontaneous and evoked brain activity, which allows for spatiotemporal localization of neuronal activity [10].

The source localization of seizure onset is very important for presurgical evaluation in epilepsy patients for neurophysiologists, neurologists, and neurosurgeons. Here, scalp EEG and long-term video EEG monitoring are biomarkers to localize the source of seizure where electrodes are placed by an international 10–20 system [11]. In recent times, although we have had more opportunity to use dense electrode caps (sensor net) to record EEG/ERP, the use of silver cup electrodes for the 10–20 system has not decreased. In both scalp EEG and video EEG monitoring, many technicians/neurologists are still using adhesive gel or collodion to put silver cup electrodes. Placement of all electrodes is based on underlying areas of the cerebral cortex. With the 10–20 system, the anterior temporal areas of the right and left sides are not covered by any electrodes. With this system, the source of epileptiform discharges is always missing if the source is from the anterior temporal areas (mesial temporal areas). T1 (left) and T2 (right), two extra electrodes, can localize the source of the ictal zone mainly for mesial temporal seizures more accurately than spheroidal electrodes [12]. Another investigation discovered that 59% of seizures come from bilateral (T1 and T2 both) electrodes, and 65.7% seizures arise from unilateral (T1 or T2) electrodes [13]. Therefore, the use of T1 and T2 electrodes are not ignored with the 10–20 system (Figure 4.3a). T1/T2 extra electrodes were used first by Silverman [14] before Chatrian et al. [15]. After that, T1/T2 electrodes were used more extensively to localize the source of seizure for epileptic patients. We strongly recommend that T1 and T2 electrodes should be used to avoid any misdiagnosis and mistreatment, though some neurologists are not aware about the use of these two electrodes. To put the silver cup electrodes, adhesive paste or collodion is used. However, to avoid some of the disadvantages of adhesive paste or collodion [16] for electrode placement to record scalp EEG and video EEG monitoring, more advanced dense electrode caps were used to record EEG. Nowadays, researchers use a higher number of channels with EEG acquisition systems; not only 128-channel (Figure 4.3b) but also 256-channel EEG systems [17] are used that can also cover T1/T2 electrode locations. The source localization of seizures, therefore, is now more accurate than before. Technology has given us more opportunities to locate the source of seizures (the inverse problem) with many algorithms.

The approximation for the location and distribution of the sources responsible for the electromagnetic activity inside the brain based on the potential recorded through the electrodes is one of the major problems in EEG. This problem is known as the source localization or the EEG inverse problem as the data (potentials) are given and one has to design the model from the available data. In other words, given a set of electric potentials from discrete sites on the surface of the head and the associated positions of those measurements and the geometry and conductivity of regions within head, the location and magnitude of the current sources within the brain are calculated [18].

Source modeling by using EEG signals for noninvasive localization of epileptogenic zones helps in clinical applications such as surgery in patients with partial seizures [19]. Therefore, EEG source localization has been an active area of research for decades. In the past few years, the source localization method, owing to its application in clinical applications for epileptic surgery, has been the subject of more than 150 research publications [20]. These publications are on software-based mathematical solutions for the EEG inverse problem. However, fewer than half of the publications addressed the clinical validation for investigation of focal epilepsy [21].

This chapter deals with the basic concepts behind the EEG inverse problem and the various algorithms that have been proposed so far. The existing algorithms such as minimum norm, LORETA, sLORETA, eLORETA, MUSIC, FOCUSS, WMN-LORETA, hybrid WMN, and shrinking LORETA-FOCUSS

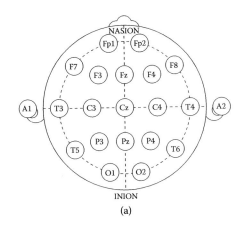

(a)

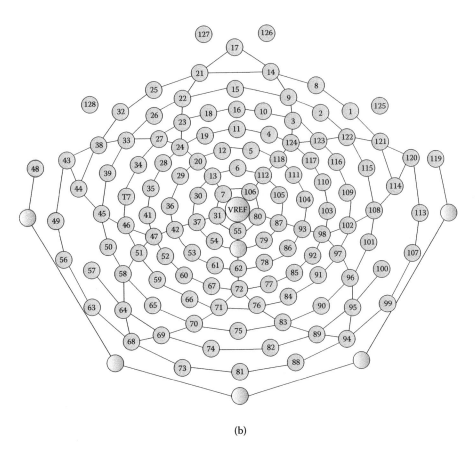

(b)

FIGURE 4.3
(a) Electrode locations shown in international 10–20 system. (b) One example of dense electrode sensor net of 128 channels.

are discussed thoroughly to give the reader a clear view of these methods. After a detailed discussion followed by the mathematical interpretation and physical meaning of each algorithm, the advantages and limitations of each algorithm are presented. Finally, tables are produced as results from the study for the various techniques employed.

The localization process involves the prediction of scalp potentials from the current sources in the brain (the forward problem) and the estimation of the location of the sources from scalp potential measurements (the inverse problem) [22]. The efforts to understand the localization problem began 40 years ago by correlating the existing body of electrophysiologic knowledge about the brain to the basic physical principles controlling the volume currents in conductive media [23–28].

4.1.1 Forward and Inverse Problem

In the physical world, if the data values are extracted/estimated from the given model with the help of physical theories applied to the model, then the problem is said to be a modelization problem, simulation problem, or forward problem [29]. This is a straightforward procedure that requires fewer computations with less error because the model with a complete description is with us. However, the inverse problem involves predicting the model with the help of available measured parameters.

In the physical world, a finite amount of data are available to reconstruct a model with infinitely many degrees of freedom. Hence, the inverse problem is not unique, and there are many models that can explain the data equally well. On the contrary, the forward problem has a unique solution. As an example taken from [29], consider measurements of the gravity field around a planet: given the distribution of mass inside the planet, we can uniquely predict the values of the gravity field around the planet (forward problem). However, there are different distributions of mass that provide exactly the same gravity field in the space outside the planet. Therefore, the inverse problem of inferring the mass distribution from observations of the gravity field has multiple solutions (in fact, an infinite number of solutions). Because of this, in the inverse problem, one needs to make explicit any a priori information on the model parameters. One also needs to be careful in the representation of the data uncertainties.

The inverse problem is nonunique in nature, which means many models can fit the data. Also, the estimated data are tainted with errors, and therefore the estimated model always differs from the true model. The model with finite degrees of freedom is termed a discrete model, and the model with continuous data and infinite degrees of freedom is termed a continuous model. The model estimation and model appraisal are different for both systems. According to Hadamard [30], if a physical problem has a solution with uniqueness and stability, then the inverse problem is assumed to be well posed; otherwise, it is termed an ill-posed inverse problem. The majority of geophysical problems are ill posed, which means they are characterized by nonuniqueness and instability. Therefore, it is assumed that the predicted model is just an approximation of the true model. The inversion problem consists of two steps, that is, an estimation problem and an appraisal problem [31]. Let the true data be denoted by d, the true model by m, the estimated model by $\tilde{m}$, and assume that the data are tainted by error. Then one can frame the inverse problem as a combination of what is to be estimated and a relationship between the estimated and the true models, as termed appraisal (Figure 4.4). Therefore,

Inverse problem = Estimated problem + Appraisal problem.

The EEG source localization is an underdetermined ill-posed inverse problem as the number of unknown parameters is greater than number of known parameters. There exist two general approaches for localization as proposed by researchers. In the first approach, the signals are assumed to be generated by a small number of focal sources. This approach is called the equivalent current dipole (ECD). However, if all possible source locations are assumed simultaneously, the approach is known as the linear distributed approach [32]. There are many existing methods to solve the EEG inverse problem, which can be categorized according to the methodology adopted for the implementation of each method. Some of the methods are defined independently. However, other methods are either hybrid methods that mix the existing algorithms with the other algorithms for better results with fewer errors and accurate localization of sources within the brain. Figure 4.5 shows the simplified EEG source localization model with EEG data on the left hand side and forward and inverse modelling at the right hand side.

Figure 4.6 shows the categorization of the various inverse methods of source localization using EEG signals.

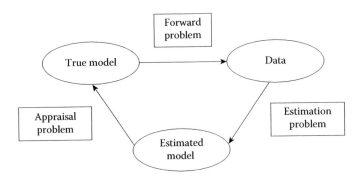

FIGURE 4.4
Inversion for a physical model. (From Snieder, R., and J. Trampert, *Inverse problems in geophysics*, in Wavefield inversion, edited by A. Wirgin, pp. 119–190, Springer Verlag, New York, 1999.)

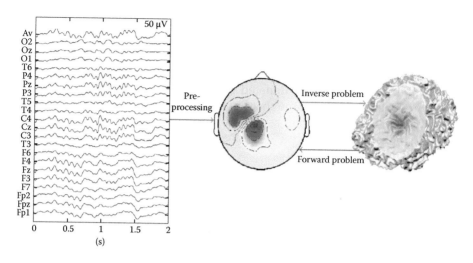

FIGURE 4.5

Inverse and forward model for EEG source localization. (From Wendel, K. et al., *Computational Intelligence and Neuroscience*, Hindawi Publishing Corporation, New York, 2009.)

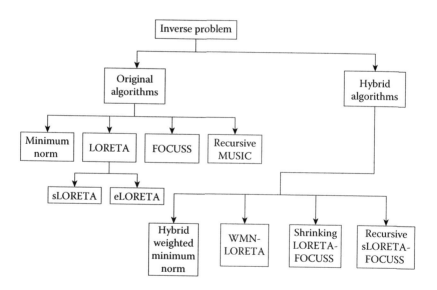

FIGURE 4.6

Flow diagram of inverse methods used for EEG source localization.

There are many solutions for the instantaneous, distributed, discrete, linear EEG inverse problem, starting with an article by Hamalainen and Ilmoniemi in 1984 titled "Interpreting Measured Magnetic Fields of The Brain: Estimates of Current Distributions" [33]. After this, different localization methods with various techniques and algorithms were developed.

These methods include minimum norm, weighted minimum norm (WMN), low resolution brain electromagnetic tomography (LORETA), standardized LORETA (sLORETA), recursive multiple signal classifier (MUSIC), recursively applied and projected MUSIC, shrinking LORETA-FOCUSS, hybrid WMN method, recursive sLORETA-FOCUSS, standardized shrinking LORETA-FOCUSS (SSLOFO), and others.

4.1.2 Localization Techniques

4.1.2.1 Minimum Norm

This solution for the distributed, discrete, and instantaneous EEG inverse problem was suggested by Hamalainen and Ilmoniemi in 1984 [33]. The EEG inverse problem is solved by proposing a linear combination of magnetometer lead fields as an estimate for the current distribution. The lead field L_i of a magnetometer at location I can be defined as

$$B_i(J) = \int L_i(r) \cdot J(r) dV \tag{4.1}$$

The linear relationship between the magnetometer readings, current distribution, and the lead fields can be expressed as follows:

$$B = LJ \tag{4.2}$$

Hence, the shortest current vector required to explain the magnetometer output can be defined by multiplying the output vector B by the pseudo-inverse of L such that

$$\hat{J} = L^+ B \tag{4.3}$$

where $L^+ = L^T(LL^T)^+$ denotes the Moore–Penrose generalized inverse [34]. The minimum norm solution was predicted for pure signals; signals contaminated by noise and smoothed noisy signals. The proposed method estimated sources with 1 cm resolution near the central sulcus and with about 4 cm resolution from the midline. However, it was suggested that the performance of the minimum norm solution would improve if some a priori information or assumption such as confining the integration area within the cortex was provided. Furthermore by increasing the number of magnetometers, better localization can be achieved.

Although the minimum norm provides good results in terms of resolution and current estimation, it fails to address the issue of deep source localization in the outermost cortex. This is because the minimum norm solution for EEG/MEG is a harmonic function, that is, $\nabla^2 J = 0$. Harmonic functions attain maximum values at the boundaries of their domain, which in this case is the outermost cortex. Upon comparison with newer techniques such as LORETA and WMN, the minimum norm solution has more localization error with the disadvantage of an inability to localize nonboundary sources [35].

4.1.2.2 Low Resolution Electromagnetic Tomography (LORETA)

Introduced and defined by Pascual-Marqui in 1994 [36], LORETA computes the current distribution throughout the full brain volume. This method assumes the simultaneous and synchronous excitation of neighboring neurons; that is, the current density at any given point on the cortex is maximally similar to the average current density of its neighbor. The discrete, 3D distributed, linear inverse solution is provided with a much better time resolution but low spatial resolution.

The generalized inverse problem for the LORETA can be defined and explained mathematically as

$$\min_J F_W \tag{4.4}$$

where

$$F_W = \left\| \Phi - KJ \right\|^2 + \alpha J^T W J \tag{4.5}$$

In this equation, the Tikhonov parameter $\alpha > 0$ is the control parameter used as a control for the relative importance of the penalty for being unfaithful to the measurements and a penalty for a large current density norm.

The solution is

$$\hat{J}_W = T_W \phi \tag{4.6}$$

The value of T can be calculated by

$$T_W = W^{-1} K^T \left(K W^{-1} K^T + \alpha H \right)^+ \tag{4.7}$$

The weight matrix $W \in R^{(3N_v) \times (3N_V)}$ is used to implement the discrete spatial Laplacian operator with the help of B. For LORETA, the weight matrix is calculated as

$$W = (\Omega \otimes I_3) B^T B (\Omega \otimes I_3) \tag{4.8}$$

where

$$\Omega_{\beta\beta} = \sqrt{\sum_{\alpha=1}^{N} k_{\alpha\beta}^T k_{\alpha\beta}} \tag{4.9}$$

for $\beta = 1...M \otimes$ defines the Kronecker product, and B is the matrix that implements the discrete spatial Laplacian operator for smoothness in the inverse solution.

In the just-defined set of equations, the inverse matrix B^{-1} implements the discrete spatial smoothing operator, d is the minimum inter-grid-point distance, and $diag(A_1 1_M)$ is a diagonal matrix whose entries are obtained from the matrix $A_1 1_M$. The set of equations explains the Laplacian operator used to implement LORETA. LORETA provides smooth and better localization for deep sources with less localization errors but with a low spatial resolution and blurred localized images of a point source with dispersion in the image. The low spatial resolution of LORETA is undesirable in some cases such as feature extraction of spatiotemporal pattern recognition, where a high resolution is needed. Also, it is shown in Ref. [36] that LORETA has a high localization capability for localizing the boundary sources: out of 819 cases, LORETA localized 383 cases with zero-localization error (47%).

4.1.2.3 Focal Underdetermined System Solution (FOCUSS)

This tomographic reconstruction technique for the solution of the ill-posed EEG/MEG inverse problem was proposed and explained in Ref. [37]. FOCUSS is a high-resolution nonparametric technique that uses the forward model which assigns the current to each element within a predetermined reconstruction region. It is recursive in nature, which means that the weights are iterated at each step from the solution of the previous step. The mathematical calculations for the recursive steps in FOCUSS are done with the help of the WMN method. The expression for the computation of the unknown current element I can be given as

$$I = W(GW)^+ B = WW^T G^T \left(GWW^T G^T \right)^{-1} B \tag{4.10}$$

where W is a dimensionless $n \times n$ matrix that can be altered to produce recursive schemes. It can be reconstructed by taking its diagonal elements to be the previous iterative step solution such that

$$W_k = \begin{bmatrix} I_{1K-1} & \cdots & 0 \\ \cdots & \cdots & \cdots \\ 0 & \cdots & I_{nk-1} \end{bmatrix}$$

where I_{ik-1} represents the ith element of the vector I at the $(k-1)$th iteration, and k is the index of the iteration step. The next weight matrix can be calculated by just multiplying W_{k-1} by W_k to get a new matrix. It can be observed that the diagonal elements of the weight matrix correspond to the current elements. With the help of the foregoing mathematical and theoretical concepts, simulations and comparisons are carried out between true, minimum norm, unbiased minimum norm, and FOCUSS algorithms for near-surface, mid-depth, and deep sources in Ref. [37].

The FOCUSS algorithm provides better localization capability as compared with the other algorithms and is able to handle nonuniquely defined localized energy sources. Furthermore, the FOCUSS algorithm has a better spatial resolution and is stable whenever subjected to changes.

4.1.2.4 Recursive Multiple Signal Classification (MUSIC)

This algorithm was proposed and explained in Ref. [38]. A single dipole is scanned through a grid confined to a 3D head or source volume. The forward model for the dipole at each grid point is projected against a signal subspace that has been calculated from the EEG measurements. The sources are located where the projection is best onto the signal subspace. However, one of the major problems with the MUSIC approach is the selection of the location that can provide the best projection in the practical case, as noise and errors are present in the signal subspace and forward model. The modified recursive MUSIC algorithm is the modified MUSIC algorithm as it can combat the limitations of MUSIC in terms of localizing synchronous sources through the use of the spatiotemporal independent topographies (IT) model.

The mathematical relationship relating the IT model and the signal subspace can be defined as [38]

$$F = AS^T + E \tag{4.11}$$

where E is a random error matrix comprising n time slices such that $E \equiv [e(t_1), ..., e(t_n)]$ added to the data $\tilde{F} = A(\rho, \theta)S^T$ produces F. The purpose is to determine the dipole location (rho) and orientation (theta) when F is given. Hence, the parameter estimation requires some mathematical manipulation of the true and estimated data by taking noise into consideration. Hence,

$$\left\{ \hat{\rho}, \hat{\theta} \right\}_{ls} = \arg\max \left\{ \left\| U_A^T \hat{\phi}_s \hat{\Sigma}_s \right\|_F^2 + \left\| U_A^T \hat{\phi}_e \hat{\Sigma}_e \right\|_F^2 \right\} \tag{4.12}$$

where ρ is the set of dipole locations, θ is the set of dipole orientations, U_A is the matrix whose columns are the left singular vectors of A that correspond to its nonzero singular values, contains eigenvectors such that span (ϕ_s) = span (A), (referring to signal subspace), is the weighted sum of the projections of the estimated signal subspace eigenvectors, and Σ_e is the weighted sum of the errors. The recursive MUSIC algorithm and its modifications such as RAP MUSIC [39] provide good localization with less complexity. Furthermore, the algorithm is extendable with a straightaway procedure rather than a sequential way.

4.1.2.5 Hybrid WMN

This algorithm was produced and explained in Ref. [40] as a hybrid algorithm with an initial reconstruction that was carried out with the help of the LORETA algorithm and then iterative calculations by using the FOCUSS algorithm. The LORETA algorithm is applied as it can localize the real current density distribution under the framework of the WMN method, which provides a smooth solution.

The discrete result for the inverse problem in this algorithm with the initialization of LORETA can be suggested as

$$\min_J \|BWJ\| \quad \text{Under constraint: } V = KJ \tag{4.13}$$

where V is an $N.1$ matrix comprising potential differences, J comprises current densities at M points within the brain volume, K is the lead field matrix defining the relationship between the scalps measurements and the current densities, and B is the discrete Laplacian operator. The weighted matrix W can be defined as

$$W = \Omega \otimes I \tag{4.14}$$

Here, I refers to the identity matrix, $\otimes$ denotes the Kronecker product, and Ω is an $M \times M$ diagonal matrix with the following diagonal elements:

$$\Omega_{ii} = \sqrt{\sum_{\alpha=1}^{N} k_{\alpha i}^T k_{\alpha i}} \tag{4.15}$$

Hence, the unique solution for the method can be described mathematically as

$$\hat{J} = \left(WB^T BW\right)^{-1} K^T \left(K\left(WB^T BW\right)^{-1} K^T\right)^+ V \tag{4.16}$$

where $\hat{J}$ is the estimated current density with all the parameters having the same definitions as defined earlier. A^+ is the Moore–Penrose inverse operator. If the regularization term (λ) is included in the foregoing approximation expression for the current density for more stability and less jamming, then one can rewrite the foregoing expression as

$$\hat{J} = \left(WB^T BW\right)^{-1} G^T \left(G\left(WB^T BW\right)^{-1} G^T + \lambda H\right)^+ V \tag{4.17}$$

The weighted iterative method is adapted to the forward problem to ensure the strengthening of the grid's energy in the solution space.

Hence, the weighted matrix W_K for the kth step can be calculated by the solution J_{k-1} of the $(k–1)$th step as

$$W_K = diag(J_{k-1}) \tag{4.18}$$

By the application of the WMN algorithm, the solution for the kth iteration is given as

$$J_K = W_K \left(K W_K \right)^+ V \tag{4.19}$$

The results in Ref. [40] are provided by using a four-shell (brain, CSF, skull, scalp) spherical head model with related electrical conductivities. The solution space has a radius of 0.84 with 729 grid points within it. The methods provide better results with initialization of LORETA and then iteration of the weight matrix. The LORETA provides a rough estimation of sources, and then the iterations improve the accuracy of the results, yielding in-depth localization with minimized errors and good estimation.

The disadvantage of this hybrid algorithm is heavy computations and repeated iterations that result in greater accuracy but at the expense of heavy calculations and time. Owing to continuous iterations, there is also a chance of loss of information and induction of noise.

4.1.2.6 sLORETA

This method is based on the assumption of the standardization of the current density, which implies that not only the variance of the noise in the EEG measurements is taken into account but also that the biological variance in the actual signal is considered [41]. This biological variance is taken as independently and uniformly distributed across the brain, resulting in a linear imaging localization technique having exact, zero-localization error. This localization technique resembles the method provided by Dale et al. [42] in which the localization is provided on a standardization of the estimates of current density. The current density estimates are given by the minimum norm method as in Dale, with the localization inference based on standardized values of the current density estimates. The way standardization for sLORETA is performed is different from Dale's method, resulting in zero localization for the sLORETA.

The mathematical formulation for the sLORETA is given as

$$F = \left\| \phi - KJ - c1 \right\|^2 + \alpha \|J\| \tag{4.20}$$

where ϕ = electrical potentials, K is the lead field matrix, J is the current density, $\alpha \geq 0$ is a regularization parameter. This functional has to be minimized with respect to J and c, for given K, ϕ, and α. By using average reference transforms of ϕ and K, the foregoing equation can be rewritten as

$$F = \|\phi - KJ\|^2 + \alpha\|J\|$$
(4.21)

with minimum

$$\hat{J} = T\phi \text{ where } T = K^T\left[KK^T + \alpha H\right]^+.$$
(4.22)

Therefore, for the standardized estimates of current density, the variance of the estimated value of $\hat{J}$ is to be calculated. So the electric potential variance $S_\phi \in \Re^{N_E \times N_E}$ can be expressed as

$$S_\phi = KS_J K^T + S_\phi^{Noise} = KK^T + \alpha H$$
(4.23)

From the preceding equation, the variance for the estimated current density can be given as

$$S_{\hat{J}} = TS_\phi T^T = T\left(KK^T + \alpha H\right)T^T = K^T\left[KK^T + \alpha H\right]^+ K$$
(4.24)

The sLORETA linear imaging method is

$$\sigma_v = \left[S_{\hat{J}}\right]_v^{-\frac{1}{2}} \hat{j}_v$$
(4.25)

where $\left[S_{\hat{J}}\right]_v \in \Re^{3 \times 3}$ is the vth 3×3 diagonal matrix in S_J, and $\left[S_{\hat{J}}\right]_v^{-\frac{1}{2}}$ is the symmetric square root inverse. The squared norm of σ_v corresponds to the estimate of the standardized current density power as

$$\sigma_v^T \sigma_v = \hat{j}_v^T\left[S_{\hat{J}}\right]_v^{-1} \hat{j}_v$$
(4.26)

The simulations are carried out by using the Talairach human brain atlas. A total of 6430 voxels at 5 mm spatial resolution were produced under these constraints. For each dipole, there exist three unknown values, making the number of unknowns $3 \times 6430 = 19{,}290$ with 25 electrodes. Different localization methods are compared with sLORETA, which include the minimum norm and that proposed by Dale et al. [42] in terms of localization errors and spatial spread. The simulations with noise and without noise demonstrate that sLORETA has far better quality with exact localization and zero-error localization as compared with the minimum norm and Dale methods, which shows that the sLORETA is a perfect first-order localization technique.

4.1.2.7 eLORETA

There have been many useful attempts to minimize the localization error by choosing the weight matrix in a more adequate way. However, there exists one methodology that gives greater importance to the deeper sources with reduced localization error. The study carried out in Ref. [43] shows that this method achieves depth weighting with reduced localization error from 12 to 7 mm. This method was developed and recorded as a working project in the University of Zurich in March 2005 [44]. According to Pascual-Marqui [43], eLORETA is a genuine inverse solution that provides exact localization with zero error in the presence of measurement and structured biological noise. Hence, the family of linear imaging methods are parameterized by a symmetric matrix $C \in \Re^{N_E \times N_E}$ such that

$$\hat{j}_i = \left[\left(K_i^T C K_i \right)^{-1/2} K_i^T C \right] \phi \tag{4.27}$$

where $\hat{j}_i \in \Re^{3 \times 1}$ is an estimator for the calculation of neuronal activity at the ith voxel. In this study [43], the localization ability of a linear imaging method is elaborated by considering the actual source as an arbitrary point test source at the jth voxel, which assumes that

$$\phi = K_j A \tag{4.28}$$

where K_j is the lead field matrix, and $A \in \Re^{3 \times 1}$ is a vector that contains dipole moments for the sources. By making use of the foregoing equations, one can write the estimation values as

$$\left\| \hat{j}_i \right\|^2 = A^T K_j^T C K_i \left(K_i^T C K_i \right)^+ K_i^T C K_j A \tag{4.29}$$

Now, considering the case of eLORETA, the current density estimator at the ith voxel can be written as

$$\hat{j}_i = W_i^{-1} K_i^T \left(K W^{-1} K^T + \alpha H \right)^+ \phi \tag{4.30}$$

Upon comparison of the foregoing equations, one can deduce that the exact, zero-error localization can be achieved with weights satisfying the following equation:

$$W_i = \left[K_i^T \left(K W^{-1} K^T + \alpha H \right)^+ K_i \right]^{1/2} \tag{4.31}$$

The eLORETA method is standardized, which implies that its theoretical expected variance is unity. The simulations for the validation of this method were carried out under the free academic eLORETA-KEY software with data available in Ref. [43]. The results show that eLORETA is an authentic localizing method with no localization bias, which provides zero-error localization in the case of nonideal conditions, that is, the presence of structured biological noise.

4.1.2.8 WMN-LORETA

This is the hybrid algorithm suggested and explained in Ref. [45], in which the WMN method is used to initialize LORETA algorithm. For the WMN method, the forward problem can be written as

$$V = KJ = \sum_{i=1}^{3M} K_i J_i = \sum_{i=1}^{3M} \frac{K_i}{W_i} . (W_i J_i) \tag{4.32}$$

where,

$$W_i = \left(\frac{1}{N_e} \right) . \sqrt{\sum_{j=1}^{Ne} K_{ij}^2} \tag{4.33}$$

Therefore, for the inverse problem, the current density can be estimated for the WMN as

$$J_{WMN} = W^{-2} K^t \left(K W^{-2} K^t \right)^+ V \tag{4.34}$$

The LORETA calculates the forward problem by minimizing the cost function min $J^t C J$ with the constraint of $V = KJ$, where $C = [BW]^t [BW]$, B is a discrete Laplacian operator to smoothen the output, and the weight matrix W is

$$W = \Omega \otimes I \tag{4.35}$$

Here, I is the identity matrix, $\otimes$ denotes the Kronecker product, and Ω is an $M \times M$ diagonal matrix with the following diagonal elements:

$$\Omega_{ii} = \sqrt{\sum_{\alpha=1}^{N} k_{\alpha i}^T k_{\alpha i}} \tag{4.36}$$

Hence, from the foregoing derivations for the different parameters related to the LORETA solution, the current density can be predicted as

$$J_{LORETA} = (C)^{-1} K^t \left[KC^{-1} K^t \right]^+ V \tag{4.37}$$

For this hybrid algorithm, the current density is calculated by using the equation for the WMN method. This vector J_{WMN} is used to build a weight matrix with the formula

$$W_h = diag\left(J_{WMN}(i) \right) \tag{4.38}$$

With this new weight matrix, C_h is developed, which is dependent on the foregoing calculation made with the help of the WMN algorithm. Hence,

$$C_h = W_h B^t B W_h \tag{4.39}$$

Ultimately, the equation for the computation of the current density for this new hybrid WMN-LORETA method is written as

$$J_{WMN-LORETA} = (C_h)^{-1} K^t \left[K(C_h)^{-1} K^t \right]^+ V \tag{4.40}$$

This technique was demonstrated by using 138 electrodes distributed on the scalp surface with 429 sources on the cerebral volume. The simulations for WMN, LORETA, and hybrid WMN-LORETA are shown for comparison. The comparison is done in terms of the resolution matrix, that is, $R = TK$. The resolution matrix for these methods shows that the WMN-LORETA method has the nearest value to that of the identity matrix, which leads to ideal conditions for less error and greater accuracy. Also, in terms of the computational time, the suggested algorithm uses less computing time than the LORETA algorithm. Hence, the hybrid algorithm is an efficient solution of the EEG inverse problem.

4.1.2.9 Recursive sLORETA-FOCUSS

This is a hybrid algorithm that combines the features of sLORETA and FOCUSS in a recursive manner to estimate the electrical activity inside the brain. The algorithm is presented and explained in Ref. [46]. It starts with the estimation of the current density by using the sLORETA method. The current density is estimated by using the sLORETA method given by

$$J_{sLORETA} = \hat{S}_j \times J_{MNE} \tag{4.41}$$

where $\hat{S}_j$ is the variance of the estimated current density, and J_{MNE} is the current density estimation for the minimum norm method. In the next step, the weight matrix is constructed by using the following mathematical relation:

$$W_i = PW_{i-1}\left[diag\left(\hat{J}_{i-1}(1),\hat{J}_{i-1}(2),...,\hat{J}_{i-1}(3M)\right)\right] \tag{4.42}$$

where $\hat{J}_{i-1}(n)$ is the nth element of vector $\hat{J}$ at the $(I-1)$th iteration. P is the following diagonal matrix:

$$P = diag\left[1/\|K_1\|,1/\|K_2\|,...,1/\|K_{3M}\|\right] \tag{4.43}$$

This method is utilized for the calculation of the current density by using following equation:

$$\hat{J}_i = W_i W_i^t K^t \left(KW_i W_i^t K^t\right)^+ V \tag{4.44}$$

FOCUSS is a recursive method in which the weight matrix is updated each time based on the data provided by the current density estimation of the previous ith iteration. This procedure is repeated (and hence the name "recursive") to eliminate the nonactive areas of the brain. Hence, after this elimination, a new space is defined for the active area only. These steps are repeated until the so-called convergence criterion is met. Here, "convergence" means that the number of nodes in the newly defined solution space is less than the number of sensors used for measurements.

The algorithm is designed and analyzed by simulation in MATLAB® by assuming the presence of two current dipoles in the brain, and a comparison is made between various localization algorithms such as sLORETA, FOCUSS, sLORETA-FOCUSS, and recursive sLORETA-FOCUSS. According to the simulated images, sLORETA produces smooth and diffused reconstructed images for two dipoles, which shows its inability to localize the dipoles correctly. The hybrid sLORETA-FOCUSS provides exact convergence to the real dipole with no localization error. However, the problem with this hybrid algorithm is the generation of small fake sources besides the space solution. The recursive sLORETA-FOCUSS algorithm gives the best solution and provides exactly the same result as the simulated dipole. Upon making a comparison between the said four algorithms in terms of the computational time taken for a method to localize the sources, the newly designed hybrid algorithm, recursive sLORETA-FOCUSS, is more time efficient as it takes 323.72 s, in contrast to sLORETA-FOCUSS (330.45 s) and FOCUSS (494.03 s).

4.1.2.10 Shrinking LORETA-FOCUSS

This is a hybrid algorithm combining the features of LORETA and FOCUSS in which the weight matrix is iterated along with the solution space, which is also subjected to alteration during the localization. The idea is presented and explained in Ref. [47]. First, the current density is computed using the LORETA algorithm as it provides smoothness and a relatively small localization error. After this, the weight matrix and the solution space are recursively iterated. The algorithm works by estimation of the current density using the LORETA algorithm. Next, the weight matrix is calculated. After that, a smoothing operator L is introduced such that

$$L.\hat{J} = l_1^T, l_2^T, ..., l_M^T \tag{4.45}$$

Therefore, the smoothest current densities are

$$l_i = \frac{1}{s_i + 1}\left(\hat{J}_i + \sum_u \hat{J}_u\right) \tag{4.46}$$

where s_i denotes the number of neighboring nodes with the region defined by u.

The smoothing operator is used for the retention of prominent nodes in the estimated topography. The selection of prominent nodes is done by setting down a threshold value in the topography. This process is repeated until convergence occurs. The results show that the algorithm provides reconstruction of sources with relatively high spatial resolution as compared to the LORETA algorithm. The localization capability is compared with other algorithms in terms of the energy error (E_{enrg}), which is calculated as

$$E_{enrg} = 1 - \frac{\left\|\hat{J}_{max}\right\|}{\left\|J_{simu}\right\|} \tag{4.47}$$

where $\left\|\hat{J}_{max}\right\|$ is the power of maxima in the estimated current density, and $\left\|J_{simu}\right\|$ is the power of the simulated point source.

The results for the comparison of the localization ability for LORETA, LORETA-FOCUSS, and shrinking LORETA-FOCUSS are presented in Table 4.1.

4.1.2.11 Summary with Comparison Tables

After this detailed discussion of the methods described so far in the literature, it is necessary to summarize the methods. The summary can be

TABLE 4.1

Localization Ability Comparison for Three Algorithms

	LORETA	LORETA-FOCUSS	Shrinking LORETA-FOCUSS
$\bar{E}_{loc}(mm)$	13.41	2.33	0.72
$E_{max\_loc}(mm)$	59.81	38.34	35.69
$E_{loc} \leq 7\ mm$	400 nodes	2136 nodes	2307 nodes
$E_{loc} \leq 14\ mm$	1591 nodes	2330 nodes	2379 nodes
$\bar{E}_{enrg}(\%)$	96.75	8.44	0.73
$E_{max\_enrg}(\%)$	99.76	76.54	79.98
$E_{enrg} \leq 0.01\%$	0 nodes	1292 nodes	2109 nodes
$E_{enrg} \leq 1\%$	0 nodes	1729 nodes	2207 nodes

Source: Hesheng, L. et al., Shrinking LORETA-FOCUSS: A recursive approach to estimating high spatial resolution electrical activity in the brain, in *Conference Proceedings. First International IEEE EMBS Conference on Neural Engineering*, pp. 545–548, 2003.

categorized by the author, year of publication, citations, advantages, and limitations of the localization method. Table 4.2 shows that LORETA is the most popular method for source localization even though there is still the disadvantage of low spatial localization. However, the methods derived from LORETA such as sLORETA and eLORETA have better localization capability but lag in terms of resolution. The hybrid weighted minimum norm method employs a hybridization of the minimum norm method and LORETA, which involves great system complexities that ultimately increase the required computational time. Shrinking LORETA-FOCUSS is also a hybrid algorithm that takes advantage of LORETA and FOCUSS, but it has not been validated experimentally. The other methods such as ICA, MUSIC, RAP MUSIC, and RCD method perform better in terms of localization but suffer from the disadvantages of complex computations, long computational time, and loss of data.

We will now present a feature-based comparison between various source localization methods in Table 4.3. The comparisons are made by different authors: a comparison between the LORETA, Backus and Gilbert, minimum norm, and WROP methods is made by Pascual in terms of the localization error and the estimated current density, and it shows that LORETA performs better than the other algorithms. The other comparison is made between Shrinking LORETA-FOCUSS, LORETA, and LORETA-FOCUSS by He Sheng Liu et al. in terms of energy and the localization error, and it is shown that the shrinking LORETA-FOCUSS algorithm has a comparatively small localization error, resulting in better performance. The sLORETA-FOCUSS technique is compared with the individual sLORETA and hybrid LORETA-FOCUSS in terms of the localization error. The other hybrid algorithms developed by Khemakhem et al. such as WMN-LORETA and recursive

TABLE 4.2

Summary of Different Techniques for Solution of EEG Inverse Problem

Method	Author	Advantages	Limitations
Minimum norm solution [33]	M. S. Hamalainen (1984)	Provides good initial results in terms of resolution and current estimation.	Fails to address the issue of deep source localization; has more localization error as compared to LORETA, WMN, etc.; incapable of localizing nonboundary sources.
LORETA [36]	R. D. Pascual et al. (1994)	High localization capability for localizing boundary sources and deep sources. Many variations are available on this basic localization algorithm.	Has low spatial resolution with blurred images. Low spatial resolution is undesirable in feature extraction of spatiotemporal pattern recognition. Regularization increases in spatial blurring.
FOCUSS [37]	I. F. Gorodnitsky et al. (1995)	Better localization with the capability to handle nonuniquely defined localized energy sources. Provides stable outputs.	Involves large mathematical calculations and hence high computational time owing to continuous iteration of weight matrix.
Recursive MUSIC, RAP MUSIC [38,39]	J. C. Mosher and R. M. Leahy (1998, 1999)	Recursive MUSIC along with its modifications provides better estimation with low localization error.	Model estimation includes random error and noise, which cause difficulties for true signal estimation. Procedure includes weighted sum of errors and noise, which increase complexity in the algorithm.
sLORETA [41]	R. D. Pascual (2002)	Exact zero-error localization as compared with minimum norm and Dale method.	Due to low-resolution imaging method, poor performance in recovering multiple sources when the point-spread functions of sources overlap. Also, owing to instability of EEG inverse problem, sometimes regularization is employed, which increases spatial blurring of LORETA and sLORETA solutions.

Method	Author	Advantages	Disadvantages
Shrinking LORETA-FOCUSS [47]	L. Hesheng et al. (2003)	Provides better results in terms of minimized localization error as compared to LORETA and LORETA-FOCUSS algorithms.	Only evaluated on simulated data and not validated through real-time data.
Hybrid weighted minimum norm [40]	C. Y. Song et al. (2005)	Provides better estimation by using features of LORETA and WMN as iterations make the algorithm more accurate with in-depth localization and less error.	The algorithm involves large computations and repeated iterations which results in high computational time. Data could also be lost owing to continuous iterations of the weight matrix.
eLORETA [43]	R. D. Pascual (2007)	Standardized method with theoretical expected variance as unity. Authentic localization technique with zero-localization error.	The low-resolution feature of eLORETA causes blurring in the images when the space is subjected to regularizations.
WMN-LORETA [45]	R. Khemakhem et al. (2008)	Hybrid method with combined features of LORETA and WMN, which provides better resolution than LORETA and WMN alone.	The system is complex and requires more computational time. System is valid for localization in highly active regions such as somatosensory evoked potentials.
Recursive sLORETA-FOCUSS [46]	R. Khemakhem et al. (2008)	More efficient in terms of computational time and localization.	No validation provided. Results were produced on simulated data.

TABLE 4.3

Comparisons between Various Localization Algorithms

Author	Method	Compared with	Comments
R. D. Pascual-Marqui (1999)	LORETA	Backus and Gilbert, Minimum norm, WROP	Results are compared in terms of localization error and estimated current density. LORETA performs well compared to others.
Hesheng Liu et al. (2003)	Shrinking LORETA-FOCUSS	LORETA, LORETA-FOCUSS	Comparison is made by using different parameters such as energy error, maximum localization error, maximum energy error, etc. The proposed algorithm shows better results on comparison.
R. Khemakhem et al. (2007)	sLORETA-FOCUSS	sLORETA, LORETA-FOCUSS	Localization error is compared for three methods: sLORETA, LORETA-FOCUSS, and sLORETA-FOCUSS. Localization error for the sLORETA-FOCUSS is less, and it has improved ability to locate simulated sources.
R. Khemakhem et al. (2008)	Recursive sLORETA-FOCUSS	sLORETA-FOCUSS, FOCUSS	Computing time and localization ability are compared. The Recursive sLORETA-FOCUSS is well suited with less computing time and more accurate results.
R. Khemakhem et al. (2008)	WMN-LORETA	WMN, LORETA	Computational time and resolution are compared. Suggested algorithm is more time efficient.

sLORETA-FOCUSS are compared with WMN, LORETA, sLORETA-FOCUSS, and FOCUSS, respectively, with respect to the computational time, resolution, and localization capability.

Before going into the section on challenges and discussion on future studies and better implementation of the EEG inverse problem, you should analyze Table 4.4, which provides a quick overview of the methods used so far in terms of the resolution, computational time (which corresponds to processing speed), validation, and localization error. The following symbols are used in Table 4.4, which defines the relative features of different algorithms.

The methods having a low spatial resolution such as LORETA, eLORETA, and sLORETA are crossed, and methods having a better resolution as compared to others are checked. The same convention is applied to the computational time, as methods characterized by large iterations and repeated procedures (such as FOCUSS and hybrid WMN) are classified as nonfavorable methods for these features, so they are marked with

TABLE 4.4

Feature-Based Comparison of Methods

S.No.	Method	Resolution	Complexity/ Computational Time	Validation	Low Localization Error
1	Minimum norm	×	√	√	×
2.	LORETA	×	√	√	×
3	FOCUSS	×	×	√	NA
4	Recursive MUSIC	NA	×	×	NA
5	sLORETA	×	√	√	√
6	Shrinking LORETA-FOCUSS	×	×	×	√
7	Hybrid weighted MN	×	×	×	NA
8	eLORETA	×	√	√	√
9	WMN-LORETA	×	×	×	√
10	Recursive sLORETA-FOCUSS	×	√	×	√

a cross, whereas the remaining methods, being efficient in terms of the computational time, are checked. There are some methods that are just checked by using simulated data and so are classified as having no validation; hence, a cross is placed for these, and vice versa. The last parameter is the localization error, which can be regarded as the most important parameter for a quality check for any localization method. A better method is one with a smaller localization error. The localization error is compared for different algorithms relatively. "NA" in Table 4.4 means that the information is not available from the literature.

4.2 Recommendations for the Future

The EEG, an imaging technique, provides good temporal resolution for the reflection of the neuronal activity but poor spatial resolution, which results in undesirable feature whenever it is used for source localization [46]. Hence, for the solution of the EEG Inverse problem, different methods are invented and explained by different researchers aiming at exact localization with less error and high accuracy. However, the mathematical relations governing the methods, the results obtained, comparisons in terms of the computational time taken, localization ability, localization error, energy error, system complexity, and improved resolution are important parameters. It can be deduced from the foregoing discussion that for the solution of the ill-posed EEG inverse problem, the following points are noteworthy:

1. The designed algorithm should avoid the problem of not localizing the deep sources, unlike the minimum norm solution.

2. The algorithm should be developed with the ability of localizing the sources with better spatial resolution, unlike the problem posed by LORETA (and its derived methods, such as sLORETA and eLORETA), because the recovery of multiple sources is difficult with low resolution. Further, due to ill-posedness of the EEG inverse problem, one has to employ regularization methods, which cause an increase in the blurring of images in the LORETA family. So the method should be developed with higher and improved spatial resolution.

3. Repeated iterations, which is an issue with some hybrid algorithms (LORETA-FOCUSS, sLORETA-FOCUSS, etc.) cause the system to become slow as the continuous iterations in the weight matrix are time consuming. This increases the system complexity and sometimes results in the loss of information and induction of noise. So the computations should be minimized to reduce the time taken and calculations.

4. The algorithm should be validated with real-time data to confirm the results produced; the results obtained with simulated data are not good enough.

From the above discussion, it is evident that the methods applied so far have though provided good results and facilitated continuous improvement in the field of the inverse problem. Solutions are emerging, but the following issues remain: low resolution, system complexity, slow processing, result validation, solution stability, and localization error. Therefore, the inverse problem needs to be solved with the aforementioned constraints so as to resolve issues related to applied neuroscience.

The following are some results showing brain activation during stimuli taken from the NETSTATION software package. The subject was shown a visual stimulus, and the EEG was taken during experiments by using a HydroCel Geodesic 128 channel montage and a 300 Net Amplifier of EGI Inc., USA. The sampling rate for the experiments was set at 250 samples per second with the impedance between scalp and sensors set as 50 kΩ. The experiments were held at the Intelligent Neuro Signal and Medical Imaging Laboratory, Centre for Intelligent Signal and Imaging Research (CISIR), Department of Electrical and Electronic Engineering, Universiti Teknologi PETRONAS, Tronoh, Perak, Malaysia.

The MRI results of the experiments were taken for LORETA and sLORETA with a four-shell spherical model known as Sun-Stok and finite difference head modeling. Sun-Stok is an isotropic four-shell spherical head modeling scheme. The term *isotropic* means that uniform conductivity is assumed for all the four shells (i.e., brain, skull, CSF, and scalp) involved in head modeling for the forward problem solution. However, the information about the radii and conductivity for each shell is predefined for this head modeling scheme. However, finite difference method (FDM) is an effective tool for solving electromagnetic boundary value problems. Both Sun-Stok and FDM are used for the solution of the forward problem, which further leads to an approximate location of the active source by using the inverse methods. Hence, LORETA and sLORETA are used with both head modeling schemes to produce results for comparison between various localization methods. The results were produced for localization comparison between various methods with results taken from the same time interval. These results show the activation in various regions of the brain by using LORETA and sLORETA in conjunction with standardized head modeling methods. The corresponding intensity level is also shown in nA for every method. In Figure 4.7, where LORETA is used as the inversion technique, activation is shown in the occipital lobe. For the second case (Figure 4.8), with the same subject but a different inversion technique (sLORETA), the activation is shown in the occipital lobe. Unlike the previous result for LORETA, sLORETA exhibits a higher intensity level (5.35 nA) compared with the intensity level of LORETA (0.828 nA). Figure 4.9 shows the activation results for LORETA with finite difference head modeling. It again shows activation in the occipital lobe with an intensity

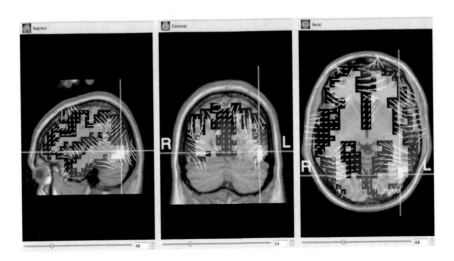

FIGURE 4.7
Activation map generated with NETSTATION for Sun-Stok-LORETA.

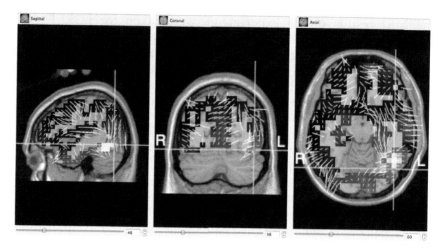

FIGURE 4.8
Activation map generated with using NETSTATION for FDM-LORETA.

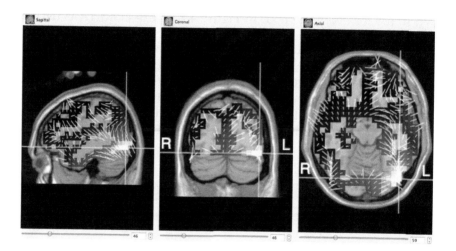

FIGURE 4.9
Activation map generated with NETSTATION for Sun-Stok-sLORETA.

level of 15.78 nA. The result for sLORETA with finite difference head modeling is shown in Figure 4.10. This shows activation in the occipital lobe with the maximum intensity level (~280 nA). Therefore, FDM-sLORETA exhibits the maximum intensity value for activation in the occipital lobe with the same time interval. It can be concluded that sLORETA with FDM head modeling can produce promising results. However, other numerical schemes (such as the finite element method (FEM) and the boundary

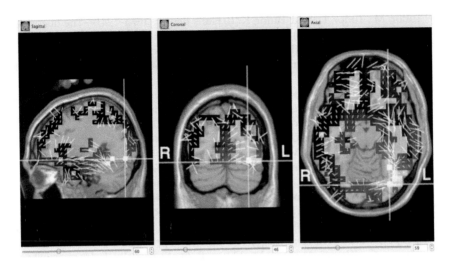

FIGURE 4.10
Activation map generated with NETSTATION for FDM-sLORETA.

element method (BEM)) can be applied in conjunction with LORETA, sLORETA, MUSIC, RAP MUSIC, and so on, to get better results in terms of less localization error and less computational complexity.

References

1. Aaronson, S. Anatomy of brain border. 2013, available at: http://universal-healthcarela.com/pic-anatomy-of-brain-border/
2. Rajapakse, J. C., Cichocki, A., and Sanchez A, V. D., Independent component analysis and beyond in brain imaging: EEG, MEG, fMRI, and PET, in *Proceedings of the 9th International Conference on Neural Information Processing, 2002, ICONIP '02*, Vol. 1, pp. 404–412, Nov 18–22, 2002.
3. Phillips, J. W., Leahy, R. M., Mosher, J. C., and Timsari, B., Imaging neural activity using MEG and EEG, *IEEE Engineering in Medicine and Biology Magazine*, Vol. 16, No. 3, pp. 34–42, 1997.
4. Modarreszadeh, M. and Schmidt, R. N., Wireless, 32-channel, EEG and epilepsy monitoring system, in *Engineering in Medicine and Biology Society, 1997. Proceedings of the 19th Annual International Conference of the IEEE*, Vol. 3, pp. 1157–1160, Oct 30–Nov 2, 1997.
5. Hauk, O. Linear distributed source analysis of EEG and MEG: Theory, implementation and application to studies on language production. *Dissertation*, Universitat Konstaz, Germany, 2000.

6. Tudor, M., Tudor, L., and Tudor, K. I., Hans Berger (1873–1941—The history of electroencephalography), *Acta Medica Croatica*, Vol. 59, pp. 307–313, 2005.

7. Yamada, T., The objective and perspective of recording electrical activity form the central nervous system, *Rinshō Shinkeigaku*, Vol. 35, pp. 1323–1331, 1995.

8. Files, B. An introduction to EEG: Neuroimaging workshop, 2011, available at: https://ngp.usc.edu/files/2013/06/BenFiles_An_introduction_to_EEG.pdf

9. Sanei, S., *EEG Signal Processing*, John Wiley, UK, 2007.

10. Wendel, K., Väisänen, O., Malmivuo, J., G. G. Nevzat., Vanrumste, B., and Durka, P., et al., EEG/MEG source imaging: Methods, challenges and open issues, *Computational Intelligence and Neuroscience*, Vol. 1, pp. 1–12, 2009.

11. Rosenow, F. and Luders, H., Presurgical evaluation of epilepsy, *Brain: A Journal of Neurology*, Vol. 124, No. Pt. 9, pp. 1683–1700, 2001.

12. Mintzer, S., Nicholl, J. S., Stern, J. M., and Engel, J., Relative utility of sphenoidal and temporal surface electrodes for localization of ictal onset in temporal lobe epilepsy, *Clinical Neurophysiology*, Vol. 113, No. 6, pp. 911–916, 2002.

13. Alarcón, G., Kissani, N., Dad, M., Elwes, R. D., Ekanayake, J., Hennessy, M. J. et al., Lateralizing and localizing values of ictal onset recorded on the scalp: Evidence from simultaneous recordings with intracranial foramen ovale electrodes, *Epilepsia*, Vol. 42, No. 11, pp. 1426–1437, 2001.

14. Silverman, D., The anterior temporal electrode and the ten-twenty system, *Electroencephalography and Clinical Neurophysiology*, Vol. 12, pp. 735–737, 1960.

15. Chatrian, G. E., Lettich, E., and Nelson, P. L., Modified nomenclature for the "10%" electrode system, *Journal of Clinical Neurophysiology*, Vol. 5, No. 2, pp. 183–186, 1988.

16. Falco, C., Sebastiano, F., Cacciola, L., Orabona, F., Ponticelli, R., Stirpe, P., and Di Gennaro, G., Scalp electrode placement by EC2 adhesive paste in long-term video-EEG monitoring, *Clinical Neurophysiology*, Vol. 116, No. 8, pp. 1771–1773, 2005.

17. Pflieger, M. E. and Sands, S. F., Abstract, in *Proceedings of the Second International Conference on Functional Mapping of the Human Brain, Neuroimage*, 3, Academic Press, New York, June 17–21, 1996, p. S10.

18. Zhukov, L., Weinstein, D., and Johnson, C., Independent component analysis for EEG source localization, *IEEE Engineering in Medicine and Biology Magazine*, Vol. 19, pp. 87–96, 2000.

19. Ebersole, J. S., *EEG Voltage Topography and Dipole Source Modeling of Epileptiform Potentials*, Lippincott Williams and Wilkins, Philadelphia, pp. 732–752.

20. Plummer, C., Simon Harvey, A., and Cook, M., EEG source localization in focal epilepsy: Where are we now? *Epilepsia*, Vol. 49, pp. 1–18, 2007.

21. Fuchs, M., Ford, M. R., Sands, S., and Lew, H. L., Overview of dipole source localization, *Physical Medicine and Rehabilitation Clinics of North America*, Vol. 15, pp. 251–262, 2004.

22. Koles, Z. J., Trends in EEG source localization, *Electroencephalography and Clinical Neurophysiology*, Vol. 106, pp. 127–137, 1998.

23. Brazier, M. A., A study of the electrical fields at the surface of the head, *Electroencephalography and Clinical Neurophysiology*, Suppl. 2, pp. 38–52, 1949.

24. Shaw, M. R., Potential distribution analysis II: A theoretical consideration of its significance in terms of electrical field theory, *Electroencephalography and Clinical Neurophysiology*, Vol. 7, pp. 285–292, 1955.

25. Plonsey, R., *Bioelectric Phenomena*, McGraw-Hill, New York, 1969, pp. 304–308.

26. Schneider, M. R., A multistage process for computing virtual dipolar sources of EEG discharges from surface information, *IEEE Transactions on Biomedical Engineering*, Vol. BME-19, No. 1, pp. 1–12, 1972.

27. Henderson, C. J., Butler, S. R., and Glass, A., The localization of the equivalent dipoles of EEG sources by the application of electric field theory, *Electroencephalography and Clinical Neurophysiology*, Vol. 39, pp. 117–130, 1975.

28. Nunez, P. L., *Electric Fields of the Brain: The Neurophysics of EEG*, Oxford University Press, New York, 1981.

29. Tarantola, A., *Inverse Problem Theory and Methods for Model Parameter Estimation*, Society for Industrial and Applied Mathematics, University City Science Center, Philadelphia, PA, 2005.

30. Zhdanov, M. S., *Geophysical Inverse Theory and Regularization Problems*, Elsevier, The Netherlands, 2002.

31. Snieder, R., and J. Trampert., *Inverse problems in geophysics*, in Wavefield inversion, edited by A. Wirgin, pp. 119–190, Springer Verlag, New York, 1999.

32. Pascual-Marqui, R. D., Theory of the EEG Inverse Problem, in *Quantitative EEG Analysis: Methods and Clinical Applications*, edited by Tong, S., and Thakor, N. V., pp. 121–140, Artech House, Boston, 2009.

33. Hämäläinen, M. S., and Ilmoniemi, R. J., *Interpreting measured magnetic fields of the brain: Estimates of current distributions.* Tech. Rep. TKK-F-A559, Helsinki University of Technology, Espoo, 1984.

34. Söderström, T. and Stewart, G. W., On the numerical properties of an iterative method for computing the Moore-Penrose generalized inverse, *Society for Industrial and Applied Mathematics Journal on Numerical Analysis*, Vol. 11, No. 1, pp. 61–74, 2006.

35. Pascual-Marqui, R. D., Review of methods for solving the EEG inverse problem, *IJBEM*, Vol. 1, pp. 75–86, 1999.

36. Pascual, R. D., Michel, C. M., and Lehmann, D., Low resolution electromagnetic tomography: A new method for localizing electrical activity in the brain, *International Journal of Psychophysiology*, Vol. 18, pp. 49–65, 1994.

37. Gorodnitsky, I. F., George, J. S., and Rao, B. D., Neuromagnetic source imaging with FOCUSS: A recursive weighted minimum norm algorithm, *Electroencephalography and Clinical Neurophysiology*, Vol. 95, No. 4, pp. 231–251, 1995.

38. Mosher, J. C. and Leahy, R. M., Recursive MUSIC: A framework for EEG and MEG source localization, *IEEE Transactions on Biomedical Engineering*, Vol. 45, pp. 1342–1354, 1998.

39. Mosher, J. C. and Leahy, R. M., Source localization using recursively applied and projected (RAP) MUSIC, *IEEE Transactions on Signal Processing*, Vol. 47, No. 2, pp. 332–340, 1999.

40. Song, C. Y., Wu, Q., and Zhuang, T. G., Hybrid weighted minimum norm method A new method based LORETA to solve EEG inverse problem, in *27th Annual International Conference of the Engineering in Medicine and Biology Society, 2005, IEEE-EMBS 2005*, (January 17–18), pp. 1079–1082.

41. Pascual-Marqui, R. D., Standardized low resolution brain electromagnetic tomography (sLORETA), *Technical Details, Methods and Findings in Experimental and Clinical Pharmacology*, Vol. 24D, pp. 5–12, 2002.

42. Dale, A. M., Liu, A. K., Fischl, B. R., Buckner, R. L., Belliveau, J. W., and Lewine, J. D., Dynamic statistical parametric mapping: Combining fMRI and MEG for high resolution imaging of cortical activity, *Neuron*, Vol. 26, pp. 55–67, 2000.

43. Pascual-Marqui, R. D., Discrete, 3D distributed linear imaging methods of electric neuronal activity. Part 1: Exact, zero error localization, October 17, 2007.
44. KEY Institute for Brain-Mind Research, University Hospital of Psychiatry, Zurich, Switzerland, 2014, available at: http://www.uzh.ch/keyinst/
45. Khemakhem, R., Zouch, W., Taleb-Ahmed, A., and Ben Hamida, A., A new combining approach to localizing the EEG activity in the brain: WMN and LORETA solution, in *International Conference on BioMedical Engineering and Informatics, 2008, BMEI 2008.* May 27–30, 2008, pp. 821–824.
46. Khemakhem, R., Ben Hamida, A., Ahmed-taleb, A., and Derambure, P., New hybrid method for the 3D reconstruction of neuronal activity in the brain, in *15th International Conference on Systems, Signals and Image Processing, 2008, IWSSIP 2008,* pp. 405, 408, June 25–28, 2008.
47. He Sheng, L., Yang, F., Gao, X., and Gao, S., Shrinking LORETA-FOCUSS: A recursive approach to estimating high spatial resolution electrical activity in the brain, in *Conference Proceedings. First International IEEE EMBS Conference on Neural Engineering,* pp. 545–548, March 20–22, 2003.

5

Epilepsy Detection and Monitoring

Arslan Shahid
Universiti Teknologi PETRONAS

John K.J. Tharakan
Universiti Sains Malaysia

CONTENTS

5.1 Introduction...123
5.2 Types of Epileptic Seizures...124
    5.2.1 Generalized Seizures...124
    5.2.2 Partial Seizures 125
5.3 Electroencephalography and Epileptic Seizure Detection 125
    5.3.1 Seizure Detection Algorithms 126
        5.3.1.1 Seizure Event Detection 126
        5.3.1.2 Seizure Onset Detection 131
5.4 Development of a Seizure Detection System Using the Singular
    Values of EEG Signals 134
    5.4.1 Database 134
    5.4.2 Machine-Learning-Based Seizure Detection System
        Using SVD..135
5.5 EEG Time-Series Analysis...136
5.6 Multivariate Measures 137
5.7 Principal Component Analysis..138
5.8 Closed-Loop Seizure Prevention System 138
5.9 Summary...139
References..139

5.1 Introduction

Epilepsy is a chronic condition that is characterized by recurrent and unprovoked seizures. A seizure can be defined clinically as "intermittent paroxysmal, stereotyped disturbance of consciousness, behavior, emotion, motor function, perception, or sensation, which may occur singly or in combination and is thought to be the result of abnormal cortical neuronal

123

discharges." The exact type of seizure will depend on where it starts and then spreads within the brain. Abnormalities in synaptic transmission, neuronal excitability, voltage gated ion channels, neuronal proteins, and gene expression are all implicated in the pathogenesis of epilepsy.

According to the WHO, epilepsy is the most common primary disorder of the brain. More than 50 million people in the world are affected by epilepsy, and 80% of them are in developing countries. About 70% of epilepsy cases can be treated effectively, but 30% generally remains refractory to medical treatment and constitute a heavy burden on the quality of life of the sufferer and the caregiver and an economic burden on health delivery systems. Alternative treatment options for these patients are surgical removal of the epileptogenic tissue or vagal nerve stimulation.

5.2 Types of Epileptic Seizures

Seizures are categorized into two main groups based on the involvement of the brain region responsible for the seizure. Generalized seizures involve almost the whole brain, whereas focal seizures, also called partial seizures, start from a particular brain region and may spread from there to other parts of the brain. Generalized seizures are further classified into petit mal, tonic–clonic, myoclonic, and atonic seizures. Partial seizures are classified into simple partial and complex partial seizures as shown in Figure 5.1.

5.2.1 Generalized Seizures

- *Petit Mal:* Petit mal seizures are also referred to as absence seizures. They are represented as a loss of consciousness that usually lasts for 20 s.

- *Tonic–Clonic:* Tonic–clonic seizures are also referred to as grand mal seizures. In the tonic phase, a person may lose consciousness for a

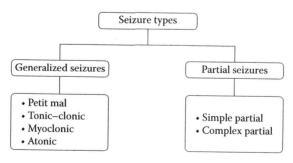

FIGURE 5.1
Summary of epileptic seizure types.

small amount of time, and in the clonic phase the body starts shaking and stretching in jerky muscular contractions.

- *Myoclonic:* Myoclonic seizures manifest as small jerks in the body.
- *Atonic:* Atonic seizures manifest as a lack of muscle control and they usually last for less than 15 s.

5.2.2 Partial Seizures

- *Simple Partial Seizures:* Simple partial seizures are also called Aura. They do not cause loss of consciousness, but they may be a warning sign of more serious seizures in the future.
- *Complex Partial Seizures:* Complex partial seizure is an epileptic seizure that is associated with unilateral cerebral hemisphere involvement and causes impairment of awareness or responsiveness (i.e., alteration of consciousness).

5.3 Electroencephalography and Epileptic Seizure Detection

Different techniques are used to detect epileptic seizures including electro-encephalography (EEG), magnetoencephalography (MEG), and functional magnetic resonance imaging (fMRI). Among them, EEG is the most widely used as it is cheaper and offers a high temporal resolution. Most of the work in epileptic seizure detection is based on EEG in conjunction with machine learning tools. The next section will discuss the most relevant studies on seizure detection using EEG signals. Figure 5.2 shows the signals of 17-EEG channels at the start of epileptic seizure.

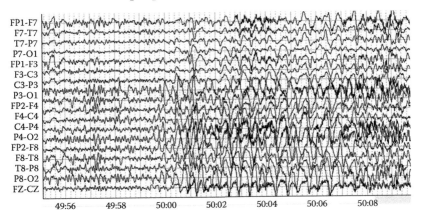

FIGURE 5.2
A 17-channel EEG showing the start of an epileptic seizure.

5.3.1 Seizure Detection Algorithms

Various studies have been done in the past two decades for the detection of epileptic seizures using EEG. Nasehi and Pourghassem [1] have surveyed several automated algorithms for the detection of epileptic seizures based on the scalp EEG and ECG signals up to 2011. They suggest that there are two types of seizure detection: epileptic seizure onset detection and epileptic seizure event detection, both of which have patient-specific or non-patient-specific applications. Seizure event detectors identify seizures with a higher percentage of accuracy but not with the shortest possible delay [2]. Seizure onset detectors identify the start of seizure with the shortest possible delay, but these algorithms are not usually as accurate as seizure event detectors [3].

Moreover, there are separate studies on seizure detection algorithms based on the intracranial EEG. Usually, it is seen that intracranial epileptic seizure detection gives better results as compared to the scalp EEG, but placement of intracranial electrodes requires surgery and can only be used in specific monitoring conditions in hospitals. Accordingly, an efficient seizure detection algorithm using the scalp EEG is required so that seizure detection for immediate epilepsy treatment can be improved. In general, seizure detection algorithms can be divided into two major classes—(i) seizure event detectors and (ii) seizure onset detectors. The two classes of seizure detectors are shown in Figure 5.3.

5.3.1.1 Seizure Event Detection

Seizure event detectors are meant for the detection of epileptic seizures in EEG signals. They usually have good accuracy but unsatisfactory latency in detecting the onset of a seizure. These algorithms are used by physicians to monitor the duration and frequency of seizures of individual patients [4]. This may allow them to prescribe better medication to suppress the disorder.

In 1992, Liu et al. [5] developed a patient-independent algorithm to detect neonatal seizures from EEG. Autocorrelation analysis was used to differentiate between the seizure activity and the normal background EEG. EEG data from 14 infants were recorded using a 12-channel EEG amplifier and divided into 30 s epochs, with 58 epochs containing seizures and 59 epochs for control EEG being selected for further analysis. No preprocessing was

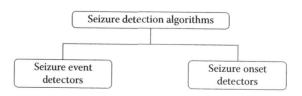

FIGURE 5.3
Types of seizure detection algorithms.

applied on the raw EEG data. Some of the control EEG epochs contained some common EEG artifacts such as eye blinks and respiratory artifacts that coincide with seizure activities and may account for false detections, but it was proved in terms of results that the developed technique is robust to these periodic and rhythmic activities as these are of low amplitude and minimal duration. Scored autocorrelation moment (SAM) analysis was performed for performance evaluation of the developed technique as it provides a measure of the periodicity of a signal. The sensitivity and specificity of the system were 84% and 98%, respectively, and 9 out of 58 test seizure epochs were falsely classified as nonseizure epochs.

In 1997, Gotman et al. [6] developed a system for neonatal seizure detection that consists of three different methods of detecting rhythmic patterns at different frequencies and different spike patterns. EEG data were taken from 43 newborns recorded at three different children's hospitals. In the first method, the EEG was divided into epochs of 10 s shifting forward with a 75% overlap. For each epoch, the frequency spectrum was calculated, and the frequency of the dominant peak was calculated. The dominant peak was defined as the peak with the largest average power in the full width half maximum of that peak. The width of the dominant peak was calculated, and if it had a small width, it was represented as a rhythmic activity. The power ratio was calculated between the power of the band of the dominant peak and the power of the same frequency band in the background spectrum. The second method consisted of multiple spike detection [7], and a seizure was declared if six spikes were detected in an epoch of 10 s. The third method considered the detection of very slow rhythmic activities that were not detected by the first method. The EEG was filtered with a 2nd order low-pass filter of cutoff frequency 1 Hz. The system gave an average seizure detection rate of 71% when all the three methods were combined. The average false detection rate was 1.7 FD/h. In another publication [8], this methodology was reapplied on a completely new data set with 54 neonatal patients, and the average seizure detection rate was 66% with 2.3 false detections per hour.

In 2004, Hassanpour et al. [9] developed a method for seizure detection based on the estimation of the distribution function of the singular vectors related to the time-frequency distribution of EEG epochs to find patterns that can differentiate between a seizure and nonseizure activity. These features were used to train a neural network to classify seizures and normal EEG activity. EEG data were recorded from eight newborns ranging in age between two days and two weeks. The EEG was recorded using a 20-channel EEG amplifier and was divided into 30-s epochs. Right and left singular values were used corresponding to the largest singular value for finding the characteristics of the signals. Each EEG epoch was mapped into the time-frequency domain using the β distribution. Two of the left and two of the right singular values were computed for the matrix containing the TF representation, and the singular values were squared transform them into density functions. These density functions were further used to compute the distribution functions. Histograms were computed, and the dominant features were selected and fed into a neural network of two hidden layers. The average good detection rate (GDR) for eight patients was 92.5%, and the false detection rate (FDR) was 3.7%. The GDR was defined as the total number of true detections.

In 2004, Wilson et al. [10] developed an algorithm for seizure detection called the Reveal algorithm. It was based on the matching pursuit algorithm, which transforms EEG signals into the sum of overlapping "atoms," which are localized in time and frequency. Each atom can then be taken as the time-frequency evolution of an independent component of the EEG waveform. It divides the EEG into 2 s epochs and computes the matching pursuit decompositions. In the Reveal algorithm, only the first two atoms are generated for each decomposition, and each atom is represented by its total root mean square amplitude, its maximum amplitude excursion, and the summed duration of flat periods corresponding to the amplifier saturation. This algorithm was tested on 672 seizures from 426 patients, and the observed sensitivity was 76% with 0.11 false detections per hour.

In 2005, Alkan et al. [11] developed a seizure detector using the EEG power spectra as an input to the classifier. Different methods were applied to compute the power spectral density (PSD) of the signals, for example, multiple signal classification (MUSIC), autoregressive (AR), and periodogram approaches. Then, different classification methods were applied and compared. Eleven patients participated in the study, out of whom five were epileptic. The EEG was recorded using four channels, namely, F7-C3, F8-C4, T5-O1, and T6-O2. The recordings were annotated by two neurologists to mark the start and end of the seizures. The PSD of the frequency of the signals was first computed using the periodogram method after dividing the data into finite-sized windows. The AR method was also used for calculating the PSD. This method for spectral estimation consisted of two steps. First, the AR parameters were estimated from a given data sequence, and then the PSD was calculated from these estimates. The subspace-based MUSIC method was also applied to estimate the PSD of the signal, which was based on the eigendecomposition of the correlation matrix of the noisy signal. This method performed better than the previous methods as it is best suited to signals that contain noise and have a low SNR. Logistic regression and multilayer perceptron artificial neural networks (MPL-NN) were used for classification. MLP-NN is more robust to noise as compared to LR, and thus the overall results showed that MLP-NN when trained with the PSD estimated with the MUSIC method was more accurate in terms of sensitivity and specificity. With this configuration, the classification accuracy was 92% with a sensitivity of 90% and a specificity of 93.6%.

In 2007, Greene et al. [12] developed a seizure detection system based on EEG and ECG. ECG also contains responses related to seizure activity, and thus it is often used as a second physiological signal along with EGG to boost the performance of seizure detectors. The algorithm was evaluated in both the patient-specific and patient-independent configurations.

The results were better in the patient-specific mode compared to the patient-independent mode. The data set contained EEG recordings of 10 neonatal epilepsy patients with 633 seizures. The EEG was divided into 12 records, and the total recording duration was 154 h and 10 min. Six features were extracted from each EEG channel, namely the dominant spectral peak, power ratio, bandwidth of dominant spectral peak, nonlinear energy, spectral entropy, and line length. The dominant spectral peak, power ratio, and the bandwidth of dominant spectral peak are the same features that were used by Gotman et al. [6]. The complexity of the signal reduces during a seizure [13], and thus spectral entropy was calculated to measure the complexity. The mean nonlinear energy was also reported to be a good feature to classify seizure and nonseizure epochs [14]. The line length was calculated as it is a measure of the fractal properties of a signal and a very strong feature for seizure detection [15]. For ECG signals, the R-R interval was calculated first. The R-R interval is defined as the time in seconds between two adjacent R-wave maximums. ECG features include the mean R-R interval, standard deviation between R-R intervals, mean spectral entropy, mean change in the R-R interval, coefficient of variation, and the PSD. ECG features were calculated from 60 s nonoverlapping epochs. Both EEG and ECG features were then fused using early integration and late integration. Early integration refers to the integration of both types of features before classification, which gives the overall result at the output. Late integration involves classifying both types of features separately and then determining the probability of seizures for each single mode and then combining both the probabilities. In the patient-specific mode and with early integration fusion, the classifier has an accuracy of 86.32%, a sensitivity of 76.37%, a specificity of 88.77%, a GDR of 95.82%, and an FDR of 11.23%. With late integration fusion, the classifier has an accuracy of 84.66%, a sensitivity of 74.08%, a specificity of 86.82%, a GDR of 97.52%, and an FDR of 13.18%. In the patient-independent mode and with early integration fusion, the classifier has an accuracy of 71.51%, a sensitivity of 71.73%, a specificity of 71.43%, a GDR of 81.44%, and an FDR of 28.57%. With late integration fusion, the classifier has an accuracy of 68.89%, a sensitivity of 74.39%, a specificity of 66.95%, a GDR of 81.27%, and an FDR of 68.89%.

In 2007, Aarabi et al. [16] developed a multistage knowledge-based neonatal EEG detection system. The system consisted of multiple stages, that is, band-pass filtering and normalizing the amplitude of the EEG signals, automatic removal of artifacts, segmentation into EEG epochs, feature extraction, feature selection, channel-by-channel classification, and integration of the individual channel decisions. EEG was recorded from 10 neonates aged between 39 and 42 weeks post conception. Eleven electrodes were used for recording with the 10-20 system for a bipolar montage. Data were recorded at a sampling rate of 256 Hz and was passed through a band-pass filter with frequencies between 0.2 and 70 Hz. A total of 110 h of EEG were recorded. Because the skull thickness and the skin impedance affect the neonatal EEG [17], the data were first equalized by dividing it by the mean standard deviation of amplitude of 1 min of data without any seizures or artifacts. At the artifact removal stage, eye blinks and eyeball movements were removed by correlation-based template matching. Segments with constant amplitudes were considered as saturation artifacts, and all segments with zero derivatives were removed. ECG artifacts were removed by summed squared error-based template matching [18]. Electrode movement artifacts and EMG muscle artifacts were removed by thresholding. The segmentation stage was divided into two steps. First, the EEG was divided into 10 s epochs with 75% overlap. Second, these epochs where further subdivided into three subsegments to separate out the slow and rapid EEG activities. Various features were extracted from these segments, including time-domain features, frequency-domain features, AR coefficients, wavelet features, and cepstral features. The feature vector was 275-dimensional for each segment. Mutual information-based forward feature selection (MIFFS) [19] is a method based on relevance and redundancy and was used for feature selection to deduce the feature vector's dimensionality.* A back propagation neural network with three hidden layers was used as a classifier. Overall, the training and test performance were measured for the classifier, giving a sensitivity of 74%, a specificity of 70.1%, and an FDR of 1.55/h.

In 2011, Temko et al. [20] developed a neonatal seizure detection system based on support vector machines (SVMs). Two postprocessing steps were proposed to increase the temporal precision of the system. The data set used contained EEG recordings of 267.9 h from 17 neonatal epileptic patients. The recordings contained 705 seizures in all. The first step was the preprocessing in which EEG was downsampled from 256 Hz to 32 Hz. The EEG was then divided into epochs of 8 s with 4 s overlap. A total of 55 different features were extracted from the frequency domain, time domain, and information theory (entropy measures). It was observed during the classification stage that feature selection to reduce the number of features does not significantly reduce the processing time, so all the 55 features were used for classification. At the classification stage, leave-one-out cross-validation method was used to assess the performance of the system. In the postprocessing stage, the classifier output for the input channel was converted into probabilistic values, filtered, and then the threshold was applied along with multichannel decision making to improve the performance of the system. After that, the outputs from all the channels were combined to make

---

* The final feature vector contained AR coefficients a3, a6, a12, a13; the relative spectral power in delta, theta, and alpha bands; zero crossings of wavelet coefficients; kurtosis of wavelet coefficients; relative energy of wavelet coefficients; mean and coefficient of variation of the first derivative of the signal; the coefficient of variation of amplitudes and curvatures of slow waves; the coefficient of variation of durations and curvatures of rapid waves; the mean of the rise time to the fall time of slow and rapid waves; the coefficient of variation of durations; slopes and curvatures of spike-like waves; mean left-side amplitude/mean right-side amplitude of slow waves; and first and second Hjorth coefficients.

TABLE 5.1

Summary of Seizure Event Detectors

Algorithm	Sensitivity (%)	Specificity (%)	GDR (%)	FDR (%)	FDR/h	Data Set Used
Liu, 1992	84	98	—	—	—	14 patients with 58 seizures
Gotman, 1997	71	—	—	—	1.7	43 patients
Hassanpour, 2004	—	—	92.5	3.7	—	8 patients with 64 seizures
Wilson, 2004	76	—	—	—	0.11	426 patients with 672 seizures
Alkan, 2005	90	93.6	—	—	—	11 patients
Greene, 2007	76.37	88.77	95.82	11.23	—	10 patients with 633 seizures
Aarabi, 2007	74	70.1	—	—	1.55	10 patients
Temko, 2011	89	—	—	—	1.0	17 patients with 705 seizures

it a multichannel system. Results showed that the algorithm detected 89% of seizures with one false detection per hour, 96% with two false detections per hour, and 100% with four false detections per hour.

A summary of all the foregoing epileptic seizure event detection algorithms is given in Table 5.1.

5.3.1.2 Seizure Onset Detection

In 2005, Saab and Gotman [21] developed a system for seizure onset detection based on the probability of seizure in small EEG epochs. A 5-level wavelet transform using the Daubechies-4 wavelet was performed for EEG epochs of 2 s for each channel. Three characterizing measures were derived from the wavelet coefficients, namely, the relative average amplitude, the relative scale energy, and the coefficient of variation of amplitude. In order to estimate the conditional probability of a seizure or nonseizure event, Bayes' formula was applied for probabilistic analysis. The system was evaluated on 652 h of continuous EEG recordings of 28 patients with 126 seizures. Training and testing of the data were done on separate subsets of the data. The system showed a sensitivity of 77.9%, a false detection rate of 0.86 false detections per hour, and an average latency of 9.8 s. After applying a tuning mechanism for the artifact rejection, the sensitivity decreased to 76% and the latency increased to 10 s, but the false detection rate improved to 0.34 false detections per hour.

In 2005, Grewal and Gotman [22] developed a clinical seizure warning system for intracerebral EEGs. The technique applied consisted of data filtering in different bands, spectral feature extraction, and spatiotemporal analysis. EEGs were first filtered into seven frequency bands using the band-pass

Chebyshev IIR filter. The EEG was divided into 4 s epochs for each frequency band, and features were extracted from these epochs, including the relative energy, relative amplitude, and coefficient of variation of amplitude. Furthermore, the coefficient of variation of amplitude and the relative amplitude were calculated from the frequency band containing the first harmonic frequency of each epoch to find the harmonic characteristics of the signal. The feature vector finally contained five features. The data set consisted of EEG recordings of 407 h from 19 patients with 152 seizures. Different subsets were used for training and testing using the probabilistic analysis. Bayes' theorem was used for classifying epochs based on the probabilities of the feature values. The system when evaluated gave a sensitivity of 89.4%, a false detection rate of 0.22 false detections per hour, and a delay time of 17.1 s.

In 2009, Shoeb [23] developed a machine learning algorithm based on EEG and ECG to detect the onset of epileptic seizures in pediatric patients. The algorithm was based on the spectral and spatial features of the EEG signals. The data set used contained 916 h of EEG recordings of 23 pediatric patients and 1 adult patient. The total data set contained 173 seizures. Spectral and spatial features were used to find the rhythmic activity in the signals as the rhythmic activity associated with the onset of a seizure is composed of many spectral components. A sliding window length of 2 s was selected, and a filter bank of 8 filters was used, which helped to better discriminate between seizures and nonseizures. For time evolution, the feature vector was stacked with a concatenation of feature vectors from three nonoverlapping 2 s epochs so that the feature vector captures the relation of an epoch relating to the epochs in the recent past. An SVM was used for classification using the radial basis kernel function. Training of the classifier was done with 20 s from each seizure and a 24 h normal EEG for each patient. When only EEG features were used to train the classifier, an overall 96% of the seizures were detected by the classifier with a mean latency of 4.6 s. The average false detection rate was 2 per 24 h period. When the EEG and the heart rate variability from the ECG were used together for 10 seizures of a specific patient, it was observed that the mean latency was reduced to 2.7 s, and all the seizures were detected correctly with five false detections per 24 h period.

In 2011, Kharbouch et al. [24] developed a seizure onset detection system using intracranial EEG. Spectral, spatial, and temporal features were extracted from iEEG and fed to a classifier to detect the onset of the seizure. The method was based on combining the time evolution of spectral properties of brain activities into a single feature vector for classification. The data set used in this study consisted of 875 h of continuous iEEG recordings from 10 patients with 67 seizures and sampled at 500 Hz. Ninety-seven percent of the 67 test seizures were detected by this method with an average detection delay of 5 s and a false detection rate of 0.6 false detections per 24 h period. The average sensitivity was 95.8%.

In 2011, Nasehi and Pourghassem [25] developed a seizure onset detector using dynamic cascade feed-forward neural networks. The discrete Fourier transform was used to calculate the spectrum energy of each epoch, which was used further as a feature for binary classification between seizure and nonseizure events. A three-layer dynamic cascade feed-forward neural network was used for classification as it is efficient for determining non-linear decision boundaries. Moreover, it is faster than multilayer percep-tron and offers a smaller latency for onset detection. Eighty percent of the data was used for training the classifier, and 20% was used for testing. The CHB-MIT scalp EEG database was used in this study, which contains continuously recorded 844 h of EEG recordings from 23 patients and with 173 seizures. The overall sensitivity and specificity were calculated as 98% and 91%, respectively. The average latency was 3.5 s. Later, Nasehi and Pourghassem [26] used a neural network based on an improved par-ticle swarm optimization (IPSONN) to determine an optimal nonlinear decision boundary. IPSO searches for a better solution to the problem of weight assignment in a neural network such as the implementation of the modified evolutionary direction operator (MEDO). This technique further improved the results. With IPSONN, the average sensitivity was 98% with a latency of 3 s. However, the false detection rate of 3 false detections per 24 h was not a good outcome.

In 2012, Khan et al. [27] developed a seizure detection algorithm based on kurtosis, skewness, and the normalized coefficient of variation of ampli-tude. The EEG was divided into nonoverlapping epochs of 1 s, and fea-tures were extracted from these epochs. Skewness measures asymmetry in the probability distribution of data, whereas kurtosis is a measure of "peakedness" [28]. A background window of 25 s was taken to normalize epoch features. Linear classification was applied for labeling the data as epileptic or nonepileptic. Ten patients' data out of 24, with 55 seizures, from the CHB-MIT scalp EEG database was used in this study, and the sensitiv-ity was 100% with 1.1 false detections per hour. The mean latency of the system was 3.2 s.

In 2012, Rabbi and Fazel-Rezai [29] developed an iEEG seizure onset detec-tion system based on fuzzy logic. The data were first divided into epochs of 2.5 s with a 0.5 s overlap. Saturation artifacts and electrode movement artifacts were removed, and the data were filtered to remove high-frequency noise, low-frequency artifacts, and power line noise. Four features were extracted from each epoch: average amplitude, rhythmicity, the dominant frequency, and entropy. A three-stage fuzzy-rule-based system comprising a feature combiner, a spatial combiner, and a final decision-making system was used for onset detection. False detections caused by artifacts are rejected at the postprocessing step. The Freiburg database was used to evaluate the system with 112.5 h of iEEG recordings from 20 patients and with 56 sei-zures. The system showed a sensitivity of 95.8% with 0.26 false detections per hour and a detection latency of 15.8 s.

The foregoing epileptic seizure event detection algorithms are summarized in Table 5.2.

TABLE 5.2

Summary of Seizure Onset Detectors

Algorithm	Sensitivity (%)	Specificity (%)	GDR (%)	FDR/h	Latency (sec)	Data Set Used
Saab, 2005	77.9	—	—	0.86	9.8	28 patients with 126 seizures
Grewal, 2005	89.4	—	—	0.22	17.1	19 patients with 152 seizures
Shoeb, 2009	96	—	—	0.083	4.6	23 patients with 173 seizures
Kharbouch, 2011	95.8	—	97	0.025	5	10 patients with 67 seizures
Nasehi, 2011	98	91	—	0.125	3	23 patients with 173 seizures
Khan, 2012	100	—	—	1.1	3.2	10 patients with 55 seizures
Fazle Rabbi, 2012	95.8	—	—	0.26	15.8	20 patients with 56 seizures

5.4 Development of a Seizure Detection System Using the Singular Values of EEG Signals

In this section, the development of a seizure event detection algorithm based on the singular values of EEG signals is proposed. Singular values contain very important information about the oriented energy of a signal and are very sensitive to sudden changes in the signal. This property of singular values is used for epileptic seizure detection using EEG signals [30]. The subsequent sections will discuss in detail the methodology used for the development of such an algorithm.

5.4.1 Database

EEG data for 24 pediatric patients with 198 seizures used in this study were acquired from the PhysioNet Online EEG database [31]. The data were recorded from pediatric patients at the epilepsy monitoring unit of Boston Children's Hospital. The EEG was captured using an 18-channel bipolar montage and sampled at 256 samples/s.

The EEG data are divided into small epochs of 8 s each that are shifted forward with a 50% overlap. Singular value decomposition (SVD) is applied on each window, and the largest r-singular values are obtained. The second norms of these values are plotted starting some time before the seizure, during the seizure, and some time after the seizure. Such a plot is shown in Figure 5.4.

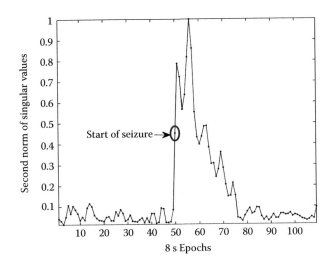

FIGURE 5.4
Second norm of singular values of 8 s EEG epochs showing the start of a seizure.

It can be clearly observed from Figure 5.4 that singular values (SVs) of EEG signals show an increasing trend during seizures. This is because SVs are very sensitive to the energy changes in a signal, and thus the increased energy of EEG signals during an epileptic attack can be detected using SVs. This finding can be made the basis for the development of a seizure detection algorithm using some machine learning tools.

5.4.2 Machine-Learning-Based Seizure Detection System Using SVD

In the past few years, many studies have used machine learning techniques to develop automated epileptic seizure detection systems. A similar method is proposed here using an SVM and using singular values of EEG signals as features to detect epileptic activity from EEG recordings.

An SVM is an efficient tool for nonlinear binary classification problems as it projects the nonlinear problem to a higher-dimensionality space where the problem can more likely be treated as a linear problem. The SVM then creates a hyperplane between the two classes, separating them with the largest possible distance. SVMs have been used in many previously developed seizure detection algorithms.

For the development of a singular-values-based seizure detection system, EEG data are first divided into epochs of 8 s, and SVD is applied on each epoch. For every epoch, a singular value matrix is generated that is then converted into a feature vector. These feature vectors are then passed to the SVM for classification. The system is trained on feature vectors obtained from an equal number of seizure and nonseizure epochs.

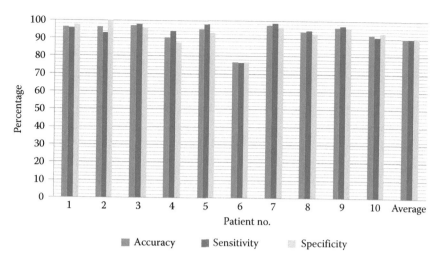

FIGURE 5.5
Evaluation parameters for singular-values-based seizure detector.

Leave-one-out cross-validation is used at the validation stage with 10 folds to achieve the highest possible accuracy.

The proposed system was evaluated in a patient-specific mode. Each time, the SVM was trained and tested on an individual patient's data. The classifier gave the confusion matrix at the output, which was used for calculating the evaluation parameters. Figure 5.5 shows the results for classification accuracy, sensitivity, and specificity achieved by this system.

The mean classification accuracy for the system is 90.14% with a sensitivity and specificity of 91.22% and 89.19%, respectively. This proves that the singular values when used as features to distinguish between epileptic and normal EEG shows very promising results, and this technique can be employed in clinical use as SVD is a very fast technique from the computational perspective and requires less hardware complexity.

5.5 EEG Time-Series Analysis

A time series is defined as "a sequence of data points, measured at successive uniform intervals of time." Time-series analysis has a major application in electroencephalography as the temporal resolution of EEG is in the milliseconds range, which is higher than that of CT or MRI. Time-series analysis is used in both the frequency domain and the time domain in EEG analysis. Time-series analysis is most widely used for epileptic seizure detection

TABLE 5.3

EEG Features for Epileptic Seizure Detection

Frequency-Domain Features	Time-Domain Features
Bandwidth	Line length
Peak frequency	Root mean square amplitude
Peak power	Zero crossings
Spectral edge frequency	Minima and maxima
Total spectral power	Nonlinear energy
Intensity weighted mean frequency	Hjorth parameters (activity, mobility,
Intensity weighted bandwidth	and complexity)
Wavelet energy	Autoregressive model fit
	Skewness
	Kurtosis
	Variance

Sources: A. Temko et al., *Clinical Neurophysiology*, Vol. 122, pp. 464–473, 2011; B. Greene et al., *Clinical Neurophysiology*, Vol. 119, pp. 1248–1261, 2008.

using EEG signals. Many studies have been done in the past two decades that use features extracted from the time-domain and frequency-domain representation of an EEG signal. These features when evaluated using machine-learning-based classification algorithms show a high degree of precision in indicating the changes in the signal due to epileptic activities. A list of commonly used time-domain and frequency-domain features used for seizure detection is shown in Table 5.3 [32].

5.6 Multivariate Measures

Multivariate measures take multiple EEG channels into consideration at once. Multivariate measures are used when different EEG channels have a high correlation among them. This is useful if multiple brain regions show similar activations during an epileptic seizure. Multivariate measures are further divided into a subset called bivariate measures, which consider only two EEG channels for analysis at a time. The two notable bivariate measures are the simple synchronization measure and the lag synchronization. Simple synchronization measures are used to detect synchronous activity in different brain regions. In case of an epileptic seizure, high rhythmic activity can be seen in different brain regions, which may be synchronous. This synchronization may be helpful in correlation analysis and ultimately for dimensionality reduction for fast analysis. In a similar way, if there is synchronization in different parts of the brain shifted by some time delay, then lag synchronization is used, which may be used for the same purpose as that of simple synchronization.

5.7 Principal Component Analysis

Principal component analysis (PCA) is a mathematical procedure for dimensionality reduction. It was invented by Karl Pearson in 1901 [33]. If the data are from different sources, it finds correlation between the data and transforms it into a set of linearly uncorrelated variables. These linearly uncorrelated variables are called principal components. The dimensions of the new set may be unchanged or reduced depending on the number of linearly uncorrelated variables. Technically, PCA is an orthogonal linear transformation that transforms the data to a new coordinate system where it becomes linearly uncorrelated.

Because usually multichannel EEG analysis is preferred, PCA is a very effective technique in EEG signal processing for dimensionality reduction. It is used to remove the channels that are correlating to a large extent owing to their proximity. This helps reduce the size of the data, which facilitates computational efficiency.

For epileptic seizure detection, because the seizure usually starts from a small region and spreads all over the brain eventually, most of the EEG electrodes are correlated to one another in some way. Therefore, PCA is effective for data analysis related to epileptic seizure detection for fast processing, especially for seizure onset detection.

5.8 Closed-Loop Seizure Prevention System

Closed-loop seizure prevention systems are used in continuous EEG monitoring to suppress a seizure before or as soon as a seizure starts. In seizure onset detection, the onset of a seizure is the start of the electrical seizure activity and may precede the visual symptoms in an individual. Thus, closed-loop systems are incorporated with seizure onset detectors to suppress the seizure by giving some stimulus to the patient. Shoeb [23] used a vagus nerve stimulator (VNS) along with their SVM-based seizure onset detector to suppress the seizure as soon as the onset is detected. A lot of work is being done in the field of seizure prediction, which means to predict the occurrence of an epileptic seizure before the event.

Seizure prediction algorithms when combined with closed-loop seizure prevention systems will be ideal to stop a seizure before it occurs. This requires a robust seizure prediction mechanism so that the seizures are predicted well in time with high precision and accuracy. Seizure prediction is a very promising research area in this field, and numerous studies are under way to develop a seizure prediction system that accurately predicts the occurrence of a seizure before the event.

5.9 Summary

This chapter discussed the types and symptoms of epilepsy along with a review of the best epileptic seizure detection algorithms developed in the past few years. A new algorithm for seizure detection based on the singular values of EEG signals has also been proposed and evaluated. Time-series analysis of EEG for epileptic seizure detection and univariate and multivariate measures of analysis have also been discussed along with principal component analysis for dimensionality reduction. From application point of view, a brief introduction of closed-loop seizure prevention systems has been provided.

References

1. S. Nasehi and H. Pourghassem, Seizure detection algorithms based on analysis of EEG and ECG signals: A survey, *Neurophysiology*, Vol. 44, pp. 174–186, 2012.
2. G. Varghese, M. Purcaro, J. Motelow, M. Enev, K. McNally, A. Levin, et al., Clinical use of ictal SPECT in secondarily generalized tonic–clonic seizures, *Brain*, Vol. 132, pp. 2102–2113, 2009.
3. A. Shoeb, H. Edwards, J. Connolly, B. Bourgeois, S. Ted Treves, and J. Guttag, Patient-specific seizure onset detection, *Epilepsy and Behavior*, Vol. 5, pp. 483–498, 2004.
4. B. Litt and J. Echauz, Prediction of epileptic seizures, *The Lancet Neurology*, Vol. 1, pp. 22–30, 2002.
5. A. Liu, J. Hahn, G. Heldt, and R. Coen, Detection of neonatal seizures through computerized EEG analysis, *Electroencephalography and Clinical Neurophysiology*, Vol. 82, pp. 30–37, 1992.
6. J. Gotman, D. Flanagan, J. Zhang, and B. Rosenblatt, Automatic seizure detection in the newborn: Methods and initial evaluation, *Electroencephalography and Clinical Neurophysiology*, Vol. 103, pp. 356–362, 1997.
7. J. Gotman, J. Ives, and P. Gloor, Automatic recognition of inter-ictal epileptic activity in prolonged EEG recordings, *Electroencephalography and Clinical Neurophysiology*, Vol. 46, pp. 510–520, 1979.
8. J. Gotman, D. Flanagan, B. Rosenblatt, A. Bye, and E. Mizrahi, Evaluation of an automatic seizure detection method for the newborn EEG, *Electroencephalography and Clinical Neurophysiology*, Vol. 103, pp. 363–369, 1997.
9. H. Hassanpour, M. Mesbah, and B. Boashash, Time-frequency feature extraction of newborn EEG seizure using SVD-based techniques, *EURASIP Journal on Applied Signal Processing*, Vol. 16, pp. 2544–2554, 2004.
10. S. B. Wilson, M. L. Scheuer, R. G. Emerson, and A. J. Gabor, Seizure detection: Evaluation of the Reveal algorithm, *Clinical Neurophysiology*, Vol. 115, pp. 2280–2291, 2004.
11. A. Alkan, E. Koklukaya, and A. Subasi, Automatic seizure detection in EEG using logistic regression and artificial neural network, *Journal of Neuroscience Methods*, Vol. 148, pp. 167–176, 2005.

12. B. R. Greene, G. B. Boylan, R. B. Reilly, P. de Chazal, and S. Connolly, Combination of EEG and ECG for improved automatic neonatal seizure detection, *Clinical Neurophysiology*, Vol. 118, pp. 1348–1359, 2007.
13. P. Celka and P. Colditz, A computer-aided detection of EEG seizures in infants: A singular-spectrum approach and performance comparison, *IEEE Transactions on Biomedical Engineering*, Vol. 49, pp. 455–462, 2002.
14. M. D'Alessandro, R. Esteller, G. Vachtsevanos, A. Hinson, J. Echauz, and B. Litt, Epileptic seizure prediction using hybrid feature selection over multiple intracranial EEG electrode contacts: A report of four patients, *IEEE Transactions on Biomedical Engineering*, Vol. 50, pp. 603–615, 2003.
15. R. Esteller, J. Echauz, T. Tcheng, B. Litt, and B. Pless, Line length: An efficient feature for seizure onset detection, in *Proceedings of the 23rd Annual International Conference of the IEEE*, 2001, pp. 1707–1710.
16. A. Aarabi, R. Grebe, and F. Wallois, A multistage knowledge-based system for EEG seizure detection in newborn infants, *Clinical Neurophysiology*, Vol. 118, pp. 2781–2797, 2007.
17. N. Roche-Labarbe, A. Aarabi, G. Kongolo, C. Gondry-Jouet, M. Dümpelmann, R. Grebe, et al., High-resolution electroencephalography and source localization in neonates, *Human Brain Mapping*, Vol. 29, pp. 167–176, 2008.
18. I. N. Bankman, K. O. Johnson, and W. Schneider, Optimal detection, classification, and superposition resolution in neural waveform recordings, *IEEE Transactions on Biomedical Engineering*, Vol. 40, pp. 836–841, 1993.
19. M. A. Hall and L. A. Smith, Feature subset selection: A correlation based filter approach, in *Proceedings of the 1997 International Conference on Neural Information Processing and Intelligent Information System*, 1997, pp. 855–858.
20. A. Temko, E. Thomas, W. Marnane, G. Lightbody, and G. Boylan, EEG-based neonatal seizure detection with support vector machines, *Clinical Neurophysiology*, Vol. 122, pp. 464–473, 2011.
21. M. Saab and J. Gotman, A system to detect the onset of epileptic seizures in scalp EEG, *Clinical Neurophysiology*, Vol. 116, pp. 427–442, 2005.
22. S. Grewal and J. Gotman, An automatic warning system for epileptic seizures recorded on intracerebral EEGs, *Clinical Neurophysiology*, Vol. 116, pp. 2460–2472, 2005.
23. A. H. Shoeb, *Application of Machine Learning to Epileptic Seizure Onset Detection and Treatment*, Massachusetts Institute of Technology, CA, 2009.
24. A. Kharbouch, A. Shoeb, J. Guttag, and S. S. Cash, An algorithm for seizure onset detection using intracranial EEG, *Epilepsy and Behavior*, Vol. 22, pp. S29–S35, 2011.
25. S. Nasehi and H. Pourghassem, Epileptic seizure onset detection algorithm using dynamic cascade feed-forward neural networks, in *2011 International Conference on Intelligent Computation and Bio-Medical Instrumentation (ICBMI)*, December 14–17, 2011, pp. 196–199.
26. S. Nasehi and H. Pourghassem, Patient-specific epileptic seizure onset detection algorithm based on spectral features and IPSONN Classifier, in *IEEE 3rd International Conference on Signal Processing Systems (ICSPS)*, August 27–28, 2011, pp. 217–222.
27. Y. Khan, O. Farooq, and P. Sharma, Automatic detection of seizure onset in pediatric EEG, *International Journal of Embedded Systems and Applications (IJESA)*, Vol. 2, pp. 81–89, 2012.

28. M. Bedeeuzzaman, O. Farooq, and Y. U. Khan, Dispersion measures and entropy for seizure detection, in *2011 IEEE International Conference on Acoustics, Speech and Signal Processing (ICASSP)*, May 22–27, 2011, pp. 673–676.

29. A. F. Rabbi and R. Fazel-Rezai, A fuzzy logic system for seizure onset detection in intracranial EEG, *Computational Intelligence and Neuroscience*, Vol. 2012, p. 1, 2012.

30. A. Shahid, N. Kamel, A. S. Malik, and M. A. Jatoi, Epileptic seizure detection using the singular values of EEG signals, in *International Conference on Complex Medical Engineering (ICME)*, May 25–28, 2013, pp. 652–655.

31. A. L. Goldberger, L. A. N. Amaral, L. Glass, J. M. Hausdorff, P. C. Ivanov, R. G. Mark, et al., PhysioBank, PhysioToolkit, and PhysioNet: Components of a new research resource for complex physiologic signals, *Circulation*, Vol. 101, pp. e215–e220, 2000.

32. B. Greene, S. Faul, W. Marnane, G. Lightbody, I. Korotchikova, and G. Boylan, A comparison of quantitative EEG features for neonatal seizure detection, *Clinical Neurophysiology*, Vol. 119, pp. 1248–1261, 2008.

33. K. Pearson, LIII. On lines and planes of closest fit to systems of points in space, *The London, Edinburgh, and Dublin Philosophical Magazine and Journal of Science*, Vol. 2, pp. 559–572, 1901.

6

Neurological Injury Monitoring Using qEEG

Waqas Rasheed and Tong-Boon Tang
Universiti Teknologi PETRONAS

CONTENTS

6.1   Human Brain Dysfunctions 143
    6.1.1   Injury 144
    6.1.2   Neurodegenerative Disorder.. 144
6.2   Assessment Criteria of Neuronal Injury 145
6.3   Concussion.. 147
6.4   Post-Concussion Syndrome... 147
6.5   Neurotherapy 148
6.6   Example: Sports Concussion and Neurotherapy 148
6.7   Abnormal Brain Activity Monitoring... 149
    6.7.1   qEEG Parameters 149
    6.7.2   Monitoring Mechanism 150
6.8   qEEG in Human Behavior Studies after Neuronal Injury.......... 150
6.9   Effect of Human Diversity on the Nature of qEEG Signals........ 151
6.10 Analysis Techniques.. 152
    6.10.1 Interpolation Technique.. 152
    6.10.2 Entropy 153
        6.10.2.1 Information Entropy... 153
        6.10.2.2 Mutual Entropy 153
        6.10.2.3 Approximate Entropy.. 153
6.11 Summary.. 154
References.. 154

6.1 Human Brain Dysfunctions

The human brain is the most resilient and pliable structure in the human body as well as in nature. The brain is the source of human emotions, perceptions, thoughts, memories, and behaviors that collectively define a personality. The brain comprises hundreds of billions of neurons or nerve cells. These microscopic elements employ a chemical messaging mechanism to regulate electrical activity throughout the brain. These activities define emotions, perceptions, and thoughts. However, the intricate mechanism of how a brain forms as a mind has not been understood as yet. It must also be noted that after developed inside the mother's womb, nerve cells cannot recover from any forms of injury, but just degenerate.

The brain undergoes several types of dysfunctions caused by internal or external stimuli. The cause may be an injury or a neurodegenerative disease.

6.1.1 Injury

Any trauma (caused by external stimuli) that results in dysfunction or bruising of scalp, skull tissue, or the brain itself is categorized as a head injury. When the external force damages the brain, it normally undergoes a temporary or permanent impairment, which is commonly known as *mild cognitive impairment* (MCI) at a preliminary state. However, if this impairment worsens, it is termed as a *traumatic brain injury* (TBI) [1]. This impairment after intracranial neurological insult involves structural, functional, or both forms of insults. The structural insult physiologically damages the brain, whereas functional disruption of neuronal network connectivity delays or prevents (temporarily or prolonged) the regular brain activities from performing regular tasks.

The type of brain injury that occurs owing to internal infection or stroke is known as a nontraumatic brain injury (nTBI). This injury is due to internal stimuli, and may also trigger certain particular dysfunctions, such as epilepsy or hemorrhage.

6.1.2 Neurodegenerative Disorder

Brain cells degenerate with age. However, several diseases, such as Alzheimer's, Huntington's, and Parkinson's diseases, result in progressive loss of structural and functional connectivity of neurons. Neurons die, which eventually disrupts the structural and functional connectivity. Therapeutic advances have been partially improving the brain's functional network activity in order to improve the effects of impairment through a process called rehabilitation.

Encephalopathy, a general term used for brain dysfunction, is often observed to be associated with *delirium*. Delirium is a temporary and acutely instable state of mind that evolves with some physical or mental illness. Studies show that delirium can be distinguished from dementia using qEEG. Delirious subjects are differentiated from controls by measuring the theta rhythm, relative power in the delta band, and activity in slow waves as compared to alpha waves. On the contrary, encephalopathy subjects are differentiated from controls by estimating the relative power in alpha waves using qEEG [2–4].

6.2 Assessment Criteria of Neuronal Injury

The initial neurological examination is based on the patient's history of memory, thinking, and attention. Physical and cognitive responses are evaluated by short screening tests followed by subjective judgment. Diagnosis and monitoring of neurodegenerative disorders are performed by conducting a range of diagnostic tests, which define the type of impairment associated with a particular neuronal dysfunction (caused by injury or disease). Mini mental status examination (MMSE) is a common dementia screening test that assesses reading, writing, orientation, and memory. Neuropsychological testing is also executed to evaluate memory recall and reasoning. A periodic assessment reveals the rate of change in neurophysiological dysfunction. Advances in technology have ensured highly accurate, measured cognitive screening with 100% sensitivity. A standardized examination is devised for neurophysiological impairment as shown in Table 6.1 [5]. However, more expert testing is required in the case of physical disabilities, because the standard tests involve speaking, writing, seeing, or hearing.

The Glasgow Coma Scale (GCS) [6,7] score is most commonly used in the clinical classification of the severity of neuronal dysfunction due to injury, out of the several most commonly schemes as shown in Table 6.2. The GCS is based on a physical assessment of eye, verbal, and motor responses.

A human brain is always experiencing chemical activity throughout a person's life, even when he or she is sleeping or doing nothing. Usually, multiple low-frequency coherent networks are active when the brain is idle. The default mode network (DMN) has been the major point of focus for dementia evaluations via functional neuroimaging modalities. The DMN is primarily composed of the posterior cingulate, medial temporal regions, lateral inferior parietal regions, and medial frontal cortex. For instance, the posterior cingulate and medial temporal regions are the first to be affected in Alzheimer's disease, because they are found to be the most closely related to memory functions. Certain regional deficits established in various studies are shown in Table 6.3 [5].

Laboratory tests are also performed in order to confirm if the symptoms are only due to some physical problems, such as a deficiency of vitamin B-12 or a malfunctioning thyroid gland—and not dementia. A psychiatric evaluation also helps to eliminate depression or a similar psychological condition as the cause of the mental condition.

Clinicians evaluate network abnormalities by gaining insight into functional connectivity using functional neuroimaging modalities such as functional magnetic resonance imaging (fMRI), computed tomography (CT), diffusion tensor imaging (DTI), positron emission tomography (PET), single-photon emission computed tomography (SPECT), and electroencephalography (EEG). EEG is relatively new, and it is used to get insight into cortical processing. EEG is a noninvasive and low-cost biomarker for visualizing structural and

TABLE 6.1

Clinical Evaluation Tests

Functionality	Standard Evaluations
Screening Measures	• Addenbrooke's Cognitive Examination (ACE-R) • Frontal Assessment Battery (FAB)
Premorbid Functions	• National Adult Reading Test • Wechsler Test of Adult Reading
Intelligence/General Ability	• Wechsler Adult Intelligence Scale (WAIS-III and IV)
Memory Functions	• Wechsler Memory Scale (WMS-III and IV) • Adult Memory and Information Processing Battery • Rivermead Behavioral Memory Test • Doors and People test • Rey-Osterrieth Complex Figure • Recognition Memory Test (Words and Faces) • List Learning Tests (Rey Auditory Verbal Learning and California Verbal Learning test) • Autobiographical Memory Interview
Executive Functions and Attention	• Behavioral Assessment of Dysexecutive Syndrome (BADS) • Delis–Kaplan Executive Function System (D-KEFS) • Wisconsin Card Sorting Test • Trail Making Test • Stroop Test • Hayling and Brixton tests • Verbal Fluency (semantic categories and FAS test) • Test of Everyday Attention
Language Functions	• Confrontation Naming Tests (Boston or Graded Naming test) • Word–Picture Matching Peabody Vocabulary Test • Pyramids and Palm Trees Test • Test for Reception of Grammar (TROG) • Token Test • Boston Diagnostic Aphasia Examination (expression/comprehension sub-tests)
Visuoperceptual/Spatial Functions	• Visual and Object Space Perception Battery • Benton Line Orientation Test • Birmingham Object Recognition Battery • Behavioral Inattention Test

Source: Bokde, A. and Meaney, J., *Neurodegenerative Disorders*, O. Hardiman and C. P. Doherty, Eds. London: Springer, pp. 17–42, 2011.

TABLE 6.2

Clinical Schemes of Severity Assessment of Brain's Neuronal Dysfunction

	Glasgow Coma Scale (GCS) [6,7]	Post-Traumatic Amnesia (PTA) [8]	Loss of Consciousness (LOC) [9]
Mild	13–15	< 1 day	0–30 min
Moderate	9–13	>1 & <7 days	>30 min & <24 h
Severe	3–8	>7 days	>24 h

TABLE 6.3

Regional Deficits

Etiology of Dementia	Regional Deficits (Hypoperfusion or Hypometabolism)
Alzheimer's Disease	• Parietal, temporal, and posterior cingulated cortices are affected early • Sparing of primary sensorimotor and visual cortex subcortical nuclei and cerebellum • Changes start out asymmetric but become bilateral and confluent
Vascular Dementia	Changes affect cortical, subcortical areas, and cerebellum
Frontotemporal Dementia	• Frontal, anterior, and mesotemporal areas affected earlier and with greater severity than parietal lateral temporal cortex • Sparing of sensorimotor and visual cortex
Huntington's Disease	Caudate and lentiform nucleus affected early
Lewy Body Dementia	Similar to Alzheimer's disease (AD) but less sparing of occipital cortex

Source: Bokde, A. and Meaney, J., *Neurodegenerative Disorders*, O. Hardiman and C. P. Doherty, Eds. London: Springer, pp. 17–42, 2011.

metabolic abnormalities. One such abnormality may be excessive or deficient connectivity between brain areas.

Sometimes, the GCS is not estimated correctly after a TBI, or hospital records may be missing. At times, the accuracy of the acute TBI measures is uncertain. Discriminant scores based on relative power, total power, coherence, and asymmetry in the phase and amplitude of EEG variables describe the neurological basis for a patient's condition. The EEG severity index may also provide an insight during the recovery process and the kind of treatment that may improve neurological and/or neuropsychological functioning in the future [10].

6.3 Concussion

A temporary loss of regular brain functionality after a mild TBI is called concussion; however, mild TBI and concussion are used interchangeably in the literature. The symptoms usually disappear under a month's time; however, prolonged dawdling of these symptoms is associated with complications, which are mostly physical, cognitive, and emotional. Repeated or prolonged concussion increases the risk of dementia. Common types of concussion described in the literature are related to military and sports injuries.

6.4 Post-Concussion Syndrome

A set of residual symptoms that persist for more than a year's time after the injury, usually overlooked owing to a lack of clinical or neuroimaging evidence after a mild TBI, is termed as post-concussion syndrome (PCS) or shell shock. PCS subjects often suffer from memory issues, social phobia, and headaches, which are usually treated with cognitive rehabilitation and psychological assistance. However, such treatment is not effective in PCS [11].

PCS is undetectable by standard clinical evaluation procedures because the damage is very slight. Usually, clinicians conduct tests related to memory, attention, and sports concussion as described in Table 6.1. Doctors can also prescribe a CT scan or an MRI. However, quantitative EEG is highly sensitive for diagnosis and prognosis of PCS, and it is used throughout *neurotherapy* in order to recover from PCS.

6.5 Neurotherapy

Studies suggest that thalamocortical dysrhythmia (lack of synchronization in the neurons of the thalamus) is the basis of psychiatric disorders [12]. Neurotherapy, also known as biofeedback, is a computer-assisted painless treatment method that trains brain waves in order to improve memory, attention, and cognitive functions. Mood disorders are also observed in many cases, followed by anxiety, concentration, dizziness, or sleep disorders.

6.6 Example: Sports Concussion and Neurotherapy

Studies suggest that football is responsible for a relatively higher rate of concussion as compared with other sports; similarly, girls are more susceptible to concussion as compared with boys within the same sports domain [13]. Young athletes heal soon, but sometimes an athlete may suffer repeated concussion when the earlier concussion is not fully healed, which causes complications. Athletes may develop emotional, cognitive, or physical issues if concussion persists for a long time or occurs multiple times before healing; this is known as *second-impact syndrome* (SIS). The situation is aggravated if such a concussion or some post-concussion syndrome remains undetectable. This may occur when the athlete returns to the field before a complete recovery from neurological damage. Minors are vulnerable to a greater threat because their brains are still immature, and any persistent symptom hinders maturity of the particular activity [14].

Regular neuroimaging modalities, such as CT/CAT, or MRI, do not prove to be useful in such a situation, because they cannot detect damage at the axon level. Topographical mapping representation indicates evidence of concussion [11], and hence qEEG-guided neurotherapy is known to significantly restore normal functionality after mild or moderate concussion.

6.7 Abnormal Brain Activity Monitoring

Most common evaluations of functional connectivity involve functional homogeneity, differentiation, or topographic reciprocities, which are measured by coherence analysis, comodulation analysis, or both. These evaluations can be performed by using qEEG recordings. Coherence or power analysis measures the phase consistency (e.g., lag) between signals. On the contrary, comodulation analysis enumerates magnitude consistency [15]. Coherent signals have a stable phase relationship, as shown in Figure 6.1. Conversely, comodulated signals have a stable magnitude relationship irrespective of the absolute difference between them, as shown in Figure 6.2.

6.7.1 qEEG Parameters

Several approaches to quantitative analysis of EEG practiced today are predominantly based on some or all of the following parameters [16].

- Frequency or period
- Amplitude
- Phase relations
- Morphology (waveform)
- Topology
- Abundance
- Reactivity and variability (e.g., continuous, random, paroxysmal)

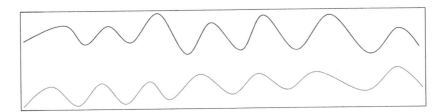

FIGURE 6.1
Example of two waves with coherent wavelength, but dissimilar amplitude.

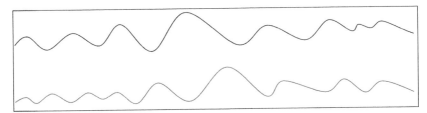

FIGURE 6.2
Example of two waves with incoherent wavelength, but similar amplitude.

6.7.2 Monitoring Mechanism

EEG voltages sensed at the scalp through the electrodes are converted to digitized frequencies during spectral analysis of EEG signals because mental processes are better reflected in the periodicities than the raw values. EEG signals are nonlinear and run the risk of being empirically uninterruptable. Nonetheless, on one level, neural coding is linearly related to perception [17], but the processing of EEG data must not be too far away from the actual recording under any approach.

Each behavioral and mental state is assumed to be reasonably homogeneous in nature during quantitative analysis of EEG signals. This assumption is termed *macrostate,* which is the basis for cognitive neuroscience. Moreover, these perceptual and cognitive operations are adequately assumed to exhibit a distinct and reliable profile of brain activity [18].

Time, frequency, and joint time–frequency analysis of EEG signals identify the features linked to neuronal disorders. Frequency analysis is specifically related to measurements of changes of the subband order, termed *spectral analysis.* Different types of neuronal disorders exhibit different kind of spectral changes; for instance, when a subject suffers a TBI, there is a significant decrease in the overall brain signal strength; conversely, epileptic subjects show very high activity at the focal point. Changes in the EEG signal due to injury is treated as a random process. Signal entropy calculations (6.10.2) are used to assess this randomness or nonstationary (or unpredictable) nature of the EEG recording of a person with a brain injury.

Studies show that psychological, behavioral, psychiatric, and neurological disorders exhibit abnormal patterns of spectral activity [19]. Neurotherapists train subjects suffering from a particular neural disorder to correct abnormal patterns of neural activity, for instance, through rehabilitation exercises. Mental rehabilitation is the process where patients suffering from neurodegenerative disorders learn how to restore their mental health. The goal is to improve learning through psychological therapies in order to achieve self-regulation of cerebral mechanisms. Neurotherapeutics are equipped with the tools that tend to improve functional connectivity inside the brain, and the behavior improves accordingly. The improvement in the functional connectivity is identified by qEEG assessment in order to monitor the recovery, and hence serves as a promising monitoring tool for the diagnosis and prognosis of neurodegenerative disease.

6.8 qEEG in Human Behavior Studies after Neuronal Injury

Behavioral analysis from brain activity using qEEG can be categorized as spontaneous and evoked. Spontaneous activity is a relaxed mental state in the absence of any task execution (e.g., relaxed wakefulness with eyes continuously opened or closed), whereas evoked activity involves a state of task execution (e.g., mathematical processing, reading). These mental states are distinct and uniform in nature, where distinct mental operations present distinct EEG and biochemical profiles. Clinicians focus on training and limiting brain activity based on a population norm; otherwise, the activity is interpreted as evidence of an abnormal neurophysiological state [20].

6.9 Effect of Human Diversity on the Nature of qEEG Signals

The degree of neuronal maturity (or immaturity) plays a vital role in evaluating a population of various ages, ethnicities, handedness (left or right), and gender. For instance, a low alpha rhythm is observed in children and is negligible in newborns, as shown in Table 6.4 [21]. It is observed that an alpha rhythm of 10 Hz is achieved when the child reaches the age of 11 years.

Children suffering from attention deficit hyperactivity disorder (ADHD) exhibit an immature frequency of the dominant thalamocortical rhythm [22]. On the contrary, a similar frequency indicates a different type of mental disorder in adults. The dominant (or peak) frequency in an EEG profile refers to the frequency range carrying the most energy in the spectrum. The peak frequency, as it is sometimes called, may exhibit topographic variability, with higher peak frequencies toward the back of the head and lower peak frequencies toward the front [23]. Furthermore, during closed eye EEG recordings, healthy adults usually exhibit a peak frequency between 8 and 12 Hz [24]. Moreover, laterality patterns associated with left-handedness are different from those associated with right-handedness [25,26]. Left-handed individuals may show abnormal hemispheric specialization such as speech functions in the right hemisphere [25,27]. Furthermore, assessment tests to discover handedness (e.g., the Edinburgh Handedness Inventory [28]) have been discovered in addition to just evaluating writing samples from each hand.

Males reveal stricter functional segregation and hence greater functional asymmetry than females [29–34].

TABLE 6.4

Alpha Rhythm and Sleep-Spindle Frequencies Observed in Preteen Children

Rhythm	Newborn	Infant	Toddler	Preschooler	Preteen
Alpha	Absent	4–6 Hz	5–8 Hz	7–9 Hz	9–10 Hz
Sleep Spindle	Absent	12–14 Hz	12–14 Hz	12–14 Hz	12–14 Hz

Source: Niedermeyer, E., *Electroencephalography: Basic Principles, Clinical Applications and Related Fields*, E. Niedermeyer and F. Lopes da Silva, Eds. Baltimore: Urban and Schwarzenberg, 1987, pp. 133–157.

6.10 Analysis Techniques

Data obtained from EEG recordings contain time (e.g., voltage amplitude, time lag) and frequency (e.g., magnitude, phase) information. Analytically, frequency analysis has been more studied than time analysis of EEG data. Frequency analysis slashes EEG data to a manageable number of coefficients. Information obtained via spectral analysis of EEG can be presented in a wide range of layouts (e.g., numerical tables, histograms, line graphs, brain maps). The most renowned layout is brain maps, which convert digital signals into values on a colored scale.

Generally, mean spectral magnitude of coherence or power for multiple frequency bands are computed and reported for subjective analysis and network diagnosis for functional connectivity variations. Evaluation of the reported information relative to existing or observed information (e.g., means, percentage change, and statistical database comparisons) is used to support subjective conclusions and recommend rehabilitation training.

The information reported only carries computations executed on values recorded at the particular positions on the scalp where the electrodes are set. The spatial accuracy can be improved by increasing the number of electrodes or by interpolation. However, interpolation cannot reconstruct the original values in the region between the electrodes. Therefore, the best way to be certain about the actual potential is to use as many electrodes as possible.

6.10.1 Interpolation Technique

In order to morph details between the scalp sensors, several interpolation techniques are used. A mathematical technique to spawn data points in the region surrounding the electrode sensors is called interpolation. The distribution of electric potentials over the scalp is smooth, and therefore interpolation furnishes fairly accurate values owing to the smearing effect. The distribution of potentials on the scalp generates an electrical field modeled by several dipoles. The values of maxima and minima of these dipoles need to be positioned as precisely as possible. K-nearest neighbors, a nonparametric classification method that is also used for interpolation, measures the closest point where the value needs to be calculated. Usually, a higher order will result in a smoother interpolation, which is deemed fit for brain maps.

The accuracy of the interpolation technique is calculated by interpolating at the position of an electrode, because it can be compared with the actual potential at that point.

Quantification of the time-series characteristics can be performed by *linear* and *nonlinear* analysis. Linear analysis deals with the amplitude component of the EEG power spectrum, whereas nonlinear analysis quantifies the EEG signal entropy [35].

6.10.2 Entropy

Information gain and learning is calculated through entropy calculation. This new concept in nonlinear analysis is often associated with more randomness and less system order [36]. EEG signals carry spatiotemporal information about potentials recorded through the electrode sensors. The temporal data are available in the form of a time series. The entropy of this time series can be quantified by using the following estimators, which indicate the dynamics and significance of a study [35]:

- Information entropy
- Mutual entropy
- Approximate entropy

6.10.2.1 Information Entropy

Let m be the dimension of the phase space of one lead data, where m is larger than the Hausdorff dimensions [37]. If the phase space is separated into l equal partitions in each dimension, then the achieved space has l^m subspaces. In such a case, each subspace carries a probability distribution $P_s(S_i)$ of the signal vector $(s = [s_1, s_2, ..., s_n], n = l^m)$. Therefore, the information entropy can be expressed as $H(S) = \sum_{i=1}^{n} P_s(S_i) \log P_s(S_i)$, where $n = l^m$.

6.10.2.2 Mutual Entropy

Mutual entropy measures the statistical dependencies between two EEG time series for linear and nonlinear analyses altogether. If Q and S are two subsystems with a dimension of m and partition l, then the joint entropy of the two is calculated with $H(QS) = \sum_{i=1}^{m} P_{sq}(S_i q_i) \log [P_{sq}(S_i q_i)]$, where $M = m^{2m}$. The mutual entropy of subsystems S to Q can be calculated as $ME(QS) = H(S) + H(Q) - H(QS)$. A lower $ME(QS)$ output reflects lower information gain and vice versa.

6.10.2.3 Approximate Entropy

A relatively new estimator, known as approximate entropy (ApEn), describes the regularity of a system, where larger values indicate a higher complexity [38,39]. Very limited and noisy data are well analyzed by using ApEn owing to its scale-invariant and model-independent nature.

Consider an EEG epoch series $\{x_i\}$, where $i = 1, 2, ..., N$, from a series of m vectors. The series X_i is given by $X_i = [x(i), x(i+1), ..., x(i+m-1)]$, where $i = 1, 2, ..., N-m+1$, and series X_j is given by $X_j = [x(j), x(j+1), ..., x(j+m-1)]$, where $j = 1, 2, ..., N-m+1$. The displacement between X_i and X_j is expressed as $d(X_i, X_j) = \max_k |x(i+k-1) - x(j+k-1)|$, where $k = 1, 2, ..., m$. Count the number of X_j for every X_i, where $I \neq j$ and displacement $d(X_i, X_j) \leq r$, where $r = N^m(i)$.

If $C_r^m(i) = N^m(i)/(N-m+1)$, then after taking the natural log and averaging it over i, $\varphi^m(r) = \dfrac{1}{N-m+1} \displaystyle\sum_{i=1}^{N-m+1} \ln C_r^m(i)$. If the dimension is increased by 1, then $ApEN(m,r,N) = \varphi^m(r) - \varphi^{m+1}(r)$, which is clearly insensitive to noise.

6.11 Summary

This chapter discusses the common human brain dysfunctions and focuses on the diagnosis, prognosis, and treatment of head injury. It provides an insight into the clinical processes and standard evaluation criteria involved in assessing the severity of brain injury and the associated neuroimaging modalities. Concussion and neurotherapy techniques are reviewed with an example of sports-related concussion. The importance of qEEG is also highlighted here, shedding some light on the commonly exercised parameters and their monitoring mechanism. The effects of neuronal injury on human behavior and variation in population are elucidated. Information on analysis techniques is presented, followed by a description of different entropy estimation techniques on EEG measurement pairs.

References

1. Phelps, C., *Traumatic Injuries of the Brain and Its Membranes*, New York: Appleton, 1897.
2. Jacobson, S., Leuchter, A., and Walter, D., Conventional and quantitative EEG in the diagnosis of delirium among the elderly, *Journal of Neurology, Neurosurgery, and Psychiatry*, 56(2), 153–158, 1993.
3. Brenner, R. P., Utility of EEG in delirium: Past views and current practice, *International Psychogeriatrics/IPA*, 3(2), 211–229, 1991.
4. Thomas, C., Hestermann, U., Walther, S., Pfueller, U., Hack, M., Oster, P., et al., Prolonged activation EEG differentiates dementia with and without delirium in frail elderly patients, *Journal of Neurology, Neurosurgery, and Psychiatry*, 79(2), 119–125, 2008.
5. Bokde, A. and Meaney, J., Advances in diagnostics for neurodegenerative disorders, in *Neurodegenerative Disorders*, O. Hardiman and C. P. Doherty, Eds. London: Springer, pp. 17–42, 2011.
6. Jones, C., Glasgow coma scale, *The American Journal of Nursing*, 79(9), 1551–1557, 2001.
7. Teasdale, G. and Jennett, B., Assessment of coma and impaired consciousness: A practical scale, *The Lancet*, 2(7872), 81–84, 1974.

8. Russell, W. R. and Smith, A., Post-traumatic amnesia in closed head injury, *Archives of Neurology*, 5(1), 4–17, 1961.

9. Eden, K. and Turner, J. W. A., Loss of consciousness in different types of head injury, *Proceedings of the Royal Society of Medicine*, 34(11), 685–691, 1941.

10. Thatcher, R. W., An EEG severity index of traumatic brain injury, *Journal of Neuropsychiatry*, 13(1), 77–87, 2001.

11. Duff, J., The usefulness of quantitative EEG (QEEG) and neurotherapy in the assessment and treatment of post-concussion syndrome, *Clinical EEG and Neuroscience: Official Journal of the EEG and Clinical Neuroscience Society (ENCS)*, 35(4), 198–209, 2004.

12. Can, R., Are thalamocortical rhythms the Rosetta Stone of a subset of neurological disorders? *Nature Medicine*, 5(12), 1349–1351, 1999.

13. Halstead, M. E. and Walter, K. D., American Academy of Pediatrics: Clinical report—Sport-related concussion in children and adolescents, *Pediatrics*, 126(3), 597–615, 2010.

14. Guskiewicz, K. and Bruce, S., National Athletic Trainers' Association position statement: Management of sport-related concussion, *Journal of Athletic Training*, 39(3), 280–297, 2004.

15. Angel, A., Processing of sensory information, *Progress in Neurobiology*, 9(1–2), 1–122, 1977.

16. Brazier, M. A. B., Cobb, W. A., Fischgold, H., Gastaut, H., Gloor, P., Hess, R., et al., Preliminary proposal for an EEG terminology by the Terminology Committee of the International Federation for Electroencephalography and Clinical Neurophysiology, *Electroencephalography and Clinical Neurophysiology*, 13(4), 646–650, 1961.

17. Johnson, K. O., Hsiao, S. S., and Blake, D. T., Linearity as the basic law of psychophysics: Evidence from studies of the neural mechanisms of roughness magnitude estimation, in *Somesthesis and the Neurobiology of the Somatosensory Cortex*, Franzén, O., Johansson, R., and Terenius, L., eds. Basel: Birkhauser Verlag, pp. 213–228, 1996.

18. Gevins, A. S., Analysis of the electromagnetic signals of the human brain: Milestones, obstacles, and goals, *IEEE Transactions on Bio-Medical Engineering*, 31(12), 833–850, 1984.

19. Hughes, J. R. and John, E. R., Conventional and quantitative electroencephalography in psychiatry, *The Journal of Neuropsychiatry and Clinical Neurosciences*, 11(2), 190–208, 1999.

20. Saxby, E. and Peniston, E. G., Alpha-theta brainwave neurofeedback training: An effective treatment for male and female alcoholics with depressive symptoms, *Journal of Clinical Psychology*, 51(5), 685–693, 1995.

21. Niedermeyer, E., Maturation of the EEG: Development of waking and sleep patterns, in *Electroencephalography: Basic Principles, Clinical Applications and Related Fields*, E. Niedermeyer and F. Lopes da Silva, Eds. Baltimore: Urban and Schwarzenberg, pp. 133–157, 1987.

22. Fernández, T., Harmony, T., Rodríguez, M., Reyes, A., Marosi, E., and Bernal, J., Test-retest reliability of EEG spectral parameters during cognitive tasks: I. Absolute and relative power, *The International Journal of Neuroscience*, 68(3–4), 255–261, 1993.

23. Gratton, G., Villa, A. E., Fabiani, M., Colombis, G., Palin, E., Bolcioni, G., et al., Functional correlates of a three-component spatial model of the alpha rhythm, *Brain Research*, 582(1), 159–162, 1992.

24. Nunez, P. L., *Electric Fields of the Brain: The Neurophysics of EEG*, New York: Oxford University Press, 1981.

25. Shepherd, R., EEG correlates of sustained attention: Hemispheric and sex differences, *Current Psychological Research*, 2(1–3), 1–19, 1982.

26. Provins, K. A. and Cunliffe, P., The relationship between E.E.G. activity and handedness, *Cortex*, 8(2), 136–146, 1972.

27. Rasmussen, T. and Milner, B., The role of early left-brain injury in determining lateralization of cerebral speech functions, *Annals of the New York Academy of Sciences*, 299, 355–369, 1977.

28. Oldfield, R. C., The assessment and analysis of handedness: The Edinburgh inventory, *Neuropsychologia*, 9(1), 97–113, 1971.

29. McGlone, J., Sex differences in functional brain asymmetry, *Cortex: A Journal Devoted to the Study of the Nervous System and Behavior*, 14(1), 122–128, 1978.

30. McGlone, J., Sex differences in human brain asymmetry: A critical survey, *Behavioral and Brain Sciences*, 3(2), 215, 2010.

31. Trotman, S. C. A. and Hammond, G. R., Sex differences in task-dependent EEG asymmetries, *Psychophysiology*, 16(5), 429–431, 1979.

32. Tucker, D. M., Sex differences in hemispheric specialization for synthetic visuo-spatial functions, *Neuropsychologia*, 14(4), 447–454, 1976.

33. Flor-Henry, P. and Koles, Z., EEG characteristics of normal subjects: A comparison of men and women and of dextrals and sinistrals, *Research Communications in Psychology*, 7(1), 21–38, 1982.

34. Lake, D. A. and Bryden, M. P., Handedness and sex differences in hemispheric asymmetry, *Brain and Language*, 3(2), 266–282, 1976.

35. Wan, B., Ming, D., Qi, H., Xue, Z., Yin, Y., Zhou, Z., et al., Linear and nonlinear quantitative EEG analysis, *IEEE Engineering in Medicine and Biology Magazine: The Quarterly Magazine of the Engineering in Medicine and Biology Society*, 27(5), 58–63, 2008.

36. Jeong, J., Chae, J. H., Kim, S. Y., and Han, S. H., Nonlinear dynamic analysis of the EEG in patients with Alzheimer's disease and vascular dementia, *Journal of Clinical Neurophysiology: Official Publication of the American Electroencephalographic Society*, 18(1), 58–67, 2001.

37. Vastino, J. A. and Swinney, H. L., Information transport in spatial-temporal system, *Physics Review Letters*, 60(18), 1773–1776, 1988.

38. Radhakrishnan, N. and Gangadhar, B. N., Estimating regularity in epileptic seizure time-series data, *IEEE Engineering in Medicine and Biology Magazine*, 17(3), 89–94, 1998.

39. Pincus, S. M. and Keefe, D. L., Quantification of hormone pulsatility via an approximate entropy algorithm, *The American Journal of Physiology*, 262(5, pt 1), E741–54, 1992.

7

Quantitative EEG for Brain–Computer Interfaces

Anton Albajes-Eizagirre, Laura Dubreuil-Vall, David Ibáñez, Alejandro Riera, Aureli Soria-Frisch, Stephen Dunne, and Giulio Ruffini

Starlab Barcelona

CONTENTS

7.1 Introduction..157
7.2 Event-Related Potentials for BCI...158
    7.2.1 P300 as a Tool for Attention Measurement.................158
    7.2.2 Steady-State Visual Evoked Potentials.........................161
7.3 Affective Computing: Novel BCIs..162
7.4 Arousal–Valence Detection 163
7.5 A qEEG Platform for Visualizing Emotional Features.........165
7.6 Stress Detection..167
7.7 Therapeutic Usage of qEEG and BCI: NF for ADHD 168
7.8 EEG Patterns...169
7.9 NeuroSurfer: NeuroElectrics NF Application 169
7.10 Conclusions...170
References..171

7.1 Introduction

This chapter attempts to give different examples of applications of quantitative EEG (qEEG) for the realization of brain–computer interfaces (BCIs). In particular, we think that neurotechnologies for BCIs based on electroencephalography (EEG) offer advanced tools of interest for other research and application fields. EEG–BCI technologies are based on the electrical nature of brain activity. EEG–BCI captures the user state or intent via electrical signals recorded on the scalp and maps them into computer or device commands. As the reader can imagine, such a process goes through an intermediate process of quantification. We explain some of these techniques in the following chapter.

We describe in Section 7.2 two of the most classical paradigms used in BCIs for communication. The so-called P300 paradigm is often used in BCIs for spelling purposes. We detail here its relationship to attentional processes and the way it can be measured. Moreover, we discuss the so-called steady-state visual evoked potential (SSVEP) responses, which are elicited while observing a flickering light source.

Affective Computing is an emerging field within BCI that involves detecting the emotional content within the subjects' brain activity. Arousal, workload, and stress can be studied through EEG quantitative analysis. We describe in Sections 7.4 to 7.6 different developments and experimental studies that we have conducted in this field.

Neurofeedback (NF) is a type of user feedback that uses real-time displays of EEG measurements of brain activity features. qEEG has been extensively used in this therapeutic field. EEG features are extracted and displayed, allowing the user to modulate their temporal evolution. This results in neuromodulatory techniques based on NF training with therapeutic applications, which are detailed in Section 7.7 on attention deficit hyperactivity disorder (ADHD). We finally present in Section 7.10 some conclusions to this overview.

7.2 Event-Related Potentials for BCI

7.2.1 P300 as a Tool for Attention Measurement

We recall in this section the quantification of attention in mental processes as described in Ref. [1]. BCIs have been using a particular paradigm based on attentional processes denoted P300 [2,3] for communication. The main idea of the experimental paradigm is that when an external stimulus attracts the attention of a subject, a characteristic brain waveform appears [4]. A typical P300 waveform or complex is formed by a negative peak at approximately 200 ms followed by a positive one of larger amplitude at 300 ms. The peak at 300 ms, which is formed by two different components, can be used for attentional studies. In particular, the component of the P300 event-related potential, which originates in the frontal part of the brain, is associated with attention mechanisms during task processing, whereas the one found in the tempoparietal part relates to memory processes [5]. In this communication we extend the observation of peripheral muscle activity [1] with the measurement of such brain activity. We expect from this paradigm shift to get a more direct access to attentional processes as originated in the brain. This could lead to a more objective quantification of the relationship between, for example, creativity and attention.

In particular, we describe our work with the so-called rapid serial visual presentation (RSVP) paradigm [6]. In this experimental setup, a sequence of images is shown to the experimental subjects. The images are shown at a high presentation rate of between 100 and 1000 ms. The time between stimuli has an influence on the P300 amplitude as shown in Ref. [5]. In such a setup, the subject is instructed to consciously detect a particular image in the sequence. Each time this image appears, her brain activity shows the so-called P300 waveform. We describe here the results obtained with such a setup.

Three different subjects went through the RSVP of 500 natural images in sets of 10 different pictures at the rate of 10 images per second. The subjects were told to detect one particular image in each of the sets. The image to be detected is denoted as the target stimulus, and the remaining images in the set are the nontarget stimuli. Each image of a 10 set gilts as the target image five times for each subject. During the experiment, the brain activity of the subject was monitored through an EEG device using 32 different channels placed in accordance with the 10–20 standard positioning.

In order to conduct the data analysis, we have used the methodology described in Ref. [7], which uses an ensemble of support vector machines (SVMs). The first stage of this methodology includes filtering the data in the band 0.1 to 10 Hz, decimating the obtained signals, and cutting a window of 667 ms duration after stimuli presentation. We show the grand averages of the waveforms obtained on all target stimuli in Figure 7.1. We have selected

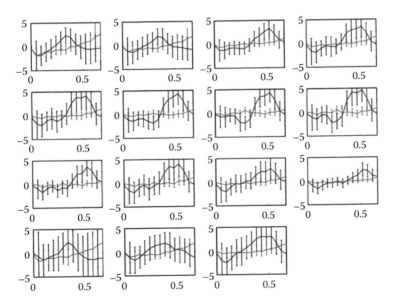

FIGURE 7.1
Grand averages with standard error bars of signals obtained with target (dark) vs. nontarget (light) stimuli in electrode positions Fp1, AF3, T7, C3, CP1, CP5, P7, P3, CP6, CP2, C4, T8, Fp2, Fz, and Cz (from top left to bottom right).

here by visual inspection the channels presenting the best response, and used them later in the classification. It is worth mentioning that the latency of the peak corresponding to the P300 appears delayed with respect to the usual value at 300 ms owing to the complexity of the stimuli used in the experiment.

After the signals have been preprocessed and reduced to 15 component feature vectors per channel, we concatenate each channel feature and deliver it to a set of 10 classifiers in order to measure the classification performance. Here, the goal is to discriminate between target and nontarget images based on the P300 waveforms. We use as performance the classification rate, that is, the percentage of right-classified stimuli. Such a classification can be used to characterize the attentional process that leads each subject to detect the presence of the target images. As in the case of the feature vectors (see Figure 7.1), the classification result will improve when averaging the detection of several results. Hence, we characterize the performance in a plot that depicts the classification rate with respect to the number of averaged classification outputs, that is, the number of times an image has to be shown to a subject in order to present a classifiable P300 waveform. The result is shown in Figure 7.2. As can be observed, the methodology gives excellent performance for two out of three subjects. The poor performance in Subject2 could already be observed in his grand average plots,

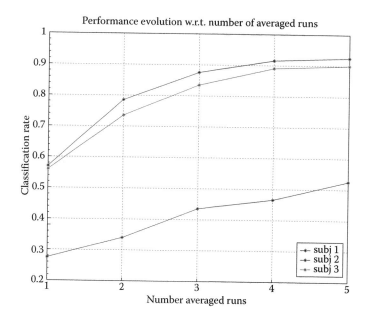

FIGURE 7.2
Classification performance of the methodology used when averaging the classification results over different runs for the three different experimental subjects (see legend).

which do not show a clear distinction between target and nontarget stimuli. Such a poor performance could be due to a lack of attention during the experimental task.

The results described in this section, that is, a characterization of the P300 wave with respect to its latency and amplitude, and its classification performance, can be used for the objective measurement of attentional psychophysical features.

7.2.2 Steady-State Visual Evoked Potentials

In recent years, there has been increased interest in using SSVEPs in BCI systems. The SSVEP approach currently provides the fastest and most reliable communication, making it the ideal BCI paradigm. SSVEP-based BCIs offer two main advantages over BCIs based on other electrophysiological sources: (1) They have a higher information transfer rate and (2) they require a shorter calibration time [8].

SSVEP is a resonance phenomenon arising mainly in the visual cortex when a person is focusing his or her visual attention on a light source flickering with a frequency from 1 to 100 Hz [9]. When an SSVEP is elicited, it is manifested as oscillatory components in the user's EEG, especially in the signal from the primary visual cortex [10]. SSVEP-based BCIs can be classified into three categories depending on the specific stimulus sequence modulation in use: time-modulated VEP (t-VEP) BCIs, frequency-modulated VEP (f-VEP) BCIs, and pseudorandom-code-modulated VEP (c-VEP) BCIs [11].

BCI systems based on f-SSVEP have been the approach most commonly used in BCI research. In f-SSVEP-based systems, each target is flashed at a unique frequency, generating a resonance response that can be measured in the EEG as oscillatory components matching the repetitive visual stimulation (RVS) frequency and its harmonics. When an SSVEP is elicited, its response is characterized by an energy increase at the RVS frequency, and its harmonics are phase-locked with the stimulus [12].

Frequency is a crucial property of the RVS. An SSVEP is elicited when a person is focusing his or her visual attention on a light source flickering with a frequency ranging from 1 to 100 Hz [9]. The SSVEP response has a strong subject and is RVS frequency dependent; therefore, for practical SSVEP-based BCI systems construction, it is recommended to tune this parameter for each subject under evaluation in order to obtain a large SSVEP response (Figure 7.3). The choice of the flickering frequency determines which cortical network synchronizes with the flickering frequency [13], and therefore the electrode or combination of electrodes in which the SSVEP response appears.

Practical SSVEP-based BCI systems in general are compound of several targets consisting of RVS sources flickering at different frequencies. When the system detects the elicited response corresponding to one of the frequencies under evaluation, it carries out the associated action. Two main types

FIGURE 7.3
User controlling an SSVEP-based BCI application with four degrees of freedom.

of SSVEP-based BCI paradigms exist: asynchronous and synchronous. Synchronous paradigms, also known as cued-paced BCI paradigms, do not consider the possibility that the user does not wish to communicate. RVS is always present, and the SSVEP response is continuously evaluated. Although these BCIs are relatively easy to use and develop, they are impractical for many in many real-world settings [14].

In asynchronous or self-paced BCI, users can interact with the BCI at their leisure, without worrying about well-defined time frames [15]. In this case, the system detects when the user wants to interact with the BCI, switching on the stimulation sources during a short time-defined period. During this period, the user focuses his or her attention on the target whose associated action is to be carried out. After the stimulation, the response of the target frequencies is evaluated, and the frequency responsible for eliciting the evoked potential is determined (Figure 7.4) [16].

7.3 Affective Computing: Novel BCIs

The effect of emotional and affective states of the user on human–computer interaction systems has been widely studied [7–19]. Moreover, the impact of affective states on BCI systems' performance has also been studied [20,21], proving the importance of incorporating mechanisms for dealing with affective states on the realization of BCIs. Affective computing methodologies and techniques have been used to develop BCI systems in which performance was improved by taking into account the affective state of the user [22,23].

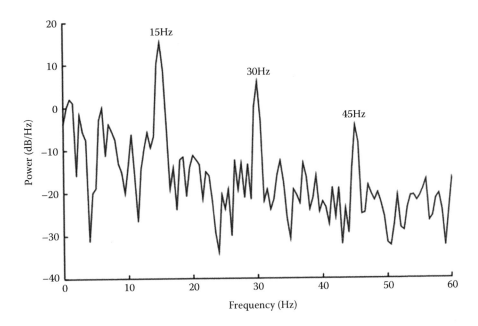

FIGURE 7.4
SSVEP power response for a 15 Hz stimulation frequency. SSVEP manifests oscillations at the stimulation frequency and higher harmonics. (From Zhu, D, et al., *Computational Intelligence and Neuroscience*, 1, 2010).

We present in the following sections our studies on emotion detection and on stress detection using EEG that can be applied to the realization of advanced BCI systems incorporating affective computing to achieve high levels of performance.

7.4 Arousal–Valence Detection

Following the dimensional approach to model emotions, we can take the valence state of the subject as the first dimensional state, and the arousal state of the subject as the second dimensional state. From the standpoint of the detection's method, these two dimensions are usually considered to be independent. Therefore, the emotional state on each of them is measured separately, and a variety of proposed detection methods are available for valence and arousal.

The detection of arousal using EEG has been widely published and different methodologies have been proposed. However, the most common approaches are those that make use of the relation between asymmetric frontal cortical activity and the valence state [24]. Studies show that in positive-valence states, the left hemisphere of the brain cortex presents higher activity compared to the right hemisphere, whereas in negative-valence states, the right hemisphere presents higher activity than the left one [25,26]. Moreover, according to recent publications, this asymmetric activation is stronger or specific to the frontal cortex [27].

In order to measure the activation of each hemisphere or region, some studies exploit the reported evidence that brain activity is inversely related to the preeminence of the alpha (8–12 Hz) frequency band of an EEG signal. In low brain activity periods, the alpha band increases its power relative to other bands, and vice versa [28].

To compute the asymmetric activation of the right and left hemisphere, the power in the alpha band is computed in the right and left hemispheres, obtaining what is commonly called alpha asymmetry: alpha power on the right hemisphere minus alpha power on the left hemisphere. Positive values of this value will respond to greater activity of the left hemisphere relative to the right hemisphere and, therefore, correspond to positive valence states. On the contrary, negative values of alpha asymmetry will correspond to negative valence states.

Following this asymmetry-based approach, we developed a valence state detector based on EEG. Placing the electrodes on the right and left hemisphere locations for the frontal and prefrontal cortex, we averaged the alpha activation of the electrodes located on the right hemisphere and the electrodes of the left hemisphere. The difference between the averaged activations of the right hemisphere and those of the left hemisphere is taken as the value corresponding to the valence state of the subject.

While for valence most of the methodologies share a clear approach based on asymmetric activation, the different methodologies to detect the arousal state from brain activity adopt a great variety of approaches. Arousal states have traditionally been detected from the reactions of the autonomic nervous system through the monitoring of the heart rate, skin temperature, or the galvanic skin response. However, more recently brain activity has also been used to detect arousal. Some publications related high mental workloads with high arousal states [29], and low mental states and boredom with low arousal [30]. Other publications related event-related reactions in different bands and locations of the brain to arousal. Variances of event-related synchronization and desynchronization in alpha (8–12 Hz) and theta (4–8 Hz) bands in different areas of the brain to different arousal states induced by pictures have been reported [31]. Gamma bands (30–65 Hz) have also been reported to present higher activation for high arousal states [32].

Following the mental-load approach, we developed an arousal state detector based on EEG. Taking advantage of the reported activation of beta bands upon higher workloads, we took the ratio between beta activation and alpha activation to compute the ratio of activation versus deactivation. This ratio corresponds to the arousal state in our detection.

7.5 A qEEG Platform for Visualizing Emotional Features

In this section, we present an EEG-based system for tracking emotions in real time, building on the application programming interface (API) for the Enobio® electrophysiological sensor*. It is based on the valence–arousal framework [33], a representation in which the arousal dimension measures how dynamic the emotional state is, and the valence is a global measure of the positive or negative feeling associated with the state.

The graphical user interface (GUI) consists of a control panel (Figure 7.5a) in which the user can configure the Enobio® setup, and an emotion visualization panel (Figure 7.5b), in which the detected emotions are shown in real time.

The control panel allows the user to configure the electrode positions that will be used to compute the valence and arousal values. Moreover the power in the alpha and beta bands will be averaged over the selected channels for each element in the valence and arousal calculation, as follows:

$$\text{Valence} = \frac{1}{I}\sum_{i=1}^{I} P_{\text{alpha}}(\text{channel}_i) - \frac{1}{J}\sum_{j=1}^{J} P_{\text{alpha}}(\text{channel}_j) \qquad (7.1)$$

$$\text{Arousal} = \frac{1}{K}\sum_{k=1}^{K} P_{\text{beta}}(\text{channel}_k) / P_{\text{alpha}}(\text{channel}_k), \qquad (7.2)$$

where P_{alpha} (channel$_i$) and P_{beta} (channel$_i$) are the powers in the alpha and beta bands in the ith channel, and I, J, and K are the number of selected channels for each element. The user can also select among four normalization methods applied to the valence and arousal values: (1) range normalization; (2) historical normalization, in which the minimum and maximum are computed only over the last L samples; (3) sigmoid normalization, where sigmoid parameters can be set up; and (4) no normalization.

The emotion visualization panel consists of a two-dimensional plot in which the valence and arousal are represented on the horizontal and vertical axes, respectively. Such a representation was proposed by Russel [34]. Every time a valence–arousal pair is computed, an emotion is represented with a circle in its corresponding position, color, and size. The position of

---

* http://neuroelectrics.com/enobio.

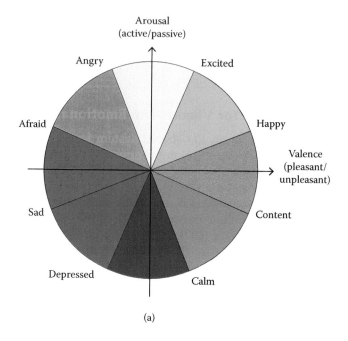

(a)

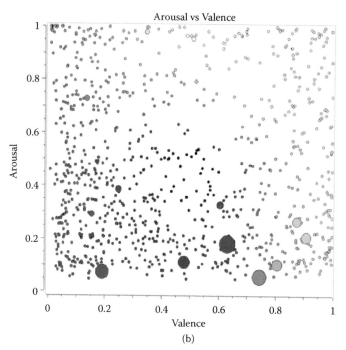

(b)

FIGURE 7.5
(a) Plutchik color scheme and (b) its usage in the visualization panel.

the circle depends on the values of the valence and arousal, and the color of the circle is set according to a color scheme derived from Plutchik [35] (Figure 7.5b). The size of the circle is determined by the recentness of the sample—the more recent the sample, the bigger the circle; as time passes, the size of the older samples becomes smaller. The resulting system outputs a new emotion every 0.4 s, taking into account that the Enobio outputs 500 samples per second and that the values of valence and arousal are computed every block of 200 samples.

7.6 Stress Detection

According to [36], psychological stress occurs when an individual perceives that the environmental demands tax or exceed his or her adaptive capacity. In this study, we focused on this type of stress and applied a protocol in which several tasks were designed to stress participants to different levels. The protocol included a baseline recording, a reading task (low stress), a Stroop test, an arithmetic task (medium stress), and finally a false blood sample test (high stress). In the first part of our study, we computed the averages over the 12 participants of our EEG features.

Following the Rel model of emotions [34], "stress" emotion is characterized by a negative valence and a positive arousal, which can be placed in a line over the plus and asterisk sectors of the diagram given in Figure 7.6. Based on the literature (for example, [37], [38], and [39]), we see that alpha asymmetry is related to the valence dimension of emotions, whereas the beta/alpha ratio is related to the arousal dimension. In our study, we have focused

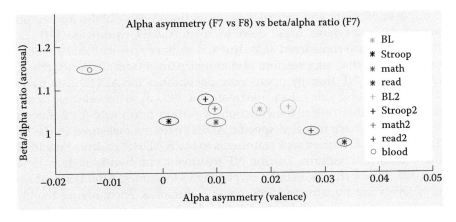

FIGURE 7.6
Average over subjects of the beta/alpha ratio and the alpha asymmetry for each of the tasks of our protocol.

on these two EEG features, and in order to extract them, we used three EEG channels (F7, F8, and Cz) as a reference.

The results shown in Figure 7.6 are as expected: our low-stress tasks have higher valence and lower arousal, whereas our high-stress tasks have the opposite characteristics. The medium-stress tasks fall in between these two extremes.

In the next step of our study and in order to study the possibility of detecting stress levels on a subject-to-subject basis, we applied machine learning techniques in order to train a classifier (linear discriminant analysis) using a leave-one-subject approach. In this case, we obtain a performance as high as 88% between the Baseline and the Fake Blood Sample task, 72% between the Stroop test and the Blood Sample, and finally 83% between the Baseline and the Stroop test.

The conclusion of this study is that EEG can be used to extract the stress level, and it is interesting to note that only three EEG channels are needed, making the system unobtrusive. Moreover, this system has the capacity to work in real time, making it very suitable for augmented reality, NF applications, and stress management therapies.

7.7 Therapeutic Usage of qEEG and BCI: NF for ADHD

ADHD is the most common psychiatric disorder in children (2–5%) [40] and is generally diagnosed in children who exhibit attention difficulties, impulsive behaviors, and extreme levels of hyperactivity. Children with ADHD frequently also exhibit a variety of physical problems such as headaches and immune system deficiencies that cause frequent illnesses.

For over 50 years and increasingly more since 1990s, Ritalin and amphetamine derivates have been used to treat ADHD worldwide [41]. Even though they are considered safe drugs, they have frequent side effects, for example, appetite suppression, abdominal pain, insomnia, headache, and anorexia [42]. NF therapy opens new possibilities for ADHD care by providing an adverse-side-effect-free treatment. EEG measures electric currents in the brain that reflect the function of certain brain activities. Since the 1970s, studies have revealed specific ADHD patterns measured in the EEG. The goal of EEG biofeedback training is to teach ADHD patients how to alter these abnormal patterns. During NF treatment, children learn how to self-regulate these patterns, usually by playing videogames. Attention and alertness levels are transformed into game commands, encouraging the child to operate at the desired levels. NF training has many therapeutic applications, including ADHD. Creative states and meditation patterns can also be found in the EEG, and the same training as in a clinical NF application can be applied.

7.8 EEG Patterns

The most common pattern found in ADHD is the excess slow wave activity in the frontal regions [42]. There are numerous other ways of looking at subtypes of ADHD based on EEG band power analysis. Excess theta activity is the most common type, as reflected in theta–beta ratios using a single channel [43]. There could also be excess alpha activity, usually in the lower alpha range (8–10 Hz). The goal of EEG biofeedback training is to teach the child how to alter these abnormal brain waves and normalize the EEG activity.

Band power meditation patterns can also be found in the EEG. Significantly increased theta power was found during nondirective meditation, while there was also a significant increase in alpha power [44]. Professionally significant enhancement of music, dance, performance, and mood has followed training with an EEG-neurofeedback protocol that increases the ratio of theta to alpha waves using auditory feedback with eyes closed [45].

7.9 NeuroSurfer: NeuroElectrics NF Application

NeuroSurfer, a flexible, general-purpose, fully configurable NF application developed by Neuroelectrics®, aims to address ADHD treatment requirements. The application has been designed to satisfy the needs of both experienced researchers and clinicians. NeuroSurfer monitors the temporal power evolution of customized frequency bands, fulfilling almost every NF band-power-based training protocol. NeuroSurfer uses Enobio® in its 8-channel version, a wearable, wireless electrophysiology sensor system for the recording of EEG, which makes it simple and quick to set up the device montage; another added advantage is that the patient need not be attached to the measuring device. NeuroSurfer performs robust band power calculations in real time. Signal processing stages include filtering, referencing, eye artifact correction, windowing, and power spectral density extraction. The calculated power feature is presented to the clinician/researcher, allowing him or her to monitor its temporal evolution and its levels is used to command the patient's video game application.

NeuroSurfer gives the possibility of automatically configuring the NF application parameters for most ADHD popular training protocols, including the ones previously described. Experienced users have also the chance of fully configuring the application setting.

During the NF training, the operator is able to follow the calculated feature evolution in real time. NeuroSurfer feature monitoring includes temporal feature evolution (Figure 7.7), power spectral density (PSD), and bar representation. qEEG (Figure 7.8) of the feature at every connected electrode is also presented. BrainSurfer also provides simple statistics of the performance

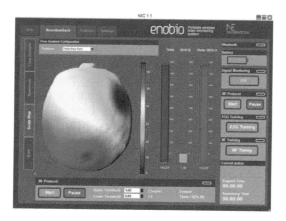

FIGURE 7.7
Attention deficit hyperactivity disorder (ADHD) feature monitoring by qEEG.

FIGURE 7.8
Attention deficit hyperactivity disorder (ADHD) temporal feature evolution monitoring.

of both training trials and the entire training session, which is useful for following the treatment evolution. Current NF clinical applications such as Neurosurfer can be easily configured to calculate meditation and creative state features, boosting the possibilities of emotional and creative NF training.

7.10 Conclusions

This chapter describes different neurotechnologies to be used for the quantification of brain activity. All types of techniques take as their starting point the electrical nature of such activity. BCIs based on the detection of event-related potentials such as the P300 or the SSVEPs described here and the detection of affects can be used for the objective measurement of cognitive processes. Different parameters of the P300 waveform and its detection can be used to quantify attentional processes. SSVEPs used the reaction to flickering stimulation sources as reflected in the brain activity. Moreover, we have described different technologies for the measurement of arousal, valence, and stress. All these interface paradigms rely on quantification techniques to be exploited in other research and application fields.

Moreover, NF, which is based on the real-time application of quantitative techniques to brain activity, is used to modify the dynamics of brain activity. The presented NF application can be tuned to monitor different brain rhythms in real time. This measurement is fed to the control loop of the visual feedback mechanism. The modulatory action of these technologies in therapy has not been exploited fully. We expect the following years to witness the full flowering of these technologies.

References

1. Friedman, R. S., Fishbach, A., Förster, J., and Werth, L., Attentional priming effects on creativity, *Creativity Research Journal*, 15(2–3), 277–286, 2003.
2. Fazel-Rezai, R., Allison, B. Z., Guger, C., Sellers, E. W., Kleih, S. C., and Kübler, A., P300 brain computer interface: Current challenges and emerging trends, *Frontiers in Neuroengineering*, 5(14), 14, 2012.
3. Farwell, L. and Donchin, E., Talking off the top of your head: Toward a mental prosthesis utilizing event-related brain potentials, *Electroencephalography and Clinical Neurophysiology*, 70, 510–523, 1988.
4. Gray, H. M., Ambady, N., Lowenthal, W. T., and Deldin, P., P300 as an index of attention to self-relevant stimuli, *Journal of Experimental Social Psychology*, 40(2), 216–224, 2004.
5. Polich, J., Updating P300: An integrative theory of P3a and P3b, *Clinical Neurophysiology: Official Journal of the International Federation of Clinical Neurophysiology*, 118(10), 2128, 2007.
6. Acqualagna, L., Treder, M. S., Schreuder, M., and Blankertz, B., A novel brain-computer interface based on the rapid serial visual presentation paradigm, in *Conference Proceedings of IEEE Engineering in Medicine and Biology Society*, 1, pp. 2686–2689, 2010.
7. Rakotomamonjy, A. and Guigue, V., BCI competition III: Dataset II-ensemble of SVMs for BCI P300 speller, *IEEE Transactions on Biomedical Engineering*, 55(3), 1147–1154, 2008.
8. Cheng, M., Gao, X., Gao, S., and Xu, D., Design and implementation of a brain-computer interface with high transfer rates, *IEEE Transactions on Biomedical Engineering*, 49(10), 1181–1186, 2002.
9. Dornhege, G., Millan, J. D. R., Hinterberger, T., and D. J. M., Eds., *Toward Brain-Computer Interfacing (Neural Information Processing)*, The MIT Press, 2007.

10. Veigl, C., Weiß, C., Ibanez, D., and Soria-Frisch, A., Carbone: Model-based design of novel human-computer interfaces—The assistive technology rapid integration and construction set (AsTeRICS), in *4th IEEE Biosignals and Biorobotics Conference (ISSNIP 2013)*, Rio de Janeiro, Brasil, February 18–20, 2013.

11. Bin, G. Y., Gao, X. R., Wang, Y. J., Hong, B., and Gao, S., VEP-based brain-computer interfaces: Time, frequency, and code modulations, *IEEE Computational Intelligence Magazine*, 4(4), 22–26, 2009.

12. Mason, S., Bashashati, A., Fatourechi, M., Navarro, K., and Birch, G., A comprehensive survey of brain interface technology designs, *Annals of Biomedical Engineering*, 35(2), 137–169, 2007.

13. Wang, Y., Gao, X., Hong, B., Jia, C., and Gao, S., Brain-computer interfaces based on visual evoked potentials, *IEEE Engineering in Medicine and Biology Magazine*, 27(5), 64–71, 2008.

14. Graimann, B., Allison, B., and Pfurtscheller, G., Brain–computer interfaces: A gentle introduction, in *Brain-Computer Interfaces*, Springer, Berlin, pp. 1–27, 2010.

15. Mason, S. G. and Birch, G. E., A brain-controlled switch for asynchronous control applications, *IEEE Transactions on Biomedical Engineering*, 47(10), 1297–1307, 2000.

16. Zhu, D., Bieger, J., Molina, G. G., and Aarts, R. M., A survey of stimulation methods used in SSVEP-based BCIs, *Computational Intelligence and Neuroscience*, 1, 2010.

17. Partala, T. and Surakka, V., The effects of affective interventions in human–computer interaction, *Interacting with Computers*, 16(2), 295–309, 2004.

18. Brave, S. and Nass, C., Emotion in human-computer interaction, in *The Human-Computer Interaction Handbook: Fundamentals, Evolving Technologies and Emerging Applications*, pp. 81–96, 2002.

19. Cowie, R., Douglas-Cowie, E., Tsapatsoulis, N., Votsis, G., Kollias, S., Fellenz, W., et al., Emotion recognition in human-computer interaction, *Signal Processing Magazine, IEEE*, 18(1), 32–80, 2001.

20. Birbaumer, N., Breaking the silence: Brain–computer interfaces (BCI) for communication and motor control, *Psychophysiology*, 43(6), 517–532, 2006.

21. Curran, E. A. and Stokes, M. J., Learning to control brain activity: A review of the production and control of EEG components for driving brain–computer interface (BCI) systems, *Brain and Cognition*, 51(3), 326–336, 2003.

22. Garcia-Molina, G., Tsoneva, T., and Nijholt, A., Emotional brain–computer interfaces, *International Journal of Autonomous and Adaptive Communications Systems*, 6(1), 9–25, 2013.

23. Chanel, G., Emotion assessment for affective computing based on brain and peripheral signals, University of Geneva, 2009.

24. Harmon-Jones, E., Gable, P. A., and Peterson, C. K., The role of asymmetric frontal cortical activity in emotion-related phenomena: A review and update, *Biological Psychology*, 84(3), 451–462, 2010.

25. Sackeim, H. A., Greenberg, M. S., Weiman, A. L., Gur, R. C., Hungerbuhler, J. P., and Geschwind, N., Hemispheric asymmetry in the expression of positive and negative emotions: Neurologic evidence, *Archives of Neurology*, 39(4), 210, 1982.

26. Robinson, R. G. and Price, T. R., Post-stroke depressive disorders: A follow-up study of 103 patients, *Stroke*, 13(5), 635–641, 1982.

27. Schutter, D. J., van Honk, J., d'Alfonso, A. A., Postma, A., and de Haan, E. H., Effects of slow rTMS at the right dorsolateral prefrontal cortex on EEG asymmetry and mood, *Neuroreport*, 12(3), 445–447, 2001.

28. Cook, I. A., O'Hara, R., Uijtdehaage, S. H., Mandelkern, M., and Leuchter, A. F., Assessing the accuracy of topographic EEG mapping for determining local brain function, *Electroencephalography and Clinical Neurophysiology*, 107(6), 408–414, 1998.

29. Brookhuis, K. A. and De Waard, D., Assessment of drivers' workload: Performance, subjective and physiological indices, in *Stress, Workload and Fatigue: Theory, Research and Practice*, P. Hancock and P. Desmond, Eds., Lawrence Erlbaum, NJ, pp. 321–333, 2001.

30. Kroemer, K. and Kroemer-Elbert, E., *Ergonomics: How to Design for Ease and Efficiency*, Prentice Hall, 2001.

31. Aftanas, L. I., Varlamov, A. A., Pavlov, S. V., Makhnev, V. P., and Reva, N. V., Time-dependent cortical asymmetries induced by emotional arousal: EEG analysis of event-related synchronization and desynchronization in individually defined frequency bands, *International Journal of Psychophysiology*, 44(1), 67–82, 2002.

32. Keil, A., Müller, M. M., Gruber, T., Wienbruch, C., Stolarova, M., and Elbert, T., Effects of emotional arousal in the cerebral hemispheres: A study of oscillatory brain activity and event-related potentials, *Clinical Neurophysiology*, 112(11), 2057–2068, 2001.

33. Lang, P. J., The emotion probe: Studies of motivation and attention, *American Psychologist*, 50(5), 372–385, 1995.

34. Russel, J. A., A circumplex model of affect, *Journal of Personality and Social Psychology*, 39(6), 1161–1178, 1980.

35. Plutchik, R., *The Psychology and Biology of Emotion*, HarperCollins, New York, 1994.

36. Cohen, S., Janicki-Deverts, D., and Miller, G. E., Psychological stress and disease, *JAMA: The Journal of the American Medical Association*, 298(14), 1685–1687, 2007.

37. Gotlib, I. H., Ranganath, C., and Rosenfeld, J. P., Frontal EEG alpha asymmetry, depression, and cognitive functioning, *Cognition and Emotion*, 12(3), 449–478, 1998.

38. Lewis, R. S., Weekes, N. Y., and Wang, T. H., The effect of a naturalistic stressor on frontal eeg asymmetry, stress, and health, *Biological Psychology*, 75(3), 239–247, 2007.

39. Zhang, Q. and Lee, M., Fuzzy-gist for 4-emotion recognition in natural scene images, pp. 1–8, 2010.

40. American Psychiatric Association, *Diagnostic and Statistical Manual of Mental Disorders* (4th ed.), Washington, DC, 1994.

41. Van den Bergh, W., Neurofeedback (EEG-Biofeedback) (BMED Press LLC), in *Neurofeedback and State Regulation in ADHD: A Therapy without Medication*, Corpus Christi, TX, 2010.

42. Thompson, L. and Thompson, M., QEEG and neurofeedback for assessment and effective intervention, in *Quantitative EEG and Neurofeedback: Advance Theory and Applications*, 2nd ed., Elsevier, 2009.

43. Monastra, V., Lubar, J. F., Linden, M., VanDeusen, P., Green, G., Wing, W., et al., Assessing attention deficit hyperactivity disorder via quantitative electroencephalography: An initial validation study, *Neuropsychology*, 13, 424–433, 1999.

44. Lagopoulos, J., Xu, J., Rasmussen, I., Vik, A., Malhi, G. S., Eliassen, C. F., et al., Increased theta and alpha EEG activity during nondirective meditation, *The Journal of Alternative and Complementary Medicine*, 15(11), 1187–1192, 2009.

45. Gruzelier, J., A theory of alpha/theta neurofeedback, creative performance enhancement, long distance functional connectivity and psychological integration, *Cognitive Processing*, 10, 101–109, 2009.

8

EEG and qEEG in Psychiatry

Likun Xia and Ahmad Rauf Subhani

Universiti Teknologi PETRONAS

CONTENTS

8.1   Introduction 176
8.2   Psychiatric Disorders 176
    8.2.1   Stress 177
    8.2.2   Depression and Subtypes 177
        8.2.2.1   Nonmelancholic Depression 177
        8.2.2.2   Melancholic Depression 177
        8.2.2.3   Psychotic Depression 178
        8.2.2.4   Atypical Depression 178
        8.2.2.5   Causes of Depression 178
        8.2.2.6   Diagnosis of Depression 178
8.3   Clinical Treatment Methods 179
    8.3.1   Stress 179
    8.3.2   Depression 179
8.4   EEG in Psychiatry 180
    8.4.1   Stress Characterization 180
    8.4.2   Depression 183
        8.4.2.1   Discriminant Functions (Depression) 183
        8.4.2.2   Depression Detection 184
        8.4.2.3   Differential Diagnostic Classification 185
        8.4.2.4   Prediction of Treatment Response for Depression 186
8.5   Case Study of Mental Stress 187
    8.5.1   Experiment and Subjects 187
    8.5.2   Data Acquisition and Analysis 188
    8.5.3   Results 189
        8.5.3.1   Performance Results 189
        8.5.3.2   EEG Results 190
        8.5.3.3   HR Results 190
8.6   Conclusion 191
References 191

8.1 Introduction

Conventional EEG techniques are based on visual inspection of patterns from EEG records. This involves observation of changes in wave shapes, frequencies, and transitions of states that may not be found in healthy subjects. During the last few years, more robust techniques based on signal digitization and processing have been proposed, including quantitative electroencephalography (qEEG).

qEEG covers techniques involving computer algorithms and methods for quantification, analysis, diagnosis, discrimination of clinical disease subtypes, and prediction of treatment response of medicine. It requires digital recording of EEG as an unavoidable preliminary step, known as digital EEG (DEEG). It is a well-established methodology and the inevitable first step before qEEG analysis can be applied. DEEG includes computer techniques for the digital storage of EEG signals. It provides flexibility in terms of signal display, change of filters settings, and so on. qEEG substantially boosts the capacity of conventional EEG techniques. Both techniques complement each other: conventional EEG is used to aid the diagnosis of a disease, whereas qEEG precisely discriminates individuals by analyzing EEG features into clinically significant subgroups.

qEEG has been suggested in clinical cases for discriminating patients into different subgroups, and it is also being used as a biomarker to aid clinicians in the selection of a treatment plan. Unfortunately, so far the outcomes have not been significant despite a huge body of research evidence. This is mainly because individual has a unique EEG signal/pattern, that makes it difficult to be measured based on a certain biomarker. Development of such a biomarker to generalize the signals is a challenge and involves advanced techniques. However, it has been recognized that the qEEG biomarker is critical and will be playing an important role in clinical cases.

This chapter introduces the development of a number of biomarkers using conventional and qEEG techniques for stress, unipolar depression, and alcohol addiction. A case study on stress assessment is included and discussed.

8.2 Psychiatric Disorders

A major application of EEG lies in psychiatric disorders, including depression and schizophrenia. They may result from social interactions, separation of partners, job dissatisfaction, or unemployment. One of the common factors that make individuals vulnerable to any psychiatric disorder is their mental health or their inability to tackle such situations.

8.2.1 Stress

Hans Selye is considered a pioneer in describing the role of stress in a general adaptation syndrome (GAS) [1]. He defined stress as the transition from a calm to an excited state for the sake of preserving the integrity of the organism [2]. According to cognitive activation theory, stress is a physiological and psychological response to threatening situations that need adjustment in homeostatic imbalance caused by a general "alarm." Formally, the alarm occurs when there is a discrepancy between what it should be and what it is [3].

8.2.2 Depression and Subtypes

Depression is a common experience in daily life. For example, we sometimes feel "depressed" about a friend's cold shoulder, misunderstandings in our marriage, tussles with teenage children, and sometimes we feel "down" for no reason. Depression is specific to the following symptoms:

1. The mood state is severe.
2. It lasts for 2 weeks or more.
3. It interferes with the individual ability to function.

8.2.2.1 Nonmelancholic Depression

It is a more commonly existing nonbiological depression mainly due to severe daily life events. It normally has the following symptoms:

1. A depressed mood for more than 2 weeks
2. Social impairment (e.g., difficulty in dealing with work or relationship)

8.2.2.2 Melancholic Depression

It is a less common form of biological depression with the following characteristics. It is normally treated with physical treatments such as antidepressants.

1. A more severe depression than the case with nonmelancholic depression, with a lack of pleasure and difficulty in being cheered up
2. Psychomotor disturbance (e.g., low energy, poor concentration, slow or agitated movements)

8.2.2.3 Psychotic Depression

It is a less common type as compared to melancholic and nonmelancholic depression. It has a low remission rate and is normally dealt with physical treatment such as antidepressants. It has the following symptoms:

1. An even more severely depressed mood than the case with either melancholic or nonmelancholic depression
2. More severe *psychomotor disturbance* than the case with melancholic depression
3. *Psychotic* symptoms (either delusions or hallucinations, with delusions being more common) and overvalued guilt ruminations

8.2.2.4 Atypical Depression

It has different symptoms as compared to other subtypes. For example, instead of a weight loss, weight gain; instead of appetite loss, appetite increase; instead of insomnia, increased sleep time. Additional features are as follows:

1. Being able to be cheered up by pleasant events
2. Significant weight gain or increase in appetite (especially with comfort foods)
3. Excessive sleeping (hypersomnia)
4. Heaviness in the arms and legs
5. A long-standing sensitivity to interpersonal rejection; the individual is quick to feel that others are rejecting him or her.

8.2.2.5 Causes of Depression

Depression may be caused by multiple factors including biological, nonbiological, and a combination of the two. Nonbiological causes may involve life events such as death of a loved one or the loss of a job, the first birth of a baby, postnatal effects, and so on. This is categorized under melancholic depression. Biological effects may be considered as biological osteoporosis of mental systems and are mainly included in either the nonmelancholic or psychotic subtypes.

8.2.2.6 Diagnosis of Depression

Depression is diagnosed based on standard clinical procedures. An individual with a potential depression problem is screened by a professional, usually a psychiatrist, using the DSM-IV criteria in a psychiatry clinic. The screening is normally based on standard questions, and the response can lead to diagnostic conclusions. Some other clinical tools are also required to check the severity of depression, for example:

1. Beck Depression Inventory-II (BDI-II) Scale
2. Hospital Anxiety and Depression Scale (HADS)
3. Hamilton Depression (HAM-D) Scale
4. Clinical Global Impression (CGI) Scale

8.3 Clinical Treatment Methods

8.3.1 Stress

Stress response can be measured and evaluated in terms of perceptual, behavioral, and physical responses. One way to quantize mental stress in psychological studies is through questionnaires. Several well-established questionnaires have been employed in clinical and psychological applications for stress evaluation, for example, the Perceived Stress Scale (PSS) [4], the Life Events and Coping Inventory (LECI) [5], and the Stress Response Inventory (SRI) [6]. The PSS, designed for adults, determines the extent to which conditions are perceived as stress, based on experience during events in the duration of one month. It quantifies adult life in unpredictable, uncontrollable, and overloaded levels. The LECI comprises 125 questions to measure the degree to which children experience stress in association with life events. The SRI consists of 39 items and produces scores for seven factors of tension, violence, somatization, rage, depression, exhaustion, and frustration [6].

It is also designed for adults, but the difference from the PSS is that it assesses mental and physical symptoms associated with psychological stress. Unfortunately, the questionnaires mostly rely on symptoms that appear after suffering from mental stress. The personal experiences in the past and the subject's capacity to feel and explain the behavioral changes vary from one subject to another, which makes the methods very subjective. Furthermore, these methods hardly provide insight into physiological variability. As a result, it is difficult to detect physiological changes at the early stages.

8.3.2 Depression

In general, depression can be treated either psychologically or physically, or both types of treatment can be simultaneously applied. Physical treatment includes antidepressants and electroconvulsive therapy (ECT). Psychological treatment is mainly based on verbal counseling by a psychiatrist or a psychologist.

Specifically, depression that is more biological in origin (melancholic depression and psychotic depression) is more likely to need physical treatments rather than psychological treatments alone. Nonmelancholic depression can be treated equally effective with physical treatments (antidepressants) or with psychological treatments.

8.4 EEG in Psychiatry

8.4.1 Stress Characterization

The effect of a real-life stressor on frontal-EEG asymmetry and negative health is investigated in Ref. [7]. Forty-nine subjects are tested during periods of low and high examination stress. Low examination stress period is the week when no test or assignment is pending for the students. The high examination stress period occurs in the week when students have at least three exams or assignment pending. Spielberger State Anxiety Inventory and Cohen's Perceived Stress Scale are used for a subjective assessment of stress. Subjects report higher stress during the high examination stress period. The asymmetry in the alpha band of EEG is measured during the stress tests. For the EEG recording, the subject took part in eight random blocks of 1 min of "eyes-open" and "eyes-closed." The results show a relative shift in the alpha asymmetry from the left to the right hemisphere. In other words, during the low examination period, the left frontal activity is greater, whereas a comparatively larger right frontal activity occurs during the high examination period.

A system is designed to categorize the subject's tasks into three difficulty levels in Ref. [8]. This study aims to establish an artificial neural network (ANN) for an individual operator and focuses on determining the parameters that would be used with an ANN. The experimental design consists of a simulation that is divided into 5 min intervals of relaxation and low and high stages of difficulty. EEG, ECG, EOG, and respiration are measured during the experiment, and an ANN is used for classification between conditions. The extracted features to train the ANN are EEG spectral features, inter-beat intervals from ECG signal, eye blinking rate and length of blink interval, and respiration rate. Initially, a training session of 5 min segments is implemented to test the algorithm; thereafter, the algorithm is run constantly to detect stress in real time. The classification accuracy of this system is 84.9%, 82%, and 86% for baseline, medium, and high stress conditions, respectively.

A thorough study is performed in Ref. [9] to analyze the EEG in order to identify mental states in which five subjects would have to go through five tasks. Every task is repeated five times, and each time, 10 s of data are recorded. This is the core investigation of the topic of identifying mental states. The tasks offered to perform in this study are: baseline, in which no activity is performed; arithmetic problems, in which subjects are shown nontrivial arithmetic multiplications to solve with static and quiet posture; and geometric figure rotation, whereby a drawing of a complex figure is exposed in front of the subject. After 30 s, the figure is removed, and the subject is instructed to visualize the rotated drawing; mental letter composing is tested, whereby the subject is asked to perceptually compose a letter to a friend or relative, without making a noise; and visual counting is tested, whereby the subject is instructed to think of an imaginary blackboard and imagine a sequence of numbers being written on it. This study did not address exactly the mental stress studies, but experiments implemented for the purpose of mental state identification followed the same procedure. Another study is conducted in Ref. [10] in order to classify mental tasks using EEG signals. Two experiments are performed to fulfill this purpose. This research did not directly address mental stress. However, the first experiment is performed in a similar manner as in mental stress studies. Eight subjects participated in this research. In the first experiment, three tasks are presented: "rest," "mental arithmetic," in which the subject is needed to mentally multiply a single digit number by a three-digit number; and "mental rotation," in which the subject is required to imagine particular objects. EEG signals are recorded from two parietal locations. The extracted features are in terms of the signal power in frequency bands of delta, theta, alpha, low-beta, high-beta, and gamma for each channel; the phase coherence in every band across channels; and the power difference in every band between the two channels. Additionally, the mean spectral power, peak frequency, peak frequency magnitude, mean phase angle, mean sample value, zero-crossing rate, number of samples above zero, and the mean spectral power difference between two input channels are also calculated in every frequency band. These features are applied to a Bayesian network. The classification accuracy is 83.8% between arithmetic and rotational tasks, 86.5% for rest versus arithmetic task, and 82.9% for rest versus rotational task. However, the accuracy decreases to 68.3% when all three tasks are classified.

Another type of study performed to assess stress uses emotional methods. The contribution of these studies is to identify different emotional states. Most of these studies are related to maintaining concentration in a simulation environment. A method proposed in 2011 [11] tries to analyze and understand stress by categorizing emotions into its four basic types: happy, calm, sad, and fearful. Four subjects participated in this study. The EEG is recorded while displaying International Affective Picture System (IAPS) images as the stimulus in order to discriminate the four emotional states. Kernel density estimation (KDE) is employed as the feature extraction method. The extracted features are then classified using a multilayer perceptrons (MLP). The classification accuracy reached maxima of 73.21%, 87.5%, 75%, and 96.43% for happy, calm, sad, and fearful classes for a single subject when the MLP is trained and tested with a combination of valence and arousal. The accuracy for the same subject is 92.68%, 55.36%, 89.29%, and 89.29% for the four discrete emotions, respectively, when valence and arousal are separately used to train and test the MLP. However, the accuracy of the system for all the subjects is 71.69%, 60.74%, 71.84%, and 65.94% for the four discrete emotions, respectively, using 5-fold cross-validation.

A study on feeling/emotion (happiness, rage, sadness, and relaxation) based on EEG data were described in Ref. [12]. In order to induce this feeling, music, television, video, and puzzle games are used. EEG data are recorded after the subject reports that this feeling is induced. Power spectra of EEG data, especially alpha, beta, and gamma bands, are used as features. The wavelet transform is used on 8 s of data for this purpose. Subsequently, principle component analysis (PCA) is applied for dimensionality reduction. Ultimately, a feed-forward neural network is applied for feelings classification. The classifications results are 54.5%, 67.7%, 59%, and 62.9% for happiness, rage, sadness, and relaxation. However, another observation is that happiness and rage exhibit large variance of amplitude, whereas sorrow and relaxation exhibit a small variance. So this observation is approached in another way; happiness and rage are considered as one class, and sadness and relaxation are considered another class. Whenever happiness or rage is given as input, the prediction accuracy of being happy or enraged increases (73.2% and 94.9% for happiness and rage, respectively). When sadness or relaxation is given as input, the prediction accuracy of being sad and relaxed also increases (89.2% and 97.2% for sadness and relaxation, respectively).

An EEG-signal-based analysis method is proposed to inspect a driver's mental response to traffic lights [13]. Driving simulation is performed in a virtual reality environment. This study includes nonparametric weighted feature extraction (NWFE), PCA, and linear discriminant analysis (LDA). PCA and LDA are also used for dimensionality reduction and projecting the data into a feature space spanned by their eigenvectors. The projected data are classified using a k-nearest neighbor classifier (KNNC) and a naive Bayes classifier (NBC). The experiment is conducted with six subjects for data collection. The classification results show that NWFE+NBC can achieve the best classification accuracy of 77%.

A study described in Ref. [14] involves two experiments. The first is for assessing stress among nine participating subjects. In order to induce stress, three difficulty levels are proposed: initial level, medium level, and hard level besides the rest condition at the start. Playing video games is the stress stimulus. EEG signals are recorded during the experiment, and the power spectral density (PSD) and inter-electrode coherence are extracted from the signals. The second experiment examines the influence of emotion on the EEG; the information might be used to design an interface for disabled persons to interpret emotions. Instead of exploring the neural mechanisms of emotions, this part of the study focused on finding EEG measures that would allow good discrimination between various emotional phases, and possibly, a quantification of emotional strength. Sound, pictures, and a combination of both are presented as the stimulus to 20 subjects who participated in the experiment. Similar to the first experiment, selected features are PSD and inter-electrode coherence. An ANN is used as a classifier on these features.

8.4.2 Depression

Numerous qEEG studies have found increased alpha and/or theta power in a high percentage of depressed patients [15–22]. Antidepressants reduce alpha activity [18,23–26], suggesting normalization of these deviant qEEG features (in contrast to the increased alpha caused by neuroleptics) [27–29]. Interhemispheric asymmetry, especially in anterior regions, has been reported repeatedly [30–34], as having decreased coherence [16,35,36].

Both EEG and qEEG studies report that a high proportion of patients with mood disorders display abnormal brain electrical activity. EEG studies report that small sharp spikes and paroxysmal events are often found, especially on the right hemisphere, and that studies on abnormal sleep are common. There is broad consensus in qEEG studies that increase in alpha or theta power, as well as asymmetry and hypocoherence in anterior regions, appear most often in unipolar depressed patients.

8.4.2.1 *Discriminant Functions (Depression)*

Conventional EEG studies have found a substantial proportion (typically 20% to 40%) of depression patients to have EEG abnormalities, with several characteristic and controversial patterns, as described in Ref. [37]. In their review of the literature, Holschneider and Leuchter [38], taking the opposite viewpoint, note that the majority of conventional EEGs are normal in depression, and that abnormalities are generally mild, such as a slowing of the posterior dominant rhythm. From this standpoint, they argue that a patient with severe cognitive impairment and a normal or nearly normal EEG may be suffering from a pseudo-dementia of depression, whereas a similarly impaired patient with severe EEG slowing is likely to be suffering from another disease process, such as Alzheimer's disease. However, this distinction is not likely to be seen in early stages of Alzheimer's disease, when the EEG is normal or only mildly abnormal, generally showing posterior slowing. Holschneider and Leuchter make the point that abnormal EEGs predict functional decline regardless of the diagnostic group. They also argue that EEG may be more useful than neuropsychological tests for identifying pseudo-dementia since motivational and attentional problems are less likely to interfere with testing. For all of these reasons Holschneider and Leuchter maintain that "although an abnormal EEG in a depressed patient is not specific for dementia, it does identify the patients at greatest risk for functional decline, and therefore is a useful part of the evaluation." Unfortunately, no indications of the accuracies (e.g., sensitivity, specificity) of these statements are presented, but the authors' view probably reflects the informed clinical consensus when EEGs are visually analyzed.

qEEG studies of depression yield widely varying results depending primarily on the analytic technique employed. A decade of studies reviewed by Pollock and Schneider [21] revealed increased alpha and beta power in slightly more than half, which in principle might allow discrimination of depression from dementia with its decreased alpha and beta. Univariate approaches using single qEEG feature fail to classify depressed patients in a clinically useful manner. In contrast, by considering several variables simultaneously, multivariate approaches appear to offer the ability to classify mood disorder patients in ways that are clinically useful. Hughes and John [37] note that numerous qEEG studies have reported increased power in the theta or alpha band and decreased coherence and asymmetry over frontal regions among unipolar depressed patients, which is essentially the opposite pattern of changes seen in schizophrenia (discussed in the following) and which may aid in the differential diagnosis of difficult cases. Similarly, unipolar and bipolar depression appears to have different patterns of qEEG changes, with schizophrenia-like alpha decreases and beta increases in the latter. Hughes and John suggest that this difference may serve to separate unipolar from bipolar patients presented in a state of depression without a prior history of mania, but this distinction may be compromised by antidepressant medication, which tends to reduce the excessive alpha among unipolar depressed patients.

8.4.2.2 Depression Detection

Using multivariate qEEG techniques, the accurate separation of depressed from healthy individuals has been demonstrated repeatedly and replicated in large samples. Prichep and John [39] attained 83% sensitivity and 89% specificity (jackknife replicated to 81% and 87%, respectively). A four-way classification (healthy, depression, alcoholism, and dementia) identified depressed individuals with 73% (jackknife replicated to 65%) sensitivity and 84% (jackknife replicated to 76%) specificity. A follow-up [16] using the same four-way classification identified depressed individuals with 72% sensitivity and 77% specificity (independently replicating to 85% and 75%, respectively). Higher accuracies are reported by John et al. [40], who found that depressed individuals could be separated from their healthy counterparts using a two-way discriminant with 83% sensitivity and 86% specificity (independently replicated at 93% and 88%, respectively). A three-way discriminant (healthy, depression, and dementia) identified depressed patients with 84% sensitivity and specificity (independently replicated at 80% and 85%, respectively), and a four-way discriminant (healthy, depression, alcoholism, and dementia) identified them with 72% sensitivity and 77% specificity (independently replicated at 85% and 75%, respectively).

These results also demonstrate the trade-off between the number of simultaneous discriminations (the number of possible diagnostic categories into which a patient might be placed) and the accuracy (sensitivity and specificity) of the discrimination. This trade-off highlights the principle that qEEG does not take the physician out of the diagnostic loop. The more the physician knows about the patient, the more alternative diagnoses can be excluded a priori, and the more accurate the qEEG discrimination can be.

8.4.2.3 Differential Diagnostic Classification

Several replicated qEEG studies of differential diagnostic classifications of depression versus other disorders have been based on four-way discriminants. Prichep and John [39] used a four-way discriminant to identify depressed patients with 73% (jackknife replicated to 65%) sensitivity, with specificities of 73% (73%) versus dementia and 74% (64%) versus alcoholism. John et al. [16,40] used a four-way discriminant with independent replication and achieved sensitivities of 72% (85%) for depression, with specificities of 79% (77%) versus dementia and 80% (90%) versus alcoholism. The latter report also used a two-way discriminant with independent replication to categorize depressed patients with 84% (88%) sensitivity and 84% (85%) specificity versus schizophrenia. Similarly, a three-way discriminant categorized depressed patients with 84% (80%) sensitivity and a specificity of 84% (71%) versus dementia. Finally, a four-way discriminant with independent replication achieved 72% (85%) sensitivity for depression with specificities of 79% (77%) for dementia and 80% (80%) for alcoholism. The crucial differentiation between unipolar and bipolar mood disorders has been assessed using multivariate techniques by Prichep and John [39], who reported jackknife replicated unipolar classification sensitivities of 87% (87%) and specificities of 90% (85%) versus bipolar patients. John et al. [16] found nearly identical unipolar sensitivities of 85% (85%) and specificities of 85% (87%) versus bipolar, and John and Prichep [41] reported independently replicated unipolar sensitivities of 84% (87%) versus bipolar specificities of 88% (94%). Prichep et al. [36] similarly found independently replicated unipolar sensitivities of 91% (76%) versus specificities of 83% (75%) for bipolar patients. These already high accuracies could be boosted further by adding quantitative evoked potential (QEP) data, giving independently replicated unipolar sensitivities of 98% (76%) and specificities of 91% (82%) versus bipolar. Some caution must be exercised, however, in generalizing results from primary depression to the secondary depression seen so often in clinical practice. Prichep et al. [42] studied qEEG characteristics of crack cocaine dependence and noted that 28 patients (54% of the total sample) had a secondary diagnosis of major depression. When a previously used depression discriminant is applied to this group, it successfully identified only 8 of the 28 patients, for a sensitivity of 29%.

8.4.2.4 Prediction of Treatment Response for Depression

One of the suggested uses of qEEG is to predict the most effective treatment for a given patient, and one of the most frequently cited papers in this regard is from Suffin and Emory [43]. These authors recorded un-medicated qEEG, treatment, and reported outcome data from 54 patients diagnosed with DSM-III-R "affective" (mood) disorders (major depression, bipolar disorder, depressive disorders not otherwise specified) and from 46 patients suffering from ADD/ADHD. Although actual treatments varied, affective disorder patients generally are treated with antidepressants, to which anticonvulsants or lithium are added in refractory cases, followed by stimulants in those cases that are still unresponsive. Attention deficit patients generally are treated initially with stimulants, then antidepressants, and finally anticonvulsants for increasingly refractory cases. Pretreatment spectral analysis revealed significantly increased alpha in some patients and increased theta in others. It also revealed hyper-coherence among some but not other patients.

The authors "heuristically" divide the data into frontal alpha excess, frontal theta excess, and "other" groups, the last of which (N = 19) is essentially dropped from further analysis or discussion. When treatment data from the remaining subjects are analyzed, patients with similar neurometric features are found to respond to the same classes of medications, despite their differing DSM-III-R diagnoses. For example, summarizing their findings they state, "The frontal theta excess group is 100% responsive to stimulants," which is an exciting finding. The ability of qEEG to predict treatment response would have immediate clinical utility and might further suggest the presence of an underlying electrophysiological taxonomy of psychiatric disorders not entirely congruent with DSM. Unfortunately, major flaws in design, analysis, and reporting plague the Suffin and Emory [43] study, and many of their conclusions appear to be overstatements of their findings. For example, the summary statement that the frontal theta excess group is 100% responsive to stimulants might lead the reader to think that all of the 21 patients found to have frontal theta excess also responded to stimulant medication. However, in the paper it is stated, "The frontal theta excess/normocoherent subgroup appeared only in the attentionally disordered clinical population. In that population it is 100% responsive to stimulants." Indeed, the data shows only seven of these patients to have responded to stimulants, and all seven had attentional disorders. This and a host of similar problems make the conclusions somewhat less exciting than they at first appear. However, the paper does illustrate the importance of identifying subgroups in making treatment predictions, and in that regard it serves as a valuable addition to the literature.

8.5 Case Study of Mental Stress

This section provides a case study of mental stress. The following sections provide a brief methodology and results of the study.

8.5.1 Experiment and Subjects

The experiment consists of six sequential phases, as shown in Figure 8.1. Phase one requires a subject to perform training by solving sample questions. Phase two involves placing sensors and checking the impedance. The third phase requires habituation for 5 min for the subjects to get used to the environment. Physiological signal recording does not start until phase four, that is, Rest 1. It performs collection of baseline data by asking subjects to look at the circle in the center of the screen for 5 min. Phase five involves performing the task phase for 20 min in which the subject is required to solve experimental tasks. The last phase (Rest 2) is the recovery phase, where physiological signals are collected by staring at a circle in the center of the screen. The minimum duration between two sessions is seven days in order to minimize the learning effect.

The experiment contains two groups of subjects: stress and nonstress (the control). In both cases, the core task is the same, that is, to solve mental arithmetic problems. The difference is that in the stress condition there is a time restriction along with stressful feedback and statements. However, that is not the case for the controlled condition.

During the experiment, two steps are performed: half the subjects appear in the stress session followed by the control session and half the subjects appear in the control session followed by the stress session. The task phase in the stress and control sessions is divided into four levels of increasing difficulty as shown in Figure 8.2. The duration of each level is 5 min.

In every level, multiple trials of the same difficulty are repeated. In the stress session, the duration to solve a trial is fixed. After the trial, a feedback is displayed in words: "correct," "incorrect," or "no response" based on

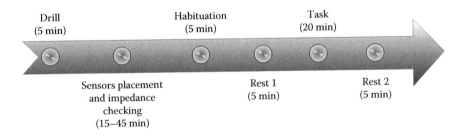

FIGURE 8.1
The process of experiment. The same process is followed for stress and control sessions.

FIGURE 8.2
Four levels of the task in the experiment.

correct/incorrect response or no response to the arithmetic question. In the stress session, the feedback display also shows the average performance at a particular level as well as the response time in order to pressurize the subject further. Moreover, in the stress session, after a certain number of trials at every level, a stressful interrupt pops up showing some stimulating statements such as "Don't guess answers," "Your performance is below average," and so on. In the control session, only "correct" or "incorrect" messages will be displayed.

Individual performance is quantified based on the task difficulty at every level using the formula shown in (8.1). It varies from 0 (min.) to 1 (max.).

$$\text{Task difficulty} = 1 - \frac{\text{Correct responses}}{\text{Total trials}}. \tag{8.1}$$

Ten healthy male subjects (age: 19–25 years) are recruited for this study based on the following criteria: their previous medical record, for example, they must have no head injury and must not be on any medication that might increase cardiac activation. The selected subjects need to fast for at least 2 h before the start of the experiment. Each subject also needs to sign an informed consent agreeing to participate and will be given an honorarium of RM40 for his or her contribution.

8.5.2 Data Acquisition and Analysis

EEG and ECG data are simultaneously measured during the experiment. The data are acquired using an Electrical Geodesic Inc. (EGI) Net Amps 300 amplifier and Net Station 4.4.5 acquisition software. EEG data is recorded using the 129 electrodes cap with reference at the Cz location at a sampling rate of 500 samples per second. The reference of offline signals is changed to the average mastoid. The impedance of all the electrodes is kept below 50 kΩ. Two Ag/AgCl surface electrodes are patched onto the bottom of the neck to measure the ECG at 500 samples per second using the same system. ECG signals are later downsampled to 200 samples per second before heart rate variability (HRV) analysis.

EEG data is manually cleaned to remove artifacts. After that, data for 1 min is selected to keep the test-retest reliability of data at 95% [44]. Heart rate (HR) analysis is performed on the 5 min blocks of Rest 1 and Rest 2 and task levels (levels 1, 2, 3, and 4), respectively. The paired *t*-test is applied on the results of the ratio $Fz(\theta)/Pz(\alpha)$, the HR, and performance in order to measure significance. It is applied between the ratio $Fz(\theta)/Pz(\alpha)$ and the HR in every level under both stress and control conditions with respect to the Rest 1 conditions. The same levels in stress and control are also tested for the ratio $Fz(\theta)/Pz(\alpha)$, the HR, and performance. The details of calculating the ratio $Fz(\theta)/Pz(\alpha)$ and HR can be found in Ref. [45].

8.5.3 Results

The experiment results are provided in the following three sections. The results are analyzed in stress vs. control conditions.

8.5.3.1 Performance Results

Figure 8.3 shows the task difficulty at every task level of the stress and control sessions. It is evident from the figure that difficulty for the stress session is considerably higher than for control session for all the levels. Under the stress conditions, the difficulty of solving tasks continuously increases with every level (0.42, 0.5, 0.59, and 0.76). Under the control condition the difficulty at levels 2 and 3 varies slightly, but at other levels it increases continuously (0.25, 0.35, 0.33, and 0.53). The subjects are less able to solve trials correctly under stress conditions as compared to the control conditions. This difference in task difficulty under stress and control is significant ($p < 0.003$ in all levels), which shows the poor performance and trouble faced under stress conditions.

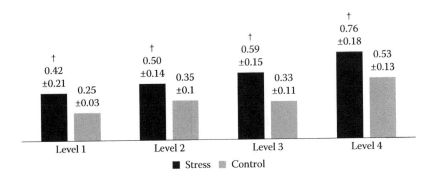

FIGURE 8.3
Task difficulty at levels 1, 2, 3, and 4 for stress and control; values are arranged as mean ± std. † indicates the significance between stress and control conditions.

8.5.3.2 EEG Results

Figure 8.4 presents results of the ratio $Fz(\theta)/Pz(\alpha)$ for all the subjects during Rest 1, for every task level, and during Rest 2. The ratios show a significant difference in values for tasks as compared to Rest 1 ($p < 0.013$). The ratio also indicates significant difference between stress and control sessions at levels 1, 2, and 3 ($p < 0.042$). The ratio in Rest 1 is very precise for both the sessions (1.26 and 1.3 in stress and control, respectively), which implies that the baseline for both conditions is almost identical. In the stress session, the first three levels of stress continuously increase (2.04, 2.12, and 2.31 at levels 1, 2, and 3). However, at level 4, the ratio slightly decreases (2.25). This reduction indicates that subjects may have lost interest in attempting to solve questions because of the restricted time and extreme difficulty at level 4. Therefore, θ in Fz is reduced in power, which is reflected in the ratio at level 4 of stress. In the control session, the ratio shows a similar pattern as shown by the task difficulty. In the first two levels, the ratios increase (1.69 and 1.92 in levels 1 and 2, respectively), then decrease for level 3 (1.84), and then increase for level 4 (2.05). The ratio during Rest 2 in both conditions revert back toward the baseline level (1.36 and 1.31 for stress and control, respectively), which indicates that the subject recovers after the experiment.

8.5.3.3 HR Results

Figure 8.5 illustrates the results of HR at all levels in the stress and control sessions. It is seen that the HR increases significantly for all task levels in stress and control sessions with respect to individual Rest 1 ($p < 0.05$ at all levels of stress and control). However, the HR fails to display any significance between stress and control conditions.

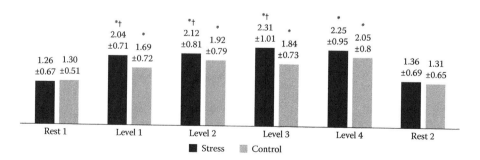

FIGURE 8.4

$Fz(\theta)/Pz(\alpha)$ values at levels 1, 2, 3, and 4 for stress and control; values are arranged as mean ± std. * indicates significance between stress and control conditions with respect to Rest 1. † indicates the significance between stress and control conditions.

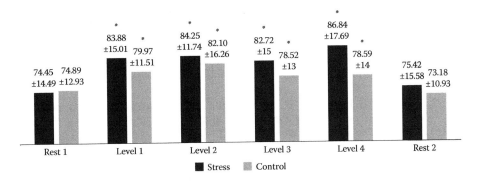

FIGURE 8.5
Heart rate (HR) at levels 1, 2, 3, and 4 for stress and control; values are arranged as mean ± std. * indicates the significance between stress and control conditions with respect to Rest 1.

8.6 Conclusion

This chapter presents a number of biomarkers using conventional and qEEG techniques for stress, unipolar depression, and alcohol addiction. The importance of qEEG techniques is highlighted in the literature. A case study is introduced with stress assessment. Both EEG and qEEG studies report that a high proportion of patients with stress, depression, and alcohol addiction display abnormal brain electrical activity. There is a broad consensus in qEEG studies that an increase in alpha or theta power, as well as asymmetry and hypo-coherence in anterior regions, appeared most often in depressed patients. Therefore, qEEG can be potentially used as a diagnosis or discriminant tool for these conditions.

References

1. Selye, H., A syndrome produced by diverse nocuous agents, *Nature*, 138, 32, 1936.
2. Selye, H., *Selye's Guide to Stress Research*, New York: Van Nostrand Reinhold, 1980.
3. Ursin, H. and Eriksen, H., The cognitive activation theory of stress, *Psychoneuroendocrinology*, 29, 567–592, 2004.
4. Cohen, S., Tyrrell, D. A. J., and Smith, A. P., Negative life events, perceived stress, negative affect, and susceptibility to the common cold, *Journal of Personality and Social Psychology*, 64, 131–140, 1993.
5. Dise-Lewis, J. E., The life events and coping inventory: An assessment of stress in children, *Psychosomatic Medicine*, 50, 484–499, 1988.

6. Koh, K. B., Park, J. K., Kim, C. H., and Cho, S., Development of the stress response inventory and its application in clinical practice, *Psychosomatic Medicine*, 63, 668–678, 2001.

7. Lewis, R. S., Weekes, N. Y., and Wang, T. H., The effect of a naturalistic stressor on frontal EEG asymmetry, stress, and health, *Biological Psychology*, 75(3), 239–247, 2007.

8. Wilson, G. F., Lambert, J. D., and Russell, C. A., Performance enhancement with real-time physiologically controlled adaptive aiding, in *Proceedings of the XIVth Triennial Congress of the International Ergonomics Association and 44th Annual Meeting of the Human Factors and Ergonomics Association, Ergonomics for the New Millennium*, San Diego, CA, pp. 61–64, July 29–August, 2000.

9. Keirn, Z. A. and Aunon, J. I., A new mode of communication between man and his surroundings, *IEEE Transactions on Biomedical Engineering*, 37, 1209–1214, 1990.

10. Johnny, C. L. and Tan, D. S., Using a low-cost electroencephalograph for task classification in HCI research, in *UIST 2006 Proceeding of the 19th Annual ACM Symposium on User Interface Software and Technology*, Montreux, pp. 81–90, October 15–18, 2006.

11. Rahnuma, K. S., Wahab, A., Kamaruddin, N., and Majid, H., EEG analysis for understanding stress based on affective model basis function, Singapore, pp. 592–597, June 14–17, 2011.

12. Ishino, K. and Hagiwara, M., A feeling estimation system using a simple electroencephalograph, in *IEEE International Conference on Systems, Man and Cybernetics*, Washington, DC, pp. 4204–4209, October 5–8, 2003.

13. Chung, I. F., Lin, C. T., Lin, K. L., Ko, L. W., Liang, S. F., and Kuo, B. C., Nonparametric single-trial EEG feature extraction and classification of driver's cognitive responses, *EURASIP Journal on Advances in Signal Processing*, 1–10, 2008.

14. Choppin, A., EEG-based human interface for disabled individuals: Emotion expression with neural networks, Master's thesis, Department of Information Processing, Interdisciplinary Graduate School of Science and Engineering, Tokyo Institute of Technology, 2000.

15. Alper, K., Quantitative EEG and evoked potentials in adult psychiatry, in J. Panksepp (ed.), *Advances in Biological Psychiatry*, vol. 1, JAI Press, pp. 65–112, 1995.

16. John, E., Prichep, L., and Fridman, J., Neurometrics: Computer assisted differential diagnosis of brain dysfunctions, *Science*, 293, 162–169, 1988.

17. Monakhov, K. and Perris, C., Neurophysiological correlates of depressive symptomatology, *Neuropsychobiology*, 6, 268–279, 1980.

18. Itil, T., The discovery of antidepressant drugs by computer analyzed human cerebral bio-electrical potentials (CEEG), *Progress in Neurobiology*, 20, 185–249, 1983.

19. Nystrom, C., Matousek, M., and Hallstrom, T., Relationships between EEG and clinical characteristics in major depressive disorder, *Acta Psychiatrica Scandinavica*, 73, 390–394, 1986.

20. Knott, V. and Lapierre, Y., Computerized EEG correlates of depression and antidepressant treatment, *Progress in Neuro-Psychopharmacology and Biological Psychiatry*, 11, 213–221, 1987.

21. Pollock, V. and Schneider, L., Quantitative, waking EEG research on depression, *Biological Psychiatry*, 27, 757–780, 1990.

22. Nieber, D. and Schlegel, S., Relationships between psychomotor retardation and EEG power spectrum in major depression, *Biological Psychiatry*, 25, 20–23, 1992.
23. Saletu, B. and Grunberger, J., Classification and determination of cerebral bioavailability of fluoxetine: Pharmaco-EEG and psychometric analyses, *Clinical Psychiatry*, 46, 45–52, 1985.
24. Itil, T., Itil, K., and Mukherjee, S., A dose-finding study with sertraline, a new 5-HT reuptake blocking antidepressant using quantitative pharmaco-EEG and dynamic brain mapping, *Integrative Psychiatry*, 7, 29–38, 1989.
25. McClelland, G., Raptopoulos, P., and Jackson, D., The effect of paroxetine on the quantitative EEG, *Acta Psychiatrica Scandinavica*, 350, 50–52, 1989.
26. Saletu, B., Grunberger, J., and Anderer, P., Pharmacodynamics of venlafaxine evaluated by EEG brain mapping, psychometry and psychophysiology, *British Journal of Clinical Pharmacology*, 33, 589–601, 1992.
27. Galderisi, S., Maj, M., and Mucci, A., QEEG alpha 1 changes after a single dose of high-potency neuroleptics as a predictor of short-term response to treatment in schizophrenic patients, *Biological Psychiatry*, 35, 367–374, 1994.
28. Saletu, B., Kufferle, B., and Grunberger, J., Clinical, EEG mapping and psychometric studies in negative schizophrenia: Comparative trials with amisulpride and fluphenazine, *Neuropsychobiology*, 29, 125–135, 1994.
29. Schellenberg, R., Milch, W., and Schwarz, A., Quantitative EEG and BPRS data following haldol-decanoate administration in schizophrenics, *International Clinical Psychopharmacology*, 9, 17–24, 1994.
30. Schaffer, C., Davidson, R., and Saron, C., Frontal and parietal electroencephalogram asymmetry in depressed and nondepressed subjects, *Biological Psychiatry*, 18, 753–762, 1982.
31. Kemali, D., Vacca, L., and Marciano, F., CEEG findings in schizophrenics, depressives, obsessives, heroin addicts and normals, *Advances in Biological Psychiatry*, 6, 17–28, 1981.
32. Henriques, J. and Davidson, R., Regional brain electrical asymmetries discriminate between previously depressed and healthy control subjects, *Journal of Abnormal Psychology*, 99, 22–31, 1990.
33. Henriques, J. and Davidson, R., Left frontal hypoactivation in depression, *Journal of Abnormal Psychology*, 100, 535–545, 1991.
34. Allen, J., Iacono, W., and Depue, R., Regional electroencephalographic asymmetries in bipolar seasonal affective disorder before and after exposure to bright light, *Biological Psychiatry*, 33, 642–646, 1993.
35. Ford, M., Goethe, J., and Dekker, D., EEG coherence and power in the discrimination of psychiatric disorders and medication effects, *Biological Psychiatry*, 21, 1175–1188, 1986.
36. Prichep, L., John, E., and Essig-Peppard, T., Neurometric subtyping of depressive disorders, in *Plasticity and Morphology of the CNS*, eds. C. L. Cazzullo, G. Invernizzi, E. Sacchetti, and A. Vita, London: MTP Press, pp. 95–107, 1990.
37. Hughes, J. and John, E., Conventional and quantitative electroencephalography in psychiatry, *Journal of Neuropsychiatry and Clinical Neuroscience*, 11, 190–208, 1999.
38. Holschneider, D. and Leuchter, A., Clinical neurophysiology using electroencephalography in geriatric psychiatry: Neurobiologic implications and clinical utility, *Journal of Geriatric Psychiatry and Neurology*, 12, 150–164, 1999.

39. Prichep, L. and John, E., *Neurometrics: Clinical Applications, in Clinical Applications of Computer Analysis of EEG and Other Neurophysiological Variables v 2. Handbook of Electroencephalography and Clinical Neurophysiology*, Elsevier, New York, 1986.

40. John, E., Prichep, L., and Almas, M., Toward a quantitative electrophysiological classification system in psychiatry, *Biological Psychiatry*, 1, 401–406, 1991.

41. John, E. and Prichep, L., Principles of neurometric analysis of EEG and evoked potentials, in *EEG: Basic Principles, Clinical Applications, and Related Fields*, eds. E. Niedermeyer and F. B. Lopes da Silva, Williams and Wilkins, Baltimore, pp. 989–1003, 1993.

42. Prichep, L., Alper, K., and Kowalik, S., Quantitative electroencephalographic characteristics of crack cocaine dependence, *Biological Psychiatry*, 40, 986–993, 1996.

43. Suffin, S. and Emory, W., Neurometric subgroups in attentional and affective disorders and their association with pharmacotherapeutic outcome, *Clinical Electroencephalography*, 26, 76–83, 1995.

44. Budzynski, T. H., Budzynski, H. K., Evans, J. R., and Abarbanel, A., *Introduction to Quantitative EEG and Neurofeedback: Advanced Theory and Applications*, 2nd ed., Elsevier INC, 2009.

45. Subhani, A. R., Xia, L., and Saeed Malik, A., Association of autonomic nervous system and EEG scalp potential during playing 2D grand turismo 5, in *34th Annual International Conference of the IEEE Engineering in Medicine and Biology Society*, San Diego, CA, pp. 3420–3423, August 28–September 1, 2012.

9

Perspectives of M-EEG and fMRI Data Fusion

Jose M. Sanchez-Bornot

Universiti Teknologi PETRONAS

Alwani Liyana Ahmad

Universiti Sains Malaysia

CONTENTS

9.1   Introduction 195
9.2   Methods for M-EEG/fMRI Fusion 198
    9.2.1   Correlation between BOLD and M-EEG Signals in
        Particular Bands 199
    9.2.2   Extract Common Spatial Signatures for M-EEG and fMRI 200
    9.2.3   M-EEG-Informed and fMRI-Informed Analyses 201
    9.2.4   DCM-Based Generative Models 203
    9.2.5   Granger-Causality-Based Models 204
    9.2.6   Neural-Mass-Based Generative Models 205
9.3   Significance of Fusion in Various Modalities 206
9.4   Practical and Realistic M-EEG/fMRI Data Integration 208
9.5   Conclusions 212
References 213

9.1 Introduction

Recently, brain imaging techniques have been developed by numerous efforts directed at multimodal data fusion, which seeks to combine high-temporal resolution information, as can be provided by electromagnetic-based techniques (M-EEG), with high-spatial resolution, as has been traditionally achieved by the use of hemodynamic-based neuroimaging methods such as positron emission tomography (PET) and functional magnetic resonance imaging (fMRI) [1]. In particular, the EEG/fMRI fusion has been well known for more than a decade owing to the higher levels of information on brain activity that can be obtained by the simultaneous combination of these complementary modalities, and the significantly increasing capacities for carrying out these studies [2,3].

EEG is a widely used brain mapping technique because it is cheaper, portable, and irreplaceable for mapping transient brain activity as in epileptic studies. The main advantage of EEG is that it is a direct measurement of electrical activity in the brain via voltage sensors on the scalps and provides a milliseconds temporal resolution measurement of brain activity [4]. The EEG can be stated mathematically by using traditional forward models that gather together the biophysics of dynamic local field potential and their propagation across the brain volume conductor medium until the conversion applied to voltage measurements at the scalp sensors. Analogously, forward models are posed for MEG technique where sensors measure the magnetic field associated with its locally tangential electrical field. Sources of electrical activity that possibly contribute to the M-EEG signals are glial potentials, neuronal action potentials, dipolar/quadrupolar electrical fields associated with excitatory and inhibitory post-synaptic potentials, and so on. It is widely accepted that the main contribution to M-EEG signals is the stable and strong dipole created by the synchronized activity of the palisade-structured pyramidal cells, which are oriented perpendicular to the cortical surface [5]. Unfortunately, estimating the sources of M-EEG signals as surrogate of underlying bioelectric activity is an ill-posed problem given that multiple current configurations can produce similar M-EEG measurements and that deep sources are highly affected by the low signal/noise rate which is worsened by the smearing properties of the skull for the EEG, and the dampening of the magnetic field because of the quasi-radial orientation of deep magnetic sources on the MEG sensor space [6,7]. These problems are numerically evidenced by the fact that forward models usually produce kernels that are ill-conditioned [4].

In the past two decades, fMRI became the standard technique for neuroimaging studies due mainly to its noninvasiveness and the fact that it allows direct measurements in the source space. This technique emerged naturally from developments in the nuclear magnetic resonance (NMR) and magnetic resonance imaging (MRI) fields. The latter originally allowed the structural mapping of the brain by exploiting the magnetic properties of the 1H nuclei. Using similar principles, Ogawa et al. [8] developed the blood-oxygen-level-dependent (BOLD) contrast based on the paramagnetic properties of the deoxygenate hemoglobin. When the functional capabilities were further developed, fMRI became essential for cognitive and behavioral studies [9]. In summary, fMRI allows quantifying the metabolic-hemodynamic cascade associated with local neuronal activity, and thus it is a reflex of the accompanying changes in blood flow and the deoxyhemoglobin content in the surrounding tissue [10]. This allows the whole brain to be mapped by the acquisition of 3D functional volumes that are separated by few seconds down to hundreds of milliseconds. This poor temporal resolution is just a minor limitation of this technique: the principal limitation is the sluggishness of the hemodynamic cascade, which only allows for a crude representation of the rich dynamics of neuronal processes.

The basic neuroimaging unit is the voxel, which in current functional studies may have a spatial resolution of $3 \times 3 \times 3$ mm^3. According to estimates, this typical voxel contains approximately 0.9 million neurons, 3.6 km of dendrites, and 36 km of axons, with less than 3% of the space occupied by capillaries, which provide the energetic elements required for brain activity and endorse the millimetric resolution of the fMRI [3]. However, as a measure of the metabolic consumptions in the voxel unit, the fMRI measure has other important drawbacks for recording the brain activity directly: (1) its potential inability to identify different balances of excitatory and inhibitory activity and (2) the changes in the density of small vessels and neuronal elements across the voxels in the different cortical and subcortical regions. The latter variability can be observed in the fMRI changes across the different brain regions. Despite these serious limitations, the use of fMRI has enabled the mapping of important brain functions, for example, visual, auditory, sensorimotor, and some integration areas; and it is currently actively used in the study of more complex systems such as those involved in memory, language, reasoning production, and emotions processing, among others [11].

The combination of EEG and fMRI has been proved clinically necessary by many current studies, while technically Ives et al. [12] showed for first time the feasibility of conjoint recording for EEG and fMRI signals. In the beginning, cognitive experiments used conjoint but interleaved recordings in order to exclude the scanner and other artifacts from the EEG analyzed segments. Today this is no longer necessary, and we can focus more on analysis and less in artifact-rejecting tasks thanks to the development of robust artifact detection techniques in the last few years (see Ref. [13] for a historical review). Multimodal measurements, such as by fusing EEG and fMRI measurements, can help neuroscientists to understand the nature and structure of cerebral activity better by exploiting the strengths for each modality. Thus, the main objective of this combination is to improve the spatial and temporal resolutions for the particular analyses [14].

In this pursuit, we have to also take into consideration the possible confounding for EEG/fMRI fusion. These are mainly threefold: (1) the absence of local field potential (LFP) recordings, due to a closed-field geometry of neuronal synaptic processes, while we still should be able to measure the BOLD response because of the metabolic activity; (2) change of LFP activity by synchronization of neuronal processes while keeping constant the balance of metabolic consumption; then we have changes in EEG activity but stable BOLD signal; and (3) glial cell metabolism, neurotransmitter replenishment, and other processes that do not directly involve the postsynaptic process but may elicit the BOLD response [3,6]. However, we may overestimate the impacts of these confounding factors if we ignore the low-band filtering properties of the BOLD signal. For example, interneurons with star-shaped dendrites make a weak contribution to LFP, though they contribute indirectly because of their effects on synaptic processes of pyramidal cells. This can be put in another perspective: lagged responses in LFP are associated with very similar BOLD waveforms [15].

Next we review several outstanding current approaches. We personally advocate integrative models mainly because they are directly based on the biophysical modeling of M-EEG/fMRI signals. This approach has the advantage that strengths and limitations can be considered directly for each modality during the modeling; however, see Ref. [16] for dealing with these problems using some carefully designed pipelined approaches.

9.2 Methods for M-EEG/fMRI Fusion

Currently, M-EEG/fMRI approaches are classified with respect to the symmetric/asymmetric relationship of developed analyses. In the former, the modalities have equal relevancies in the analyses. That is, there is no bias of relying more on one modality, for example, estimating parameters for some particular model by assigning equal weights to the fitting cost functions for the different data [17]. Otherwise, the asymmetric approaches are subdivided into (1) M-EEG oriented, where the M-EEG modalities are the main focus of the analyses, and fMRI is only used to restrict the space of solutions [18,19]; and (2) fMRI oriented, where features are extracted from M-EEG signals, or ERP-ERF analysis, and after convolving with convenient hemodynamic response functions, the obtained time series are used as standard covariates in generalized linear models, which is a classical step in statistical parametric mapping (SPM) of SPM-based analyses [20] (see also Refs. [21,22] for further discussion).

However, in our opinion, this division does not establish a clear distinction among the developed methods. For example, in (1) the significant regions detected in fMRI analyses that are viewed as prior information for M-EEG inverse problems can be used in a Bayesian or regularization framework, where control hyperparameters and assumed prior distributions are used to modelize the available information; being estimated as part of the optimization process. Therefore, the prior information weights change gradually during the optimization, which occurs in parallel with the fitness of the data; it can be shown using primal/dual theory that both parts (cost and penalty functions) are interchangeable and are thus relatively symmetric. On the other hand, the mathematical modeling for generating M-EEG and fMRI signals simultaneously has to consider that these data have different scales, distinct signal/noise rates, and contain complementary traits of the brain spatiotemporal dynamics. The way that we capture the realism for each modality, relying more or less on some modalities in order to reflect their faithfulness with the neuronal activity, automatically impose different weights to each modality [23,24]. In our opinion, the division between symmetric and asymmetric approaches is philosophical rather than mathematical.

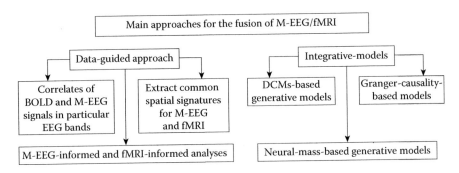

FIGURE 9.1
Main subdivisions of standard methodologies for fusing multimodal data.

We prefer here a more traditional classification based on the mathematical/computational procedure that is followed for combining the multimodal data. In Figure 9.1, we introduce a scheme that illustrates this criterion for separating the current approaches. In summary, data-guided approaches usually lead to a pipeline of procedures for extracting the common features and extracting the accompanying information. (We do not use the term *data-driven* because we believe that *data-guided* is a more flexible term and does not exclude the use of interleaved models in the analysis.) On the contrary, integrative-model approaches are based on the elaboration of realistic models that simultaneously generate the multimodal data. We may think that the former is suboptimal compared to the latter because the different modalities only contribute to the analyses separately. However, integrative-model approaches may also be biased because they do not adequately represent the complexity of brain dynamics. In the following, we review each subdivision by separate with focus on the more outstanding results. Obviously, the proposed criterion is much simpler than the previous criterion, and we think that it avoids the confusions introduced by the latter.

9.2.1 Correlation between BOLD and M-EEG Signals in Particular Bands

Mantini et al. [25] analyzed simultaneous recording of EEG and fMRI data for 15 healthy subjects in order to study resting state networks (RSNs). In particular, they selected six RSNs using temporal independent component analysis of the BOLD signal, which represented the $44 \pm 6\%$ variability of the data. For all the participants, they tested the components' spatial replicability and consistency using, respectively, the multidimensional scaling technique and comparing the extracted components for the first 2 min with those extracted for the last 2 min. On the contrary, after cleaning the EEG data of the MRI scanner and other artifacts, five waveforms were computed as the time-varying EEG power for delta, theta, alpha, beta, and gamma bands. Owing to the significant correlations among these waveforms, Mantini et al. [25]

claimed that, in particular, slow and fast rhythms are not independent, and thus a multivariate analysis is necessary to study the relationship between EEG rhythms and fluctuations in the BOLD signal. Their results are in agreement with many previous studies about RSNs and the concordance between BOLD signal fluctuation and EEG rhythms.

For example, the analysis of the spatiotemporal patterns of their estimated RSN3 corroborates the disassociation between the alpha rhythm and the BOLD signal in the occipital cortex [26,27]. However they were able to obtain a wider range of interactions including also delta, theta, and beta rhythms. A similar link is established for their results for RSN1 and the positive correlation that Laufs et al. [28,29] reported for the 17–23 Hz beta-II rhythm in the posterior cingulate, dorsomedial prefrontal gyrus, and temporoparietal junction.

9.2.2 Extract Common Spatial Signatures for M-EEG and fMRI

The currently most used approaches for extracting joint features from M-EEG and fMRI signals simultaneously are ICA-based methods. There are many proposed procedures that are well reviewed by Calhoun et al. [30]. In the beginning, they proposed the joint ICA (jICA) procedure for fusing the temporal M-EEG with the spatial fMRI information, thus ignoring their respective spatial and temporal parts [31]. For example, for a classical auditory oddball task with 23 normal subjects, they computed for each subject (1) the time-locked ERP waveform and (2) the fMRI activation maps (SPM-based ROI analysis), thus obtaining the corresponding data $\mathbf{Y}_{23 \times V}^{fMRI}$ and $\mathbf{Y}_{23 \times T}^{ERP}$, where V and T are the number of spatial and temporal points, respectively.

Next, they exploited the data variability and implemented a jICA algorithm based on the infomax principle [32]. They assumed that both the fMRI and ERP extracted from spatial/temporal waveforms can be predicted using a unique mixing matrix $\mathbf{A} \in \mathbb{R}^{N \times C}$, where N is the number of subjects ($N = 23$ in their experiment) and C is the number of jICA components to be determined by the algorithm. They proposed to estimate the latter using the minimum description length criteria [33].

The jICA method is determined by the formula

$$\mathbf{Y}^{fMRI} = \mathbf{A}\mathbf{S}^{fMRI}, \text{ and } \mathbf{Y}^{ERP} = \mathbf{A}\mathbf{S}^{ERP}. \tag{9.1}$$

The outputs of this algorithm are the jICA components' spatial sources ($\mathbf{S}^{fMRI}$) and their corresponding temporal sources ($\mathbf{S}^{ERP}$), which are estimated by computing the critical unmixing matrix of ICA-based algorithms [32]. The family of ICA-based algorithms has continued evolving in a way that resembles the vanilla cake or tutti-frutti linear model varieties, that is, for all possible combinations and tasters. Currently, these algorithms concatenate together the fMRI and M-EEG signals either spatially or temporally (depending on the analysis) and extract adaptively ICA components that "separate" the multimodal data in independent waveforms (see Figure 1 in Moosmann et al. [34]

and Figure 5 in Calhoun et al. [35]). In this analysis, the M-EEG signal is usually convolved with the hemodynamic response function (HRF) in order to alleviate the temporal differences between these modalities.

A somehow different approach is decomposing the data into two or more "cores." This can be done simultaneously for several modalities where some "equivalent cores" can be cross-linked in order to extract common information from the modalities. In this vein, Martinez-Montes et al. [36] proposed using the PARAFAC tensor decomposition and the multiway partial least squares algorithm for fusing M-EEG and fMRI data. Similar to Mantini et al. [25] and Goldman et al. [26], they computed the time-varying spectra and analyze all the frequencies instead of the banded spectra, and consider all the derivations in the analysis. As a result, they created a 3D tensor of dimensions corresponding to the number of spatial (electrodes), spectral, and temporal measurements for keeping the transformed EEG data. On the contrary, the fMRI data were collected in a 2D matrix of dimensions corresponding to the number of spatial (voxels) and temporal measurements. From the creation process, both matrices have in common the number of temporal measurements, which coincides with the number of fMRI volumes. Later, they proceed to apply PARAFAC and matrix factorization methods to both matrices, respectively, but linking the temporal dimension by a mathematical expression that aims to maximize the covariance between the temporal atoms. Similarly, but arranging the M-EEG-derived information in a standard matrix form, Correa et al. [37] proposed a multimodal canonical correlation analysis (mCCA) as a hybrid between the previous procedure and the ICA-based decompositions. The latter avoid the tensor decomposition approximation and also is more flexible, being able to include more than two modalities in the analysis (see also Correa et al. [38]).

Some of these methods have been extensively validated with simulations and the analyses of real data, which is evidence that nonparametric approaches are also capable of yielding interesting conclusions (see Correa et al. [38], which we have discussed in greater detail in the section on clinical applications). However, these methods have a serious limitation, which in essence is that they consistently ignore the volume conductor properties of the head and then establish direct connections between the M-EEG signals and the fMRI. In addition, we are of the opinion that the simulations used are nonrealistic and this, together with the lack of a reference and a generative model, makes a cautious interpretation of the extracted components necessary. Therefore, we firmly believe that our understanding of why and how these methods work must be fine-tuned with a benchmark of realistic simulations.

9.2.3 M-EEG-Informed and fMRI-Informed Analyses

These analyses can be performed on conjoint recorded EEG and fMRI data or in separate analysis that can include MEG, ECoG, and other modalities. The basic idea is to undertake preliminary analysis using one particular modality, and use the results to guide further investigation of the multimodal data. These analyses may be conducted using data that is recorded either conjointly or separately The main advantage of conjoint recording is that the modalities are similarly affected for the same environment, stimulus synchrony, and subject state. This is mandatory for continuous tracking of an ERP that also requires its localization [20,39,40], because the M-EEG and fMRI signals are affected by the experimental conditions [41]. In addition, M-EEG-informed studies for conjoint data are an indispensable tool for studying transient epileptic activity, for example, using the onsets of spikes for ROI-based analysis with the aim of detecting the epileptic focus. Benar et al. [40] also claimed that conjoint recordings are necessary for auditory paradigms where the noisy condition affects the subject's brain activity and response. On the contrary, it can also be claimed that separate recordings may allow optimizing the experiment, considering different sessions' duration, stimulus intensity, and the sequence for each modality in order to increase the signal noise rate and reduce the subject's discomfort. One obvious advantage of separate analyses is that related data obtained from different laboratories can be combined. This potential advantage is not fully exploited yet, maybe because of the lack of installed facilities for worldwide data sharing.

In an important application of asymmetric studies, Debener et al. [20] used both M-EEG-informed for fMRI analysis and fMRI-informed for M-EEG analysis to predict hemodynamic changes in the rostral cingulate zone (RCZ) that correlate with the extracted error-related negativity (ERN) independent component for a performance-monitoring task. The fMRI-informed approach was used to determine a dipole located in the RCZ that explained 90.2% of the ERN component variance. The RCZ's ROIs were selected by standard fMRI analysis contrasting the corresponding conditions; then an EEG-informed analysis allowed predicting the BOLD signal changes for this dipole, thus demonstrating the coupling between the fMRI and the single-trial ERN. Other similar studies report significant fMRI activation for single-trial changes of N1, P2, and P3 amplitudes [42] and also for changes in the P3 latency [40].

Another perspective is obtained when spatial information extracted from SPM-ROI analysis is used to constrain the space of solutions for M-EEG inverse problems. In this case, mismatches between the M-EEG and fMRI signals can produce bias in the inverse solution. However, in some situations, this can be nicely checked; for example, Debener et al. [20] tested several dipoles that predict the variability of the ERN component very poorly compared to the dipole selected in the RCZ (see the foregoing discussion). The fMRI-informed approach has changed the development of M-EEG inverse methods. Now, researchers either use the task-related activations of fMRI studies to restrict the generator locations [18–20,43] or use this information in a general Bayesian framework [17,24]. Interestingly, without requiring fMRI priors, Friston et al. [44] proposed a multiple sparse priors (MSP) algorithm based on modeling variance components corresponding to small cortical patches. This can be seen as an adaptive extension of the Bayesian modeling averaging (BMA) methodology introduced by Trujillo-Barreto et al. [45]. Henson et al. [24] compared the MSP algorithm with a similar Bayesian approach that additionally uses fMRI priors, but they did not find an improvement in the MSP results due to fMRI priors. However, MSP is critically restricted to cortical surfaces, ignoring possible deep sources that can be recovered with the fusion methodology.

9.2.4 DCM-Based Generative Models

The central idea of dynamic causal modeling (DCM) is to create a simplified but plausible mathematical model for neuronal dynamics. This should be able to produce realistic neuronal signaling by interconnecting different elements that stand for neuronal populations. These elements can be topographically identified as voxels or ROIs and thus can be simultaneously linked to generators of magneto-electroencephalographic and BOLD signals. Currently, the estimation is restricted to low-scale systems owing to limitations in the numerical algorithms; therefore, the connections can only represent long-range interactions in the brain.

The DCM equation for bilinear systems as stated in the seminal paper [46] can be written using the system of ordinary differential equations:

$$\frac{d\mathbf{x}}{dt} = F(\mathbf{x}, \mathbf{u}, \theta) = \left(\mathbf{A} + \sum_{j=1}^{m} u_j \mathbf{B}_j \right) \mathbf{x} + \mathbf{Cu}, \qquad (9.2)$$

where $\mathbf{x}(t) \in \mathbb{R}^{n \times 1}$ is a time-varying vector containing the time series for each element, $u_j(t)$ represents each of the inputs $j = 1, K, m$ that perturb the system, A represents the elements' connectivity, and C denotes the direct influence of inputs over the system elements. The bilinear term is controlled by the matrices $\mathbf{B}_j$ that modulate the second-order interactions between inputs and the elements' dynamic.

Usually, dynamic causal models (DCMs) are linked to hemodynamic state equations in different ways to disclose the hidden neuronal connectivity that is responsible for the BOLD signal measured in an fMRI experiment [47,48]. This approach is not free of difficulties, because of the sluggishness of the hemodynamic response and the consequent ill-posedness of the deconvolution problem. However, these models have shown their usefulness for interpreting real data [46,49,50]. The development of DCMs has been essentially determined for cognitive studies with controlled conditions, taking for granted the influence of an exogenous input. This may handicap the development of stochastic DCMs, which are not constrained by such assumptions. However, Daunizeau et al. [51,52] showed the potentialities of this new methodology. A much less explored but potentially much more interesting approach is to fit DCMs to recorded electrocorticogram activity, which should provide richer information about the content of neuronal processing. We find the fusion of ECoG with fMRI recordings an attractive approach that may be done conjointly or in separated studies, but using the same stimulation paradigm; for example, the brain language areas can be explored directly with intracortical recording and stimulation previously a mandatory surgical intervention [53].

9.2.5 Granger-Causality-Based Models

Developing in parallel with DCMs is a methodology based on Granger causality to represent the typical cause-effect relationship of neuronal information processing. These models are simpler in the sense that they do not rely on differential equations and are usually stated using linear interactions. At the same time, this is both an advantage and a disadvantage for studies of brain connectivity. The linearity allows granger causality can be applied to analysis of high-dimensional data, thus overcoming missing data problems by considering as much information as possible [54,55]. In particular, Garg et al. [55] estimated whole brain connectivity using the Blue Gene supercomputer and a full-brain autoregressive modeling (FARM) approach. However, in order to deal with these high dimensions, it was necessary to use Lasso/LARS regressions [56] under the valid assumption that brain connectivity is remarkably sparse. Otherwise, the intrinsic nonlinearities of neuronal processing can lead to spurious findings [57], which implies that the brain connectivity findings based on Granger causality should be considered functional rather than effective; this is the subject of ongoing discussions [57–60]. Currently, it is accepted that Granger causality allows explorations of brain-effective connectivity [60].

fMRI brain connectivity studies based on Granger causality are sustained mainly by two assumptions: (1) the HRF exhibits a small variability across the different brain regions and (2) the hemodynamic effect can be represented with a convolution without additional effects on coupling [61]. Garg et al. [55] showed with simulations that FARM may provide valuable information over the brain effective connectivity whenever this variability is comparable to the time scale of the information processing in the brain, as in the case of evoked potential studies. In addition, they hypothesized that FARM could generate useful information for studies of default mode networks, which assume a background brain state with waves that oscillate with frequencies lower than 0.1 Hz, which is supposed to sustain conscious brain dynamics. However, as a personal conclusion about this study, FARM's strengths lie in estimating the connection directionality rather than its strength, and many spurious influences may arise as result of the discretization of transient brain dynamics and the HRF-based assumptions. Smith et al. [62] discussed a set of realistic simulations that showed general limitations for these approaches.

In order to close the gap between Granger-based and dynamic causal models, several new methodologies have been proposed. An issue that has been attacked is the instantaneous associations that arise in fMRI analysis.

Independently, Faes and Nollo [63] and Gates et al. [64] proposed new autoregressive models to estimate mixed lagged and instantaneous influences. Additionally, Faes and Nollo [63] proposed an extension of partial directed coherence [65], which is a methodology that has been shown to be robust for discovering nonlinear interactions as shown by Schelter et al. [66,67]. Moreover, Gates et al. [64] developed a group-connectivity estimation strategy based on these lagged and instantaneous causal models and applied it to the simulated data created by Smith et al. [62] with an excellent recovering rate for simulated connections.

A final idea in this open discussion is the lack of theoretical convergence of previous approaches, which is a dissonant feature compared to the unified DCM analysis of brain connectivity. However, the papers reviewed in the previous paragraph display a possibly common direction to follow, though it is still not clear whether these methods effectively link the underlying neural activity with the accompanying hemodynamic response. Constraints of space and time do not allow us to continue tackling these rich discussions; however, see Refs. [60,68,69] for future avenues for investigation.

9.2.6 Neural-Mass-Based Generative Models

A more realistic mechanism for generating neuronal signals is provided by neural mass models (NMMs). They are based on simplified analyses of the cerebral cortical column that includes basically three neuronal populations widely regarded as pyramidal cells, inhibitory interneurons, and excitatory stellate cells. These models allow a realistic and simultaneous generation of M-EEG and fMRI signals, as elegantly demonstrated by Babajani and Soltanian-Zadeh ([15]; Equations 5 and 12), who used the synthetic postsynaptic activity in pyramidal cells and the sum of the absolute values for all synaptic processes, respectively (see also Refs. [22,70]). Figure 9.2 shows a diagram

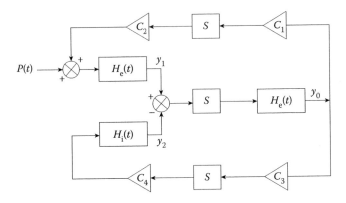

FIGURE 9.2
Jansen-Rit's neural mass model showing the interactions between pyramidal, excitatory stellate, and inhibitory interneuron populations.

of the interactions inside the columnar space for the aesthetic Jansen and Rit model [71], which can be expressed with the following set of equations:

$$\ddot{y}_0(t) = AaSigm\{y_1(t) - y_2(t)\} - 2a\dot{y}_0(t) - a^2 y_0(t)$$

$$\ddot{y}_1(t) = Aa\left(P(t) + c_2Sigm\{c_1 y_0(t)\}\right) - 2a\dot{y}_1(t) - a^2 y_1(t)$$

$$\ddot{y}_2(t) = Bb\left(c_4Sigm\{c_3 y_0(t)\}\right) - 2b\dot{y}_2(t) - b^2 y_2(t)$$

NMMs have currently several advantages and limitations. First, they are a realistic way to generate both M-EEG and fMRI signals with neurophysiological realism [72–75]. Currently, these models represent the most realistic and flexible approach for generating brain signals, which put them in a privileged position for analyzing fused M-EEG and fMRI signals. However, these models are difficult to estimate in real scenarios owing to mainly two facts: (1) only postsynaptic potentials at pyramidal cells are directly linked to electroencephalographic recordings and (2) the relationship with the BOLD signal is sluggish and unspecific. We believe that these problems have limited the role of NMMs for M-EEG and fMRI fusion applications with the benefit of closer relatives and practical DCM approaches.

9.3 Significance of Fusion in Various Modalities

The M-EEG techniques have been crucially used in the past for studying cognitive impairments and brain pathologies, either by using ERP studies, or by exploring single trials directly and detecting the transient changes that occur in the signal such as the spikes or discharges occurring for epileptic patients. The accurate estimation of the affected regions is also important, for example, for making surgical decisions in intractable epilepsies. This cannot be done simply by using M-EEG analysis, whereas the combination with fMRI-based criteria can produce the desired results. In the future, localization criteria will also be necessary for the treatment of other brain pathologies when the estimation of a target region is required for releasing a nano-drug medicine in the brain.

Currently, electrical and hemodynamic/metabolic changes can be traced with simultaneous recordings of EEG and fMRI activities. This widens the opportunities for studying the spatiotemporal patterns of brain signals [3]. Now, the fusion of EEG and fMRI is commonly used for epileptic analyses [76–78], where, additionally, the findings are usually corroborated using intracortical recordings. A good surgical outcome depends on the accurate localization of the epileptic foci and also on the location of the affected region; for example, the treatment of extratemporal lobe epilepsies is still challenging [79]. However, fusion methods are progressing rapidly, and we believe that remaining challenges will be overcome in the forthcoming years. Next, we limit the discussion to interesting applications of the fusion methodology (for additional information, we refer to some specialized reviews [30,77,80]).

Some studies have evaluated the concordance between M-EEG inverse solutions and fusion EEG-fMRI in focal epilepsy by analyzing data acquired in different sessions, based mainly on comparisons of fMRI activation versus sources estimated from M-EEG inverse problems [81–85]. For example, Daunizeau et al. [17] analyzed interictal epileptiform discharge (IED) data within a Bayesian framework approach. In their study, two EEG sections were acquired: the first was conjoint with the fMRI, and the second was for measurement of a prolonged EEG. Spikes in both EEGs were selected by an expert and served in the conjoint study for use as stimulus onsets in standard ROI analysis for detecting the corresponding activations. In the prolonged EEG, the selected spikes were averaged to create an ERP scalp map that was used simultaneously with the fMRI locations for solving the optimization procedure. With this method, they found four significant active regions that were confirmed later with electrocorticography (ECoG) recordings. However, the intracortical recordings also revealed epileptic activities in regions not detected by the method. All the estimated regions show a negative BOLD response with peak around 4–6 s, widely regarded as fMRI deactivations. The relations between IEDs and negative BOLD signals still remain unclear but are considered the result of deactivation both in the EEG and BOLD signals [86,87].

As another example, Vulliemoz et al. [88] evaluated nine patients with refractory focal epilepsy as part of a surgical procedure. The resting eyes-closed state was recorded in two sections, outside and inside the scanner, first for EEG only and then for conjoint EEG and fMRI analysis. They proceed to locate the fMRI activation similarly to the Daunizeau et al. study [17], and then solve the inverse problem for the EEG corresponding to the conjoint analysis with possible sources limited to gray matter [89]. They reported a good concordance between BOLD and EEG-based activations, and both were also in agreement with intracortical recordings.

In another perspective, Correa et al. [38] used mCCA for analysis of fMRI and EEG, and fMRI and structural MRI (sMRI) data separately for patients diagnosed with schizophrenia and a healthy group. For the first study, the data were collected conjointly for an auditory oddball experiment revealing associations of the temporal and motor areas with the N2 and P3 peaks. In particular, the P3 wave for the schizophrenia group was found to be abnormal when compared to the control group. For the second study, the data were collected for an auditory sensorimotor task, and the analysis showed a curious relationship between fMRI activations and gray matter densities. Compared to the control group, the schizophrenia group showed more activation in the motor areas and less in the temporal areas as well as less gray matter densities in both regions. Their results were consistent with standard findings for this pathology [90–94].

9.4 Practical and Realistic M-EEG/fMRI Data Integration

This last section is a little technical but is presented to show ongoing research that may allow the estimation of a brain dynamical system with a large number of interacting elements. We also present a convenient model for simulating "realistic" brain activity and methods that allow computational tractability. Next, we consider the DCM model but disregard the bilinear part, which we will call dynamic ARX (dARX) models. That is,

$$\dot{x}(t) = Ax(t) + cu(t), \tag{9.3}$$

where we assumed only one input to simplify the discussion; then $c \in \mathbb{R}^{n \times 1}$ is a vector representing the input influence over the system elements.

The following theorem and discussion show how to generate simulations based on this model by assuming that the input acts like a train of Dirac's delta.

Theorem: Let $\phi(t)$ be any fundamental solution of $\dot{x} = Ax$; $x(t_k) = x_k$. Then the solution of $\dot{x} = Ax + cu$; $x(t_k) = x_k$, where $u(t)$ is a continuous scalar function, is unique, and is given by

$$x(t) = \phi(t)\phi^{-1}(t_k)x_k + \phi(t)\int_{t_k}^{t}\phi^{-1}(\tau)cu(\tau)d\tau. \tag{9.4}$$

Given that the solution for the homogeneous system is $\phi(t) = e^{A(t-t_k)}$, then the solution for the nonhomogeneous linear system is

$$x(t) = e^{A(t-t_k)}x_k + e^{At}\int_{t_k}^{t}e^{-A\tau}cu(\tau)d\tau. \tag{9.5}$$

The proof of this theorem can be found in Coddington and Levinson's book [95]. Here, we are interested in the case where $u(t)$ represents Dirac delta impulses. Therefore, it is not continuous but measurable, and the theory also applies [95]. This can also be shown by using limit theory if we represent the impulse by a Gaussian distribution with infinitesimal variance.

Informally, we show that this numerical integration has an explicit form. For example, we can use an integration step small enough so that $u(t)$ only has a single impulse in each integration interval. Suppose now that the impulse occurs at t_0, then the previous expression is integrated as follows:

$$\mathbf{x}(t) = e^{A(t-t_k)}\mathbf{x}_k + e^{A(t-\tau_0)}\mathbf{c}\int_{t_k}^{t}u(\tau)d\tau = e^{A(t-t_k)}\mathbf{x}_k + e^{A(t-\tau_0)}\mathbf{c}. \tag{9.6}$$

In general, if we set $m_k > 1$ as the number of impulses in the semi-interval (t_k, t) and represent $u(t)$ as a sum of single impulse functions with the impulse onset occurring at $t_k \leq \tau_j < t$, $j = 1, 2, \ldots, m_k$, then the previous integration formula can be represented more generally as

$$\mathbf{x}(t) = e^{\mathbf{A}(t-t_k)}\mathbf{x}_k + \left(\sum_{j=1}^{m_k} e^{\mathbf{A}(t-\tau_j)} \right)\mathbf{c}. \qquad (9.7)$$

The latter expression helps us to simulate and estimate dynamic ARX (dARX) models for Dirac delta impulses as input.

We introduce now two situations for generating the simulations: (1) the stochastic case, where we consider that noise has an effect over the signals only in discretely equal-spaced points and (2) a more realistic situation where, additionally, measurement noise is added after the entire integration process. The noise perturbation limited to discrete points can be justified using the stochastic theory for continuous Wiener process while stating the integration step for the corresponding discrete time points.

For the simulations that follow, we consider the connectivity matrix

$$\mathbf{A} = \begin{pmatrix} 0.5 & 0.4 & 0 & 0 \\ -0.4 & -0.4 & 0 & 0 \\ 0.5 & 0 & -0.3 & -1 \\ 0 & 0 & 0.8 & -0.3 \end{pmatrix}. \qquad (9.8)$$

In Figure 9.3, we present a single simulation for each of the noisy scenarios using the previous values, which guarantees that the system will generate oscillations. In the simulations, we create 50 spikes with random onsets as exogenous input that affects the elements' dynamics with vector influences $\mathbf{c} = (3, 0, 0, 0)^T$. The stochastic noise is inserted in a similar way, but onsets are equally spaced in the time window, and the value is generated with a Gaussian distribution with a standard deviation of 0.1. At the end of the integration process, the time series are contaminated with measurement noise, also determined by a Gaussian distribution, but with the standard deviation equal to 0.1 times the maximum magnitude of the original time series.

For estimating the parameters for the dARX model, we propose essentially two variants. The first employs a standard cost function based directly on the integration formula (9.7) and is written as

$$F_k = \mathbf{x}_k - e^{\mathbf{A}(t-t_{k-1})}\mathbf{x}_{k-1} - \left(\sum_{j=1}^{m_k} e^{\mathbf{A}(t-\tau_j)} \right)\mathbf{c}, \qquad (9.9)$$

where $F_k \in \mathbb{R}^{n \times 1}$ is defined for the n elements (for each step $k = 2, 3, \ldots$). The second cost function includes the noise in the formula, and hence

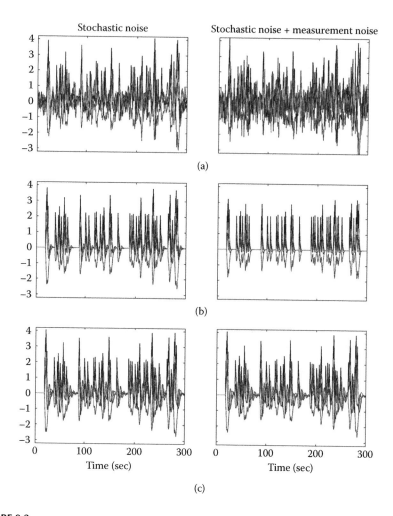

FIGURE 9.3
Simulated (a) and predicted (b and c) time series for realistic signals generated with dARX models. (b) Predicted signals using the MATLAB® lsqnonlin function with the Levenberg–Marquardt option for the optimization algorithm for the cost function (9.9). (c) Predicted signals using the same computational tools but for cost function (9.10)

the measurement errors have also to be estimated as part of the optimization. This cost function reads as

$$F_k = \left[\varepsilon_{k-1}, x_k - e^{A(t-t_{k-1})}(x_{k-1} + \varepsilon_{k-1}) - \left(\sum_{j=1}^{m_k} e^{A(t-\tau_j)} \right) c \right], \qquad (9.10)$$

where $F_k \in \mathbb{R}^{n \times 2}$ and $\varepsilon_{k-1} \in \mathbb{R}^{n \times 1}$ represent the local estimator for the errors measurements for the n elements.

In Figure 9.4, we presented the estimated coefficients versus the true coefficients for 20 repetitions of each noise scenario. While the best fit of all is achieved when we optimize using the first cost function for the "only stochastic noise" scenario, using this cost function fails almost completely when we considered the noisier scenario. On the contrary, the use of the second cost function allows stable estimators for each case. Computationally, the estimation procedure only takes a few seconds and, besides, we did not exclude the zero connections as is usually done in DCM approaches, which can be seen in the graphs of Figure 9.4. This makes us optimistic with respect to solving high-dimensional dynamical systems.

We do not present here the link with the hemodynamic response because this is currently an ongoing study; however, we are ensuing similarly to DCM models, using the balloon and hemodynamic equations [96,97].

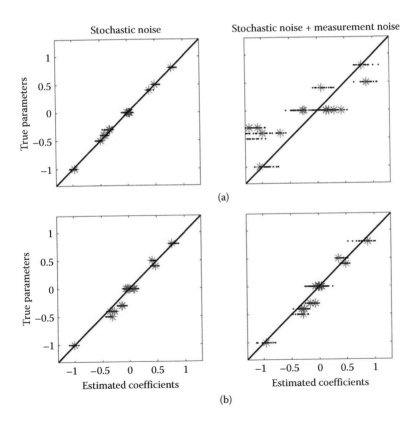

FIGURE 9.4
Parameter estimation for the proposed noisy scenarios for the first (a) and second (b) cost functions. The identity line denotes perfect correspondence between true and estimated coefficients. The black dots represent the estimated coefficients for the whole 20 simulations plotted against their corresponding true values, and the asterisks denote the individual coefficients estimated for the predicted time series shown in Figure 9.3.

We conjecture that sooner than later it should be possible to estimate high-dimensional synthetic frameworks such as those proposed by Bojak et al. [98] and Valdes-Sosa et al. [22], with subsequent application to real multimodal data analysis for directing to cognitive and clinical applications.

9.5 Conclusions

From the words of Nikos Logothetis [3]—"Research shows that the subcortical input to cortex is weak; the feedback is massive, the local connectivity reveals strong excitatory and inhibitory recurrence, and the output reflects changes in the balance between excitation and inhibition, rather than simple feedforward integration of subcortical inputs"—it may be thought that we are very far from modeling the richness of brain dynamics (maybe we are not able to achieve this with the current techniques and must wait for nanotechnology), but in the last 10 years important progress has been made in multimodal fusion methods.

The optimism expressed in the last part of this chapter is based on the possibility of characterizing the spatiotemporal functions and interactions of the brain modules that lead to the execution of the principal cognitive task, and understand why they are disrupted by the severity of particular diseases that affect the human brain. Within the intricate world of hyperactive interactions, we can say that fMRI is a measure of "something that is happening in some place," with good spatial accuracy. A similar observation is valid for the local field potential that can potentially be obtained by solving the M-EEG inverse problems or directly measured by the ECoG technique, though with the use of the M-EEG/ECoG technique this "something" can be quantified in real time, but possibly spatially displaced as in the case of EEG/MEG-based inverse solutions. Thus, as widely interpreted, M-EEG and ECoG allow meaning/content to be added to the findings made by applying fMRI, which may help to disclose the patterns of temporal dynamic and connectivity among the neuronal populations. Undeniably, neither ECoG nor EEG/MEG produces massive measures with the spatial accuracy that fMRI can provide. Therefore, a fusion of analyses and methodologies is necessary to make further discoveries.

A final conclusion is that the best current alternative in our hands is to continue developing fusion models owing to their advantages of enhanced characterizations of the brain functions and their interweaving of integration/divergence patterns. A nonevident but encouraging conclusion is that soon we will be able to estimate parameters for dynamical-systems-based inverse problems, and this will pave the way for modeling the brain in more realistic ways and thus understanding its behavior and learning capacities.

References

1. A. M. Dale and E. Halgren, Spatiotemporal mapping of brain activity by integration of multiple imaging modalities, *Curr Opin Neurobiol*, 11, 202–208, 2001.
2. T. Deneux, EEG-fMRI fusion: Adaptations of the Kalman filter for solving a high-dimensional spatio-temporal inverse problem, in *Adaptive Filtering*, Lino Garcia, Ed., InTech, pp. 233–258, 2011.
3. N. K. Logothetis, What we can do and what we cannot do with fMRI, *Nature*, 453, 869–878, 2008.
4. M. Hämäläinen, R. Hari, R. J. Ilmoniemi, J. Knuutila, and O. V. Lounasmaa, Magnetoencephalography—Theory, instrumentation, and applications to non-invasive studies of the working human brain, *Rev Mod Phys*, 65(2), 413–497, 1993.
5. P. L. Nunez, *Electric Fields of the Brain*, Oxford University Press, New York, 1981.
6. P. L. Nunez and R. B. Silberstein, On the relationship of synaptic activity to macroscopic measurements: Does co-registration of EEG with fMRI make sense? *Brain Topogr*, 13, 79–96, 2000.
7. C. M. Michel and D. Brandeis, Sources and temporal dynamics of scalp electric fields, in *Simultaneous EEG and fMRI: Recording, Analysis, and Application*, M. Ullsperger and S. Debener, Eds., Oxford University Press, Oxford, 3–19, 2010.
8. S. Ogawa, T. M. Lee, A. S. Nayak, and P. Glynn, Oxygenation-sensitive contrast in magnetic resonance image of rodent brain at high magnetic fields, *Magn Reson Med*, 14(1), 68–78, 1990.
9. B. J. Casey, M. Davidson, and B. Rosen, Functional magnetic resonance imaging: Basic principles of and application to developmental science, *Dev Sci*, 5(3), 301–309, 2002.
10. N. K. Logothetis and B. A. Wandell, Interpreting the BOLD signal, *Annu Rev Physiol*, 66, 735–769, 2004.
11. P. Ritter and A. Villringer, Simultaneous EEG–fMRI, *Neurosci Biobehav Rev*, 30, 823–838, 2006.
12. J. R. Ives, S. Warach, F. Schmitt, R. R. Edelman, and D. L. Schomer, Monitoring the patient's EEG during echo planar MRI, *Electroencephalogr Clin Neurophysiol*, 87, 417–420, 1993.
13. H. Laufs, A personalized history of EEG-fMRI integration, *Neuroimage*, 62, 1056–1067, 2012.
14. J. Daunizeau, H. Laufs, and K. Friston, EEG-fMRI information fusion: Biophysics and data analysis, in *EEG-fMRI: Physiological Basis, Technique, and Application*, C. Mulert and L. Lemieux, Eds., Springer-Verlag, Berlin, 511–526, 2010.
15. A. Babajani and H. Soltanian-Zadeh, Integrated MEG/EEG and fMRI model based on neural masses, *IEEE Trans Biomed Eng*, 53, 1794–1801, 2006.
16. M. Wibral, C. Bledowski, and G. Turi, Integration of separately recorded EEG/MEG and fMRI data, in *Simultaneous EEG and fMRI: Recording, Analysis, and Application*, M. Ullsperger and S. Debener, Eds., Oxford University Press, New York, pp. 209–234, 2010.
17. J. Daunizeau, C. Grova, G. Marrelec, J. Mattout, S. Jbabdi, M. Pelegrini-Issac, et al., Symmetrical event-related EEG/fMRI information fusion in a variational Bayesian framework, *Neuroimage*, 36, 69–87, 2007.

18. A. M. Dale, A. K. Liu, B. R. Fischl, R. L. Buckner, J. W. Belliveau, J. D. Lewine, et al., Dynamic statistical parametric mapping: Combining fMRI and MEG to produce high-resolution spatiotemporal maps of cortical activity, *Neuron*, 26, 55–67, 2000.
19. F. Babiloni, C. Babiloni, F. Carducci, C. Del Gratta, G. L. Romani, P. M. Rossini, et al., Cortical source estimate of combined high resolution EEG and fMRI data related to voluntary movements, *Methods Inf Med*, 41, 443–450, 2002.
20. S. Debener, M. Ullsperger, M. Siegel, K. Fiehler, D. Y. von Cramon, and A. K. Engel, Trial-by-trial coupling of concurrent electroencephalogram and functional magnetic resonance imaging identifies the dynamics of performance monitoring, *J Neurosci*, 25, 11730–11737, 2005.
21. R. J. Huster, S. Debener, T. Eichele, and C. S. Herrmann, Methods for simultaneous EEG-fMRI: An introductory review, *J Neurosci*, 32, 6053–6060, 2012.
22. P. A. Valdes-Sosa, J. M. Sanchez-Bornot, R. C. Sotero, Y. Iturria-Medina, Y. Aleman-Gomez, and J. Bosch-Bayard, Model driven EEG/fMRI fusion of brain oscillations, *Hum Brain Mapp*, 30(9), 2701–2721, 2009.
23. N. Trujillo-Barreto, E. Martinez-Montes, L. Melie-Garcia, and P. A. Valdes-Sosa, A symmetrical Bayesian model for fMRI and EEG/MEG neuroimage fusion, *Int J Bioelectromag*, 3(1), 2001.
24. R. Henson, E. Mouchlianitis, and K. J. Friston, MEG and EEG data fusion: Simultaneous localisation of face-evoked responses, *Neuroimage*, 47, 581–589, 2009.
25. D. Mantini, M. G. Perrucci, C. Del Gratta, G. L. Romani, and M. Corbetta, Electrophysiological signatures of resting state networks in the human brain, *Proc Natl Acad Sci U S A*, 104, 13170–13175, 2007.
26. R. I. Goldman, J. M. Stern, J. Engel, Jr., and S. Cohen, Simultaneous EEG and fMRI of the alpha rhythm, *Neuroreport*, 13, 2487–2492, 2002.
27. M. Moosmann, P. Ritter, I. Krastel, A. Brink, S. Thees, F. Blankenburg, et al., Correlates of alpha rhythm in functional magnetic resonance imaging and near infrared spectroscopy, *Neuroimage*, 20, 145–158, 2003.
28. H. Laufs, A. Kleinschmidt, A. Beyerle, E. Eger, A. Salek-Haddadi, C. Preibisch, et al., EEG-correlated fMRI of human alpha activity, *Neuroimage*, 19(4), 1463–1476, 2003.
29. H. Laufs, K. Krakow, P. Sterzer, E. Eger, A. Beyerle, A. Salek-Haddadi, et al., Electroencephalographic signatures of attentional and cognitive default modes in spontaneous brain activity fluctuations at rest, *Proc Natl Acad Sci U S A*, 100(19), 11053–11058, 2003.
30. V. D. Calhoun, J. Liu, and T. Adali, A review of group ICA for fMRI data and ICA for joint inference of imaging, genetic, and ERP data, *Neuroimage*, 45, 163–172, 2009.
31. V. D. Calhoun, T. Adali, G. D. Pearlson, and K. A. Kiehl, Neuronal chronometry of target detection: Fusion of hemodynamic and event related potential data, *Neuroimage*, 30, 544–553, 2006.
32. A. J. Bell and T. J. Sejnowski, An information maximisation approach to blind separation and blind deconvolution, *Neural Comput*, 7(6), 1129–1159, 1995.
33. M. Wax and T. Kailath, Detection of signals by information theoretic criteria, *IEEE Trans Acoust Speech Signal Process*, 33, 387–392, 1985.
34. M. Moosmann, T. Eichele, H. Nordby, K. Hugdahl, and V. D. Calhoun, Joint independent component analysis for simultaneous EEG-fMRI: Principle and simulation, *Int J Psych*, 67, 212–221, 2008.

35. V. D. Calhoun, T. Eichele, and G. Pearlson, Functional brain networks in schizophrenia: A review, *Front Hum Neurosci*, 3, 17, 2009.
36. E. Martínez-Montes, P. A. Valdes-Sosa, F. Miwakeichi, R. I. Goldman, and M. S. Cohen, Concurrent EEG/fMRI analysis by multiway partial least squares, *Neuroimage*, 22, 1023–1034, 2004.
37. N. M. Correa, T. Eichele, T. Adali, Y. O. Li, and V. D. Calhoun, Multi-set canonical correlation analysis for the fusion of concurrent single trial ERP and functional MRI, *Neuroimage*, 50, 1438–1445, 2010.
38. N. M. Correa, Y. O. Li, T. Adali, and V. D. Calhoun, Canonical correlation analysis for feature-based fusion of biomedical imaging modalities and its application to detection of associative networks in schizophrenia, *IEEE J Sel Top Signal Process*, 2(6), 998–1007, 2008.
39. F. Kruggel, C. J. Wiggins, C. S. Herrmann, and D. Y. von Cramon, Recording of the event-related potentials during functional MRI at 3.0 Tesla field strength, *Magn Reson Med*, 44, 277–282, 2000.
40. C. G. Benar, D. Schon, S. Grimault, B. Nazarian, B. Burle, M. Roth, et al., Single-trial analysis of oddball event-related potentials in simultaneous EEG-fMRI, *Hum Brain Mapp*, 28, 602–613, 2007.
41. T. Matsuda, M. Matsuura, T. Ohkubo, H. Ohkubo, Y. Atsumi, M. Tamaki, et al., Influence of arousal level for functional magnetic resonance imaging (fMRI) study: Simultaneous recording of fMRI and electroencephalogram, *Psychiatry Clin Neurosci*, 56, 289–290, 2002.
42. T. Eichele, K. Specht, M. Moosmann, M. L. Jongsma, R. Q. Quiroga, H. Nordby, et al., Assessing the spatiotemporal evolution of neuronal activation with single-trial event-related potentials and functional MRI, *Proc Natl Acad Sci U S A*, 102, 17798–17803, 2005.
43. C. Bledowski, K. K. Cohen, M. Wibral, B. Rahm, R. A. Bittner, K. Hoechstetter, et al., Mental chronometry of working memory retrieval: A combined functional magnetic resonance imaging and event-related potentials approach, *J Neurosci*, 26, 821–829, 2006.
44. K. Friston, L. Harrison, J. Daunizeau, S. Kiebel, C. Phillips, N. Trujillo-Barreto, et al., Multiple sparse priors for the M/EEG inverse problem, *Neuroimage*, 39, 1104–1120, 2008.
45. N. J. Trujillo-Barreto, E. Aubert-Vazquez, and P. A. Valdes-Sosa, Bayesian model averaging in EEG/MEG imaging, *Neuroimage*, 21, 1300–1319, 2004.
46. K. J. Friston, L. Harrison, and W. Penny, Dynamic causal modeling, *Neuroimage*, 19, 1273–1302, 2003.
47. T. Deneux and O. Faugeras, Using nonlinear models in fMRI data analysis: Model selection and activation detection, *Neuroimage*, 32, 1669–1689, 2006.
48. K. E. Stephan, N. Weiskopf, P. M. Drysdale, P. A. Robinson, and K. J. Friston, Comparing hemodynamic models with DCM, *Neuroimage*, 38, 387–401, 2007.
49. K. E. Stephan, J. C. Marshall, W. D. Penny, K. J. Friston, and G. R. Fink, Inter-hemispheric integration of visual processing during task-driven lateralization, *J Neurosci*, 27, 3512–3522, 2007.
50. S. J. Kiebel, S. Klppel, N. Weiskopf, and K. J. Friston, Dynamic causal modeling: A generative model of slice timing in fMRI, *Neuroimage*, 34, 1487–1496, 2007.
51. J. Daunizeau, L. Lemieux, A. E. Vaudano, K. J. Friston, and K. E. Stephan, An electrophysiological validation of stochastic DCM for fMRI, *Front Comput Neurosci*, 6, 103–123, 2013.

52. J. Daunizeau, K. E. Stephan, and K. J. Friston, Stochastic dynamic causal modelling of fMRI data: Should we care about neural noise? *Neuroimage*, 62(1), 464–481, 2012.
53. J. G. Ojemann, R. L. Buckner, E. Akbudak, A. Z. Snyder, J. M. Ollinger, R. C. McKinstry, et al., Functional MRI studies of word-stem completion: Reliability across laboratories and comparison to blood flow imaging with PET, *Hum Brain Mapp*, 6(4), 203–215, 1998.
54. P. A. Valdes-Sosa, J. M. Sanchez-Bornot, M. Vega-Hernández, L. Melie-García, A. Lage-Castellanos, and E. Canales-Rodríguez, Granger causality on spatial manifolds: Applications to neuroimaging, in *Handbook of Time Series Analysis: Recent Theoretical Developments and Applications*, S. Björn, M. Winterhalder, and J. Timmer, Eds., John Wiley & Sons, pp. 461–491, 2006.
55. R. Garg, G. A. Cecchi, and A. R. Rao, Full-brain auto-regressive modelling (FARM) using fMRI, *Neuroimage*, 58, 416–441, 2011.
56. B. Efron, T. Hastie, I. Johnstone, and R. Tibshirani, Least angle regression, *Ann Stat*, 32(1), 407–499, 2004.
57. K. Friston, Dynamic causal modeling and Granger causality comments on: The identification of interacting networks in the brain using fMRI: Model selection, causality and deconvolution, *Neuroimage*, 58(2–2), 303–305, 2011.
58. O. David, I. Guillemain, S. Saillet, S. Reyt, C. Deransart, C. Segebarth, et al., Identifying neural drivers with functional MRI: An electrophysiological validation, *PLoS Biol*, 6(12), 2683–2697, 2008.
59. A. Roebroeck, E. Formisano, and R. Goebel, Reply to Friston and David fMRI: Model selection, causality and deconvolution, *Neuroimage*, 58(2), 310–311, 2011.
60. P. A. Valdes-Sosa, A. Roebroeck, J. Daunizeau, and K. Friston, Effective connectivity: Influence, causality and biophysical modeling, *Neuroimage*, 58(2), 339–361, 2011.
61. K. Friston, Functional and effective connectivity: A review, *Brain Connect*, 1, 13–36, 2011.
62. S. M. Smith, K. L. Miller, G. Salimi-Khorshidi, M. Webster, C. F. Beckman, T. E. Nichols, et al., Network modeling methods for FMRI, *Neuroimage*, 54, 875–891, 2011.
63. L. Faes and G. Nollo, Extended causal modelling to assess partial directed coherence in multiple time series with significant instantaneous interactions, *Biol Cybern*, 103(5), 387–400, 2010.
64. K. M. Gates, P. C. M. Molenaar, F. G. Hillary, and S. Slobounov, Extended unified SEM approach for modeling event-related fMRI data, *Neuroimage*, 54, 1151–1158, 2011.
65. L. A. Baccala and K. Sameshima, Partial directed coherence: A new concept in neural structure determination, *Biol Cybern*, 84, 463–474, 2001.
66. B. Schelter, M. Winterhalder, M. Eichler, M. Peifer, B. Hellwig, B. Guschlbauer, et al., Testing for directed influences among neural signals using partial directed coherence, *J Neurosci Methods*, 152, 210–219, 2006.
67. B. Schelter, J. Timmer, and M. Eichler, Assessing the strength of directed influences among neural signals using renormalized partial directed coherence, *J Neurosci Methods*, 179, 121–130, 2009.
68. J. F. Smith, A. S. Pillai, K. Chen, and B. Horwitz, Identification and validation of effective connectivity networks in functional magnetic resonance imaging using switching linear dynamic systems, *Neuroimage*, 52, 1027–1040, 2010.

69. P. Hagmann, P. E. Grant, and D. A. Fair, MR connectomics: A conceptual framework for studying the developing brain, *Front Syst Neurosci*, 6, 43, 2012.

70. R. C. Sotero and N. J. Trujillo-Barreto, Biophysical model for integrating neuronal activity, EEG, fMRI and metabolism, *Neuroimage*, 39, 290–309, 2008.

71. B. H. Jansen and V. G. Rit, Electroencephalogram and visual evoked potential generation in a mathematical model of coupled cortical columns, *Biol Cybern*, 73, 357–366, 1995.

72. L. H. Zetterberg, L. Kristiansson, and K. Mossberg, Performance of a model for a local neuron population, *Biol Cybern*, 31, 15–26, 1978.

73. H. Wilson and J. Cowan, Excitatory and inhibitory interaction in localized populations of model neurons, *Biophys J*, 12, 1–23, 1972.

74. H. Wilson and J. Cowan, A mathematical theory of the functional dynamics of cortical and thalamic nervous tissue, *Kybernetik*, 13, 55–80, 1973.

75. F. H. Lopes da Silva, A. Hoeks, H. Smits, and L. H. Zetterberg, Model of brain rhythmic activity, the alpha-rhythm of the thalamus, *Kybernetik*, 15, 27–37, 1974.

76. F. Grouiller, R. C. Thornton, K. Groening, L. Spinelli, J. S. Duncan, K. Schaller, et al., With or without spikes: Localization of focal epileptic activity by simultaneous electroencephalography and functional magnetic resonance imaging, *Brain*, 134, 2867–2886, 2011.

77. R. Thornton, H. Laufs, R. Rodionov, S. Cannadathu, S. Vulliemoz, A. SalekHaddadi, et al., EEG-correlated fMRI and post-operative outcome in focal epilepsy, in *European Epilepsy Congress*, September 2008, Berlin, 2009.

78. S. Vulliemoz, L. Lemieux, J. Daunizeau, C. M. Michel, and J. S. Duncan, The combination of EEG source imaging and EEG-correlated functional MRI to map epileptic networks, *Epilepsia*, 51, 491–505, 2010.

79. R. C. Knowlton, R. A. Elgavish, A. Bartolucci, B. Ojha, N. Limdi, J. Blount, et al., Functional imaging: II. Prediction of epilepsy surgery outcome, *Ann Neurol*, 64, 35–41, 2008.

80. J. Gotman, E. Kobayashi, A. P. Bagshaw, C. G. Benar, and F. Dubeau, Combining EEG and fMRI: A multimodal tool for epilepsy research, *J Magn Reson Imaging*, 23, 906–920, 2006.

81. M. Seeck, F. Lazeyras, C. M. Michel, O. Blanke, C. A. Gericke, J. Ives, et al., Non-invasive epileptic focus localization using EEG-triggered functional MRI and electromagnetic tomography, *Electroencephalogr Clin Neurophysiol*, 106, 508–512, 1998.

82. L. Lemieux, K. Krakow, and D. R. Fish, Comparison of spike-triggered functional MRI BOLD activation and EEG dipole model localization, *Neuroimage*, 14, 1097–1104, 2001.

83. C. G. Benar, C. Grova, E. Kobayashi, A. P. Bagshaw, Y. Aghakhani, F. Dubeau, et al., EEG-fMRI of epileptic spikes: Concordance with EEG source localization and intracranial EEG, *Neuroimage*, 30, 1161–1170, 2006.

84. C. Grova, J. Daunizeau, J. M. Lina, C. G. Benar, H. Benali, and J. Gotman, Evaluation of EEG localization methods using realistic simulations of interictal spikes, *Neuroimage*, 29, 734–753, 2006.

85. C. Grova, J. Daunizeau, E. Kobayashi, A. P. Bagshaw, J. M. Lina, F. Dubeau, et al., Concordance between distributed EEG source localization and simultaneous EEG-fMRI studies of epileptic spikes, *Neuroimage*, 39, 755–774, 2008.

86. E. Kobayashi, A. P. Bagshaw, C.-G. Bénar, Y. Aghakhani, F. Andermann, F. Dubeau, et al., Temporal and extra-temporal BOLD responses to temporal lobe interictal spikes, *Epilepsia*, 47(2), 343–354, 2006.

87. B. Stefanovic, J. M. Warnking, and E. Kobayashi, Hemodynamic and metabolic responses to activation, deactivation and epileptic discharges, *Neuroimage*, 28, 205–215, 2005.

88. S. Vulliemoz, R. Rodionov, R. Thornton, D. W. Carmichael, M. Guye, S. Lhatoo, et al., BOLD correlates of interictal epileptic activity: Additional contributions from continuous EEG Source Imaging, in *15th Annual Meeting of the Organisation of Human Brain Mapping*, June 18–23, San Francisco, CA, 2009.

89. R. Grave de Peralta Menendez, S. Gonzalez Andino, G. Lantz, C. M. Michel, and T. Landis, Noninvasive localization of electromagnetic epileptic activity. I. Method descriptions and simulations, *Brain Topogr*, 14(2), 131–137, 2001.

90. W. J. Gehring, G. Grantton, M. G. Coles, and E. Donchin, Probability effects on stimulus evaluation and response processes, *J Exp Psychol Hum Percept Perform*, 18, 198–216, 1992.

91. D. E. Job, H. C. Whalley, S. McConnell, M. Glabus, E. C. Johnstone, and S. M. Lawrie, Structural gray matter differences between first-episode schizophrenics and normal controls using voxel-based morphometry, *Neuroimage*, 17, 880–889, 2002.

92. K. A. Kiehl and P. F. Liddle, An event-related functional magnetic resonance imaging study of an auditory oddball task in schizophrenia, *Schizophrenia*, 48, 159–171, 2001.

93. A. A. Stevens, P. S. Goldman-Rakic, J. C. Gore, R. K. Fulbright, and B. E. Wexler, Cortical dysfunction in schizophrenia during auditory word and tone working memory demonstrated by functional magnetic resonance imaging, *Arch Gen Psych*, 55, 1097–1103, 1998.

94. R. W. McCarley, S. F. Faux, M. E. Shenton, P. G. Nestor, and J. Adams, Event-related potentials in schizophrenia: Their biological and clinical correlates and a new model of schizophrenia pathophysiology, *Schizophr Res*, 4, 209–231, 1991.

95. E. Coddington and N. Levinson, *Theory of Ordinary Differential Equations*, McGraw-Hill, New York, 1955.

96. R. B. Buxton, E. C. Wong, and L. R. Frank, Dynamics of blood flow and oxygenation changes during brain activation: The balloon model, *Magn Reson Med*, 39, 855–864, 1998.

97. K. J. Friston, A. Mechelli, R. Turner, and C. J. Price, Nonlinear responses in fMRI: The balloon model, Volterra kernels, and other hemodynamics, *Neuroimage*, 12, 466–477, 2000.

98. I. Bojak, T. F. Oostendorp, A. T. Reid, and R. Kotter, Towards a model-based integration of registered electroencephalography/functional magnetic resonance imaging data with realistic neural population meshes, *Philos Transact Math Phys Eng Sci*, 369, 3785–3801, 2011.

10

Memory Retention and Recall Process

Hafeez Ullah Amin and Aamir Saeed Malik

Universiti Teknologi PETRONAS

CONTENTS

10.1 Introduction...219
10.2 Overview of Traditional Approaches 221
    10.2.1 Atkinson and Shiffrin Memory Model (1968)...........................221
    10.2.2 Dual Coding Theory of Memory (1971)......................................222
    10.2.3 Baddeley's Model of Working Memory (1974) 222
    10.2.4 Cognitive Load Theory (1988)..223
    10.2.5 Cognitive Theory of Multimedia Learning (1999) 223
10.3 Understanding Memory Retention and Recall...224
    10.3.1 Brain Areas Associated with Memory Functions.....................225
        10.3.1.1 Cerebral Cortex..225
        10.3.1.2 Hippocampus 226
        10.3.1.3 Amygdala...226
    10.3.2 Long-Term Potentiation and Memory Retention.....................227
    10.3.3 Factors Affecting Memory Retention and Recall.....................228
        10.3.3.1 Attention..228
        10.3.3.2 Rehearsal 228
        10.3.3.3 Sleep 229
        10.3.3.4 Exercise and Nutrition 229
        10.3.3.5 Mnemonics..230
        10.3.3.6 Testing Effect 230
        10.3.3.7 Reward..232
10.4 Experimental Design Issues...233
10.5 Summary..234
References...235

10.1 Introduction

Our knowledge is a collection of our experiences, which expands daily as we experience new things. The way we imbue our surroundings and ourselves with meaning depends on the knowledge and understanding we have, and this knowledge depends on our memorization of what we have learned.

In daily life, we take in new information and store it in our brain, maintaining it and recalling it depending on our needs. This happens because our brain has the capability of learning new skills and experiences, storing what has been learned and reusing the stored knowledge. These capabilities of storing and reusing experiences and skills are informally known as the human memory system. Everything we do or think depends on our memory, which is active every moment, receiving new information from our senses, updating existing knowledge using focus and attention, retrieving the stored experiences and skills, and planning for future activities that have not occurred yet. Thus far, neuroscientists have been expecting to find specific stores of memory in the brain and discover their exact location to know which type of memory lies where. Unfortunately, because of the great complexity of the human brain system (Figure 10.1 [1]), this concept has not been proved. However, some cognitive and mental functions are found in certain brain areas.

Generally, there are two different memory types—short-term and long-term memory—that store and access information differently, and many brain regions are involved in the process. Short-term memory retains information for a few seconds, and its capacity ranges from seven to nine items for a normal person. It tends to weaken as the individual's age increases. Long-term memory retains unlimited information for an infinite duration. The information could be personal events, temporal and spatial relations among these events, and real-world entities and their meaning such as symbols, words, and concepts. There are three fundamental memory processes: encoding, retention, and recall. Encoding allows converting the perceived information

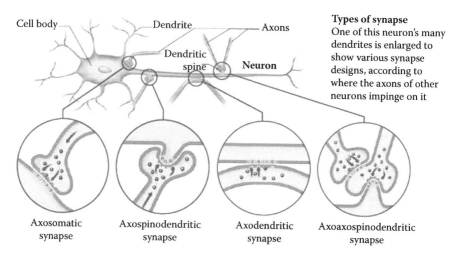

FIGURE 10.1
Neuronal communication in brain. (From Carter, R. et al. *The Human Brain Book*. London: Dorling Kindersley Publisher, 2009.)

of interest into a construct that can be retained in the brain. It is the first stage of creating a new memory in the brain. Retention is the storage of encoded information in the brain. Retrieval or recall is the re-accessing of retained events or information in the brain. A part of encoding or retention is memory consolidation, which stabilizes a memory trace after its initial formation. Neurologically, the consolidation process employs long-term potentiation; it strengthens the synapses by increasing the number of signals that are sent and received between the two neurons.

Scientists utilized different neuroimaging techniques to view these processes such as electroencephalography (EEG), functional magnetic resonance imaging (fMRI), magneto-encephalography (MEG), positron emission tomography (PET), etc. This chapter will focus on recent studies utilizing EEG for the two important memory processes: (1) retention and (2) recall, by focusing on their neurophysiological understanding and studying the factors that affect the retention and recall performance. The following sections will describe the traditional cognitive and memory theories of memory processes, the brain regions associated with memory functions, the neural understanding of retention and recall, applications and, finally, the conclusions.

10.2 Overview of Traditional Approaches

10.2.1 Atkinson and Shiffrin Memory Model (1968)

Richard Atkinson and Richard Shiffrin proposed a multi-store human memory model (see Figure 10.2 [2]) that divided the human memory into three distinct stores: sensory memory, short-term memory (STM), and long-term memory (LTM). This model has many limitations and has attracted much

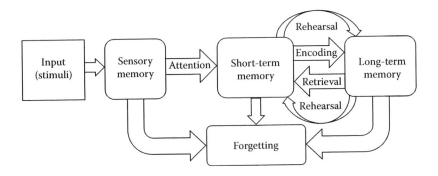

FIGURE 10.2
Atkinson and Shiffrin memory model. (From Atkinson, R. C. and Shiffrin, R. M. *The Psychology of Learning and Motivation*, 2, 89–195, 1968.)

criticism such as an ill-defined concept of the rehearsal buffer, presence of a single short-term store, and incorrect prediction of total recall probability in free recall [3]. However, it is still useful for researchers, who can propose new models based on this model.

10.2.2 Dual Coding Theory of Memory (1971)

Allan Paivio [4] developed dual coding theory in 1971. According to this theory, "human cognition is unique in that it has become specialized for dealing simultaneously with language and with nonverbal objects and events." This theory categorizes the process of intelligence and memory into two separate subsystems: verbal and nonverbal (imagery) systems. The verbal system is specialized for handling the language directly, while the imagery system is specialized for handling visual objects and events.

Although this theory was strongly criticized conceptually and was experimentally refuted, it initiated research on verbal and visual effects on memory.

10.2.3 Baddeley's Model of Working Memory (1974)

Alan Baddeley and G. Hitch proposed a working memory model to describe more accurately the concept and model of the short-term system in Atkinson and Shiffrin's memory model. This model consists of three main parts: (1) the central executive, (2) the phonological loop, and (3) the visuospatial sketchpad. It proposes the concept of two independent short-term memory storage subsystems: the phonological loop, for verbal information storage, and the visuospatial scratchpad, for visuospatial information storage (see Figure 10.3 [5]). The central executive controls both the subsystems and deals with cognitive tasks such as problem solving and mental arithmetic.

Baddeley and Hitch's original working memory model also attracted much criticism, which led to the exploration of the episodic buffer [6] in the original

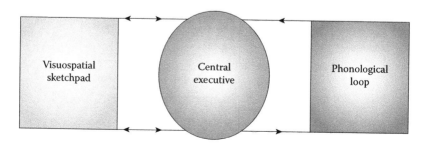

FIGURE 10.3
Baddeley and Hitch working memory model. (From Baddeley, A. D. and Hitch, G. Working memory. *The Psychology of Learning and Motivation,* 8, 47–89, 1975.)

WM model. However, this model is very critical to understand how the human mind manipulates and retains information during problem solving, reasoning, and thinking. Davidson et al. investigated cognitive control and executive functions for memory manipulation and inhibition in the visual switching task. They stated that "the mind has the ability to recollect plan and other things related to the present, future, and past" [7]. The episodic buffer organizes the arrangement of verbal sentences with the help of the phonological loop into a coherent sequence along with memory. For example, when people are involved in mutual dialogue, their brain responds by keeping track of what has been said, and their assumptions about what different speakers intended by their explanations.

10.2.4 Cognitive Load Theory (1988)

This theory took shape from the idea of working memory's limited capacity and its operations. John Sweller developed cognitive load theory [8,9] when studying problem-solving strategies. This theory is concerned with the way cognitive resources are focused and utilized while learning and solving problems. It encouraged learners to utilize working memory efficiently while solving difficult tasks. It has implications in the designing of learning material to reduce cognitive load such as multimedia content for learning.

10.2.5 Cognitive Theory of Multimedia Learning (1999)

Richard E. Mayer [10] proposed modality principles stating that when verbal and imagery content are presented together to a learner, they are processed along two distinct channels (see Figure 10.4). These principles are known as the cognitive theory of multimedia learning.

This theory combines several other concepts such as dual code theory, working memory limitation, connections between text-based and image-based

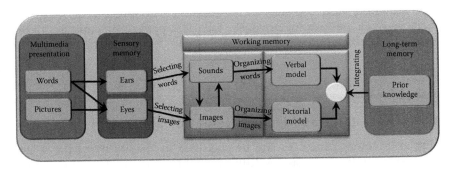

FIGURE 10.4
Cognitive theory of multimedia learning. (From Moreno, R. and Mayer, R. E. *Journal of Educational Psychology*, 91, 358, 1999.)

TABLE 10.1

Summary of Traditional Models

Model	Main Objective	Strengths	Weaknesses
Atkinson and Shiffrin Memory Model [3]	Multi-store memory system	It provides the basis for other theories. It explores the distinction between STM and LTM memory stores. It emphasizes the effect of rehearsal on memory recall.	There is a lack of emphasis on unconscious processes. Intermediate-level activation between STM and LTM is missing; absence of memory subsystems. The effect of rehearsal is overstated.
Dual-coding theory of memory [4]	Verbal and nonverbal information processing	It explains human behavior and experience in terms of verbal and imagery representation.	It does not consider the likelihood of cognition being mediated by routes other than words and images.
Baddeley's model of working memory [11]	Concept of working memory	It is applicable in everyday experience of processing information during problem solving. Rehearsal is not necessary to remember and recall all types of information.	The functions and capacity of central executive system are not clearly described and difficult to determine in practice.
Cognitive load theory [8]	Efficient use of working memory during problem solving	It identifies the methods to reduce extraneous cognitive load in learning. It initiates research on effective instructional design strategies.	When material is presented in a way that does not relate to actual performance, then transfer of learning will be more challenging.
Cognitive theory of multimedia learning [10]	Learning from multimedia content	It suggests five principles for designing multimedia instructions that lead to learning that is more effective.	It is based on multimedia technology, which has a tendency to overwhelm the brain, and needs to be designed more effectively.

representation, and transfer of information from long-term to working memory while performing any task. Table 10.1 presents a summary of these theories with their strengths and weaknesses.

10.3 Understanding Memory Retention and Recall

Human memory processes can be classified as the ability of the mind to understand, retain, and successfully recall information. The role of retention is to store encoded events and information, and the role of recall is to re-access the retained events and information in the mind in response to external stimuli. Although memory is achieved in multiple phases, recall is the only way to measure memory performance. How much information is encoded? How much information is retained? How much information is retrieved? What is the memory performance? All these questions can be answered in the memory recall phase by asking someone questions and recording their recall responses. The correct responses will indicate the *memory performance, retained information,* or amount of *encoded information.*

Memory is closely related to the learning process, which is concerned with acquiring skills or knowledge. Memory is also the representation of learning that has been acquired. Various brain parts are involved in learning and the memory formation process, and different mental and physical factors are reported that influence memory retention and recall performance. These brain areas and the factors that affect retention and recall performance are discussed in the subsequent sections.

10.3.1 Brain Areas Associated with Memory Functions

Many areas of the brain are associated with memory and memory processes, including the cerebral cortex of the frontal, parietal, and temporal lobes; the hippocampus; the amygdala; and the diencephalon (Figure 10.5).

Loss of the hippocampus leads to an inability to encode short-term memory to long-term memory. The causes of memory loss may be trauma, injury, stroke, disease, drugs, etc.

10.3.1.1 Cerebral Cortex

The outer covering of gray matter on the hemispheres is the cerebral cortex. It is a thin layer of tissue and covers the outer portion of cerebrum up to 5 mm. It consists of 19% of all the brain cells (approximately 86.1 (±8.1) billion neurons) [12],

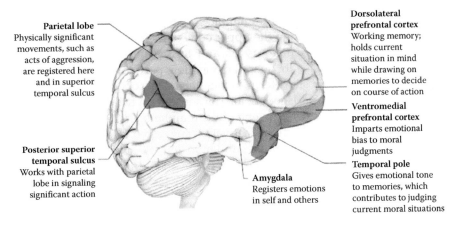

FIGURE 10.5
Brain regions associated with memory functions. (From Carter, R. et al. *The Human Brain Book.* London: Dorling Kindersley Publisher, 2009.)

which are organized into a complex neural network of synaptic connections and develops the distributed connections among the specialized brain system. It is divided into four lobes: the frontal, central, parietal, temporal, and occipital lobes, which are responsible for specific functions such as thinking, reasoning, decision making, perception, movement, touch, smell, vision, listening, attention, memory, and emotion. Some cortical regions are specialized for certain functions, and some cortices are associated with more complex functions, including working memory, emotion, attention, judgment, etc. The main functions of the frontal lobe include problem solving, reasoning, planning, memory, and decision making. It is experimentally proved that this lobe is primarily responsible for dealing with working memory and information processing in connection with other task-specific brain regions [11].

The parietal lobe is responsible for spatial orientation, pain and touch sensation, and cognitive functions. The posterior parietal cortex has been associated with memory retrieval, especially in episodic memory: personal experiences occurring at a certain time and place [13]. Several fMRI studies have investigated the intra-parietal sulci in memory encoding and retrieval, and their results reflected parietal activation in the memorization process and engagement of the brain attentional network [14]. Temporal lobes are involved in auditory, emotional, speech, language, learning, and memory related functions. High gamma activity in the left temporal cortical region has discriminated true from false memory responses [15] in memory retrieval tasks. The primary visual cortex in the occipital lobe is responsible for processing visual information received from the eyes through the thalamus. The entire cerebral cortex is linked directly or indirectly in memory processes: encoding, retention, and retrieval.

10.3.1.2 Hippocampus

The hippocampus is a part of the limbic system, which lies inside each temporal lobe. It is located along the upper edge of the parahippocampal gyrus. It interlocks with the dentate gyrus and forms the hippocampal–dentate complex. It is responsible for spatial navigation, particularly the encoding of new information from short-term to long-term memory, and the retrieval of long-term memories. Karlsgodt et al., [16] showed an fMRI study that used a verbal working memory test. In encoding, maintenance, and recall, a within-subject comparison of functional activation suggested that the hippocampus had a role in working memory similar to its role in long-term memory, implying that the hippocampus might be involved in overall encoding and recalling rather than just in long-term memory tasks.

10.3.1.3 Amygdala

The amygdala is an almond-shaped nuclei located in the deep anterior inferior medial temporal lobe near the hippocampus and holds 13 nuclei, each of which has other subsystems. It is associated with many functions in the

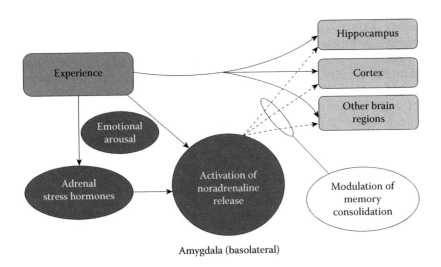

FIGURE 10.6
Amygdala activation modulates memory consolidation. (From McGaugh, J. L. *Trends in Cognitive Sciences*, 10, 345–347, 2006.)

brain, such as fear and anxiety, emotional and social impact of the environment, support for memory storage across multiple brain regions, and regulation of automatic responses. It consists of three classes of nuclei: baso-lateral, superficial, and centromedial. This classification is not consistent with all amygdala-anatomy-related studies. The input signals to the amygdala from the cortex, thalamus, hypothalamus, and brain stem include olfactory, somatosensory, gustatory, auditory, and visual information.

The amygdala itself is not a storage place for memory, but influences the consolidation and acquiring of memories in various learning situations in the brain, such as interactions with the hippocampus in emotional memory [17], involvement in memory storage in other brain systems [18], drug addiction memory [19], reward expectancy [20], and spatial learning [21]. Several other studies have reported that emotional arousal stimulates the amygdala and has modulatory effects on memory; for example, neuroimaging studies investigated the positive association between emotional memory retention and the amygdala activation during learning [22]. These studies clarify the role of the amygdala in memory consolidation (Figure 10.6) via projections to other brain memory system [23], especially in emotional memory.

10.3.2 Long-Term Potentiation and Memory Retention

In neuroscience research, it is a significant challenge to find the cellular and molecular processes involved in the process of learning and creating new memories. New information is acquired through the learning process, and memory is the route by which the acquired information is stored. Memory formation is dependent on neuronal changes at the synaptic level that allow strengthening of connections between neurons in the brain [24]. It is believed that task-dependent synaptic plasticity at the corresponding synapses during the formation of new memory is crucial and is sufficient to store information. This phenomenon of activity-dependent enhancement of synaptic transmission, which facilitates long-term information in the brain, is known as long-term potentiation (LTP). For further reading about the properties and detailed mechanism of LTP, see Ref. [25].

10.3.3 Factors Affecting Memory Retention and Recall

Cognitive research studies [26–28] have emphasized that retention and recall processes are related to one another and also are connected with other concepts such as learning, testing, capacity limit of memory, attention demand, and complexity of material. However, there are many other factors that affect retention and recall performance, such as attention, rehearsal, sleep, testing, mnemonics, exercise and nutrition, and reward. The conventional concept of learning and retrieval is that learning takes place during studying, while retrieval helps to assess the learned content.

10.3.3.1 Attention

As human memory has a limited capacity, it is crucial to determine the information of interest to be encoded and subsequently retained. Attention helps the brain to encode items selectively into memory. Recent studies have established a link between attention and memory processes. In an experiment, it was established that attention-directing cues can influence the collection of objects from visual memory [29]. Another study concluded that attention and memory cannot work separately [30], but it improves memory performance. Dividing attention may reduce the strength of retention. A recent study investigated long-term memory retention with full and divided attention. They found superior results with full attention as compared to divided attention during the memory recognition task [31]. A similar study examined the relationship between attention and memory processes and reported a strong association between attention and retention in working memory [32]. These studies established a strong association between attention and memory retention and the recall process.

10.3.3.2 Rehearsal

Rehearsal is the repeated reception, verbally or visually, of the same content, events, or information. According to the phonological loop model [33], verbal information is maintained in the phonological loop, but declines after a moment unless rehearsed. However, some studies have reported that rehearsal does not improve in the word length in a serial recognition memory task [34]. It can be concluded that rehearsal may not improve memory recall in each task.

10.3.3.3 Sleep

The association between sleep and memory has been studied for the last few decades, but it is now well established that sleep plays an important role in the memory consolidation process [35], which is a part of retention or the encoding process. There is electrophysiological and behavioral evidence to prove that sleep helps memory consolidation and brain plasticity. Frank and Benington [36] reviewed the role of sleep in memory consolidation and found that sleep promotes plastic changes in the brain. They divided the link between sleep and memory into three categories: sleep deprivation and memory consolidation, their correlation, and its effect on learning. Recently, Bell et al. [37] investigated the impact of sleep and spacing gap on long-term memory. Their results support the positive effects of sleep on long-term memory retention.

10.3.3.4 Exercise and Nutrition

Studies have reported the benefits of exercise in individuals whose cognitive performance is lowest [38]. The results showed that exercise improves memory performance in those individuals whose cognitive score in working memory was the lowest (see Figure 10.7).

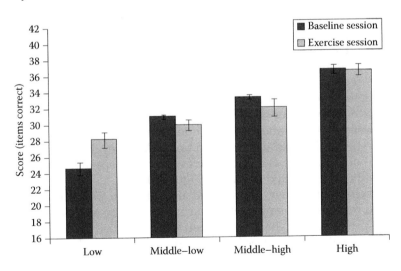

FIGURE 10.7
Memory performance for baseline and exercise sessions for four groups (low quartile, middle–low quartile, middle–high quartile, high quartile). (From Sibley, B. A. and Beilock, S. L. *Journal of Sport and Exercise Psychology*, 29, 783–791, 2007.)

The effects of nutrition and exercise on verbal memory were studied for type 2 diabetic old age participants, and positive results were reported. This study showed that exercise and nutrition may help memory for old adults at high risk of developing type 2 diabetes [39]. The reason for this finding may be that diet and exercise affect neurotropic elements and synaptic plasticity in the brain's regions, which are directly related to memory.

10.3.3.5 Mnemonics

Mnemonics is a mental technique that helps the brain remember information easily by associating the memories with some other keywords such as letters, images, or numbers. Several mnemonic techniques are available that help remember various pieces of information easily, including acronyms, acrostics, method of loci, chunking, rhymes and songs, stories, etc. [40].

10.3.3.6 Testing Effect

In educational systems, the theory of studying and testing is followed. Learning occurs during lectures, reading, and study groups. Tests have been designed to judge what has been absorbed or learned by studying. These tests are considered assessments or evaluations of learned knowledge. Researchers studied learning by trials of study (S) and a test (T). The critical supposition is that learning occurs through study phases, while a test simply measures what was learned in previous study phases (of STSTST ... order). Recently, a study of reviewed evidence opposes this conventional perception: retrieval exercises during tests have often resulted in better learning and long-term retention than has studying [41].

In related studies, Roediger and colleagues [27,41,42] investigated the link between retention and learning with repeated testing, and they proposed that the repeated recalling of retained information led to better learning and long-term retention. In one experiment, participants were given a list of items to study under two different conditions. In one condition, the list was studied 15 times and tested 5 times, while in the other condition the list was studied 5 times and tested 15 times. A retrieval task after one week showed better learning and retention results in the repeated test condition as compared to the repeated study (see Figure 10.8).

In another study, they investigated the retention performance between repeated retrieval and single retrieval conditions. The results showed a very high improvement in the repeated retrieval condition as compared to the single retrieval condition (see Figure 10.9).

These studies explained that testing/retrieval is a powerful technique of enhancing memory retention and recall performance. The implications of these are that students should test themselves repeatedly instead of studying the content frequently. However, feedback should be included in the frequent testing technique to avoid errors.

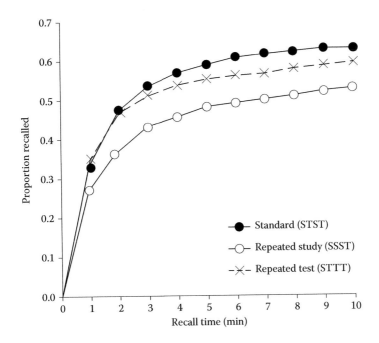

FIGURE 10.8
Retention performance of repeated recall versus repeated study. (From Karpicke, J. D. and Roediger, H. L., III. *Journal of Memory and Language,* 57, 151–162, 2007.)

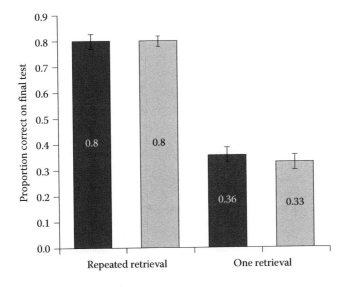

FIGURE 10.9
Correct memory recall performance of repeated versus single test. (From Karpicke, J. D. and Roediger, H. L., III. *Science,* 319, 966–968, 2008.)

10.3.3.7 Reward

Researchers investigated the process of reward and punishment stimuli and the neural understanding of reward processing in the brain. A recent study [43] reported that training under reward conditions results in substantial long-term retention, whereas training under neutral conditions showed a significant decline in memory gain. The research found reward-based training and learning more effective than neutral or punished conditions.

The effectiveness of all these techniques for memory retention and recall processes is experimentally proved. However, one technique may not give good results in all situations. Table 10.2 summarizes the factors that influence memory retention and recall performance with the underlying reasons.

TABLE 10.2

Summary of Factors Affecting Memory Retention and Recall

Factor	Affects	Reasons
Attention [30]	Full attention enhances memory recall and longer retention. Partial attention results in reduced recall performance.	When attention is divided, the limited capacity of working memory cannot accommodate the desired information and the result is partial memory consolidation, which ultimately reduces recall performance.
Rehearsal [44]	Frequent rehearsal improves recall performance.	Frequent rehearsal strengthens the synaptic connections among the neurons involved in a certain task, and stronger synaptic connections support better subsequent information recollection.
Sleep [45]	Sleep promotes brain plasticity and memory consolidation.	It remains to be determined how sleep produces neuronal changes that affect the consolidation process.
Exercise [38] and nutrition [39]	Exercise and nutrition offer benefits for mental health and affect neurotransmitters.	Nutrition and regular exercise influence neurotrophic elements and plasticity in brain regions related to memory performance. Further, for the elderly, regular exercise and balanced nutrition prevent the brain from shrinking and maintain memory.
Mnemonics [40]	Mnemonics helps encode difficult-to-remember information in such a way that it becomes much easier to recall correctly.	It strengthens network interactions and makes it easy to retain and retrieve difficult information.
Testing effect [41]	Testing allows recalling stored information from memory. It also minimizes the errors in the recall result and helps to retain correct information.	Repeated testing allows the same neurons involved initially in learning to fire repeatedly. The neuronal networks become more stable and synchronized, and recollection of stored information becomes easy.
Reward [43]	Reward-based training or learning leads to better results than nonreward learning or punishment situation.	Rewards may attract more attention during a certain task and maintain neuronal connections, which result in stable memory performance.

10.4 Experimental Design Issues

Various practical, theoretical, and ethical issues have been cited in the literature about memory research. Recently, common design issues of neurocognitive research have been highlighted [46,47]. Here, we summarize all the possible memory experimental design issues which are normally not given explicit importance, but are sensitive to the results of the experimental investigations.

Aim: When designing a memory experiment, the aim of the experimenter should be well defined. If the investigator is going to study the cognitive load, retention time, reaction time, recall performance, modality effect, short-term memory, long-term memory, recognition memory, and verbal or visual memory, the experiment must reflect the aim.

Task Simplicity: Although complex tasks have been used to examine cognitive load and brain processing speed, extreme complications in shorter time interval tasks may confuse the subject. It is better to avoid complexity from the experimental task and specify it as per memory type, subject health and age, and the nature of the memory.

Memory type: Researchers discriminate between memory types: short-term memory or working memory and long-term memory, on the basis of their capacity, retention length, and processing power. In memory experimental design, memory types are as essential as other experimental parameter specifications.

Memory modality: Visual and auditory are the two main memory modalities. The brain separately processes the information received from visual and auditory sources, and the capacities differ between visual memory and auditory memory. In designing an experiment, it should be considered that the experimental task with both modalities may have high cognitive load compared to the task with a single modality.

Specific Brain: Brain diseases such as Alzheimer's, brain tumor, depression, and Parkinson's affect different brain regions. Researchers have considered unusual tasks that focus on the affected brain part. If subjects are suffering from any brain disorder, the experimental task should be revised in consultation with area experts to handle the target subjects.

Subjects' characteristics: Individuals or volunteers who participate in the experiment are collectively called the sample. The individualities of a sample may be normal or patients, old or young, adolescent or children, and male or female. These individual differences need to be considered in task design to start the memory investigation.

Brain mapping method: Various brain-mapping methods are available such as EEG (invasive/non-invasive), fMRI, MEG, and PET. Each method has its own strengths and weaknesses. However, the choice of method depends on equipment availability, and the experimental design should be flexible to coordinate and overcome the limitation of the selected brain mapping method.

Predefined memory paradigm: Several predefined memory paradigms are available [47]. In order to use the existing available paradigm, one must be able to identify the memory paradigm that best reflects a subject's responses.

Experiment duration: The duration of the experimental task tends to vary depending on the study objectives. However, patients may feel anxiety while taking part in a prolonged experiment. This affects the patients' attention and interest in the experiment and eventually disturbs the desired consequences. The experimental task duration can be specified according to patients' health, the research aim, and the selected modality.

Consent: In memory research, the subject's consent is an important issue. The research study and designed task should meet the regional, legal, and ethical laws and should protect patients' well-being, mental privacy, and self-incrimination.

10.5 Summary

In this chapter, we have presented an overview of traditional theories of memory types and memory processes. We have focused on two important memory processes: retention and recall. The traditional theories explored short-term and long-term memory storage, working memory phenomena, verbal and nonverbal information processing, efficient use of cognitive capacity, and brain dual channels information processing of visual and verbal data. All these theories have their own distinctions. However, they are linked with each other and have important contribution in today's memory research.

Brain regions, especially the cerebral cortex, hippocampus, and amygdala have been strongly linked with memory processes. The cerebral cortex in the frontal, parietal, and temporal lobes contributes to working memory, episodic memory, and correct memory retrieval, respectively. In the medial temporal lobe, the hippocampus is the region for memory encoding and formation of new memory or transfer of information from short-term to long-term storage. The amygdala is the hub of emotional memory representation: fear, sadness, happiness, and other emotion-related events are associated with amygdala activation.

The process of long-term potentiation has a significant role in learning and new memory formation; the strong and weak memory depends on the synaptic communication network strength. Strong synaptic links have a high probability of storing new information and correct recall.

Various factors have been investigated, which have contributed to retaining more information and recall after a longer time. Many of them (such as attention, sleep, rehearsal, etc.) have been experimentally proved, such as their crucial role in utilizing the limited working memory capacity and learn more information, and transfer to long-term storage. Not all of these factors may be in support of high retention or recall. However, each of them has its own contribution in certain scenario.

In the end, various issues regarding memory experimental design were summarized. As a whole, these issues may be useful to consider in designing a memory experiment, especially for new researchers in memory research.

References

1. Carter, R., Aldridge, S., Page, M., and Parker, S., *The Human Brain Book*, London: Dorling Kindersley, 2009.
2. Atkinson, R. C. and Shiffrin, R. M., Human memory: A proposed system and its control processes, *The Psychology of Learning and Motivation*, 2, 89–195, 1968.
3. Tarnow, E., Why the Atkinson-Shiffrin model was wrong from the beginning, *WebmedCentral Neurology*, 1(10), WMC001021, 2010.
4. Paivio, A., *Imagery and Verbal Processes*, New York: Holt, Rinehart, and Winston, 1971.
5. Baddeley, A. D. and Hitch, G., Working memory, *The Psychology of Learning and Motivation*, 8, 47–89, 1975.
6. Baddeley, A. D., Allen, R. J., and Hitch, G. J., Binding in visual working memory: The role of the episodic buffer, *Neuropsychologia*, 49, 1393–1400, 2011.
7. Davidson, M. C., Amso, D., Anderson, L. C., and Diamond, A., Development of cognitive control and executive functions from 4 to 13 years: Evidence from manipulations of memory, inhibition, and task switching, *Neuropsychologia*, 44, 2037–2078, 2006.
8. Sweller, J., Cognitive load theory, learning difficulty, and instructional design, *Learning and Instruction*, 4, 295–312, 1994.
9. Sweller, J., Cognitive load during problem solving: Effects on learning, *Cognitive Science*, 12, 257–285, 1988.
10. Moreno, R. and Mayer, R. E., Cognitive principles of multimedia learning: The role of modality and contiguity, *Journal of Educational Psychology*, 91, 358, 1999.
11. Baddeley, A., Working memory: Looking back and looking forward, *Nature Reviews Neuroscience*, 4, 829–839, 2003.
12. Azevedo, F. A., Carvalho, L. R., Grinberg, L. T., Farfel, J. M., Ferretti, R. E., Leite, R. E., et al., Equal numbers of neuronal and nonneuronal cells make the human brain an isometrically scaled-up primate brain, *Journal of Comparative Neurology*, 513, 532–541, 2009.
13. Cabeza, R., Ciaramelli, E., Olson, I. R., and Moscovitch, M., The parietal cortex and episodic memory: An attentional account, *Nature Reviews Neuroscience*, 9, 613–625, 2008.
14. Rossi, S., Pasqualetti, P., Zito, G., Vecchio, F., Cappa, S. F., Miniussi, C., et al., Prefrontal and parietal cortex in human episodic memory: An interference study by repetitive transcranial magnetic stimulation, *European Journal of Neuroscience*, 23, 793–800, 2006.

15. Sederberg, P. B., Schulze-Bonhage, A., Madsen, J. R., Bromfield, E. B., Litt, B., Brandt, A., et al., Gamma oscillations distinguish true from false memories, *Psychological Science*, 18, 927–932, 2007.
16. Karlsgodt, K. H., Shirinyan, D., van Erp, T. G., Cohen, M. S., and Cannon, T. D., Hippocampal activations during encoding and retrieval in a verbal working memory paradigm, *NeuroImage*, 25, 1224–1231, 2005.
17. Phelps, E. A., Human emotion and memory: Interactions of the amygdala and hippocampal complex, *Current Opinion in Neurobiology*, 14, 198–202, 2004.
18. McGaugh, J. L., McIntyre, C. K., and Power, A. E., Amygdala modulation of memory consolidation: Interaction with other brain systems, *Neurobiology of Learning and Memory*, 78, 539–552, 2002.
19. Luo, Y.-X., Xue, Y.-X., Shen, H.-W., and Lu, L., Role of amygdala in drug memory, *Neurobiology of Learning and Memory*, 105, 159–173, 2013.
20. Holland, P. C. and Gallagher, M., Amygdala–frontal interactions and reward expectancy, *Current Opinion in Neurobiology*, 14, 148–155, 2004.
21. Roozendaal, B., Hahn, E. L., Nathan, S. V., Dominique, J.-F., and McGaugh, J. L., Glucocorticoid effects on memory retrieval require concurrent noradrenergic activity in the hippocampus and basolateral amygdala, *The Journal of Neuroscience*, 24, 8161–8169, 2004.
22. Canli, T., Desmond, J. E., Zhao, Z., and Gabrieli, J. D., Sex differences in the neural basis of emotional memories, *Proceedings of the National Academy of Sciences*, 99, 10789–10794, 2002.
23. McGaugh, J. L., Make mild moments memorable: Add a little arousal, *Trends in Cognitive Sciences*, 10, 345–347, 2006.
24. Lynch, M., Long-term potentiation and memory, *Physiological Reviews*, 84, 87–136, 2004.
25. Abraham, W. C. and Williams, J. M., Properties and mechanisms of LTP maintenance, *The Neuroscientist*, 9, 463–474, 2003.
26. Parker, A. W., Wilding, E. L., and Bussey, T. J., *The Cognitive Neuroscience of Memory: Encoding and Retrieval*, Psychology Press, New York, 2002.
27. Karpicke, J. D. and Roediger, H. L., III, Repeated retrieval during learning is the key to long-term retention, *Journal of Memory and Language*, 57, 151–162, 2007.
28. Eichenbaum, H., *The Cognitive Neuroscience of Memory: An Introduction*, Oxford University Press, New York, 2002.
29. Matsukura, M., Luck, S. J., and Vecera, S. P., Attention effects during visual short-term memory maintenance: Protection or prioritization? *Perception and Psychophysics*, 69, 1422–1434, 2007.
30. Chun, M. M. and Turk-Browne, N. B., Interactions between attention and memory, *Current Opinion in Neurobiology*, 17, 177–184, 2007.
31. Dudukovic, N. M., DuBrow, S., and Wagner, A. D., Attention during memory retrieval enhances future remembering, *Memory and Cognition*, 37, 953–961, 2009.
32. Fougnie, D., The relationship between attention and working memory, in *New Research on Short-Term Memory*, Noah B. Johansen (Ed.), Nova Science Publishers, New York, pp. 1–45, 2008.
33. Burgess, N. and Hitch, G. J., Memory for serial order: A network model of the phonological loop and its timing, *Psychological Review*, 106, 551, 1999.
34. Campoy, G., The effect of word length in short-term memory: Is rehearsal necessary? *The Quarterly Journal of Experimental Psychology*, 61, 724–734, 2008.

35. Potkin, K. T. and Bunney, W. E., Jr., Sleep improves memory: The effect of sleep on long term memory in early adolescence, *PLoS One*, 7, e42191, 2012.
36. Frank, M. G. and Benington, J. H., The role of sleep in memory consolidation and brain plasticity: Dream or reality? *The Neuroscientist*, 12, 477–488, 2006.
37. Bell, M. C., Kawadri, N., Simone, P. M., and Wiseheart, M., Long-term memory, sleep, and the spacing effect, *Memory*, 22, 276–283, 2014.
38. Sibley, B. A. and Beilock, S. L., Exercise and working memory: An individual differences investigation, *Journal of Sport and Exercise Psychology*, 29, 783–791, 2007.
39. Watson, G. S., Reger, M. A., Baker, L. D., McNeely, M. J., Fujimoto, W. Y., Kahn, S. E., et al., Effects of exercise and nutrition on memory in Japanese Americans with impaired glucose tolerance, *Diabetes Care*, 29, 135–136, 2006.
40. Bellezza, F. S., Mnemonic devices and memory schemas, in *Imagery and Related Mnemonic Processes*, Mark A. McDaniel and Michael Pressley, Springer, New York, 1987, pp. 34–55.
41. Roediger, H. L. and Butler, A. C., The critical role of retrieval practice in long-term retention, *Trends in Cognitive Sciences*, 15, 20–27, 2011.
42. Karpicke, J. D. and Roediger, H. L., III, The critical importance of retrieval for learning, *Science*, 319, 966–968, 2008.
43. Abe, M., Schambra, H., Wassermann, E. M., Luckenbaugh, D., Schweighofer, N., and Cohen, L. G., Reward improves long-term retention of a motor memory through induction of offline memory gains, *Current Biology*, 21, 557–562, 2011.
44. Svoboda, E. and Levine, B., The effects of rehearsal on the functional neuro-anatomy of episodic autobiographical and semantic remembering: A functional magnetic resonance imaging study, *The Journal of Neuroscience*, 29, 3073–3082, 2009.
45. Stickgold, R. and Walker, M. P., Memory consolidation and reconsolidation: What is the role of sleep? *Trends in Neurosciences*, 28, 408–415, 2005.
46. Amin, H. U., Malik, A. S., Badruddin, N., and Chooi, W.-T., Brain activation during cognitive tasks: An overview of EEG and fMRI studies, in *2012 IEEE EMBS Conference on Biomedical Engineering and Sciences (IECBES)*, December 17–19, pp. 950–953, 2012.
47. Amin, H. and Malik, A. S., Human memory retention and recall processes: A review of EEG and fMRI studies, *Neurosciences*, 18, 330–344, 2013.

11

Neurofeedback

Mark Llewellyn Smith
Neurofeedback Services of New York, PC

Thomas F. Collura
BrainMaster Technologies, Inc.

Jeffrey Tarrant
University of Missouri

CONTENTS

11.1 Introduction..240
11.2 Early History of the Field..241
    11.2.1 Joe Kamiya and Alpha Training..241
    11.2.2 Barry Sterman and SMR Training...242
    11.2.3 Eugene Penniston and Elmer Green: Alpha/Theta Training.....242
    11.2.4 Fehmi, Hardt, and Crane: Multichannel Alpha
           Synchrony Training..243
    11.2.5 Lubar: Theta/Beta Training 243
11.3 Description of the Neurofeedback Process...244
    11.3.1 Signal Processing 245
    11.3.2 Operant Conditioning...247
11.4 Neurofeedback Modalities: qEEG-Guided Neurofeedback 248
    11.4.1 Description of qEEG Metrics..249
    11.4.2 Databases 250
           11.4.2.1 BrainDX 250
           11.4.2.2 Applied Neuroscience/Neuroguide 251
           11.4.2.3 SKIL..251
           11.4.2.4 HBI/WinEEG 251
           11.4.2.5 BRID...251
11.5 qEEG-Based Neurofeedback-Inhibit/Enhance Training 251
    11.5.1 z-Score Training 252
    11.5.2 sLORETA/LORETA..253
11.6 Neurofeedback Modalities: Symptom-Based Approaches.................255
    11.6.1 Alpha Training...255
    11.6.2 SMR Training 255
    11.6.3 Alpha-Theta Training...255

11.6.4 Alpha Asymmetry Training..256
11.6.5 Slow Cortical Potential...256
11.6.6 Infra-Slow Fluctuation/Infra-Low Frequency..........................257
11.7 Patient Assessment 257
11.7.1 Symptom Assessment 258
11.7.2 Neuropsychological Assessments......................................258
11.7.3 Computerized Neuropsychological Batteries..........................258
11.7.4 Targeted Neuropsychological Tests 259
11.7.5 Subjective Assessment Questionnaires 259
11.7.6 qEEG Integrated Assessments 259
11.7.7 Event-Related Potential Assessment 260
11.8 Addressing Conditions with Neurofeedback...............................260
11.8.1 ADHD..260
11.8.2 Other Conditions 261
11.9 Conclusion 262
References..263

11.1 Introduction

Neurofeedback is a form of brain wave training that makes use of the principle of learning, defined as the general process by which an organism alters its behavior according to certain goals. By measuring and providing feedback related to brain wave activity, the process of neurofeedback provides an additional channel of information that increases awareness of brain behavior by creating subjective experiences that are derived from electroencephalography (EEG).

Neurofeedback has been commonly referred to as training rather than treatment owing to similarities to physical fitness training, including nonspecific intervention stratagems applied to specific disorders and the need for multiple repetitions to produce positive results. Clinicians may apply a variety of interventions that produce symptom remission for the same disorder. EEG biofeedback has proved useful with an array of psychological and medical conditions. A diversity of protocols has been used to achieve this result. It can be argued that neurofeedback does not treat anything specific, but rather EEG biofeedback optimizes the central nervous system thereby improving general function in a variety of cognitive, emotional, and homeostatic domains. It accomplishes this task by conditioning brain electrical activity through repetitive application and so is more aptly referred to as training.

Despite its birth in the halls of academia, neurofeedback as a discipline largely developed outside of the academy as a tool in clinical practice. Many of the early adopters of this intervention were not traditional mental health or medical practitioners. Some became developers and vendors of equipment and software as a bulwark against the lack of a wider community.

In this way, vendors often became leaders in the field. In many instances, these individuals filled traditional roles as therapists, teachers, and leaders of professional organizations inhabited by clinicians, academicians, and scientists in other health-related disciplines.

This early history contributed to the synergy between developers and clinicians, allowing for the rapid development of clinical tools. Where there has been interaction between clinical requirements and a manufacturer's capacity, rapid advancement has been the result. This environment has historically contributed to the field being less focused on scientific proof and more on clinical demands and market forces, adding to the field's vulnerability to criticism by traditional medicine and science.

With the production of more methodologically sound research, the field of neurofeedback is on the precipice of a much broader acceptance. The tumultuous birth of the field is inevitably giving way to the order imposed by professional standards, the demands of the growing influence of mainstream science, and the desire for a wider service delivery. Neurofeedback is now commonly performed by healthcare professionals worldwide. Professional guilds for neurofeedback and biofeedback have firmly taken root and a nascent quantitative electroencephalogram (qEEG) guild is developing. EEG biofeedback is increasingly being implemented in university programs, especially in Europe.

This chapter is intended as a general survey of neurofeedback. It will give a short history of the field and offer an introductory description of the existing clinical modalities that developed as a result. It will describe basic clinical processes and neurofeedback interventions, including biological, technical, and scientific considerations. This chapter is not intended to be exhaustive, and some areas of neurofeedback will not be included. Rather, this chapter was written to help the reader understand neurofeeback in a general way and provide a starting place for deeper investigation.

11.2 Early History of the Field

11.2.1 Joe Kamiya and Alpha Training

The roots of modern neurofeedback began with the work of Joe Kamiya at the University of Chicago in about 1962. Kamiya discovered that human subjects could be taught to consciously control alpha burst activities. Initially, this was done through verbal prompts each time an alpha burst spontaneously appeared. Later, Kamiya used a simple electronic device that would sound a tone when alpha was present in the recording. This early work suggested that feedback could be used to teach subjects what it felt like to generate increased levels of alpha [1,2]. Because this "felt state" was most typically associated with calm and relaxation, some of the earliest neurofeedback involved teaching a client to increase alpha activity as a treatment for anxiety [3].

11.2.2 Barry Sterman and SMR Training

At approximately the same time as Joe Kamiya's experiments, Barry Sterman began examining the ability of animals to control EEG activity. Sterman, a professor of neurobiology and psychiatry at the University of California, Los Angeles, taught cats to increase the sensorimotor rhythm (SMR) over the motor cortex [4]. In fact, he found that directly rewarding increases in SMR activity, which was accompanied by decreased motoric activity while remaining alert, was more successful than directly rewarding bodily stillness.

Sterman later found that the cats trained to increase SMR activity were also resistant to seizures when exposed to seizure-causing chemical agents. This finding was later replicated with monkeys and used to develop an early neurofeedback protocol for seizure disorders in human [5].

These early experiments helped lay the groundwork for the understanding that brain wave patterns were connected to states of consciousness and behavior. It was not long before the pioneers in this field began applying the basic methods of neurofeedback to conditions such as anxiety and seizure disorders.

11.2.3 Eugene Penniston and Elmer Green: Alpha/Theta Training

In the early 1960s, Elmer and Alyce Green began working with the Menninger Foundation on the development of simple biofeedback tools and autogenic training, primarily to help patients ease muscular tension. In 1965, at a meeting of the Psychophysiological Research Society, Elmer was exposed to the work of Joe Kamiya. This, in addition to other research tying brain waves to states of consciousness, eventually led the Greens to begin experimenting with theta training to enhance creativity [6]. Much of this early work was based on the observations of EEG signatures in experienced meditators, healers, and yogis. The use of neurofeedback to enhance alpha and theta brain waves was an attempt to develop some of the exceptional abilities observed in these individuals.

After taking a seminar with the Greens, Eugene Penniston began exploring specific training protocols with alcoholics in treatment at the VA Medical Center in Fort Lyon, Colorado [7]. Penniston observed that for many alcoholics, there was a decreased level of theta brain wave activity in the posterior region of the brain that was at times accompanied by a decrease in alpha activity. In conjunction with Paul Kulkosky, Penniston developed what became known as the alpha/theta protocol. This form of neurofeedback involves rewarding both alpha and theta frequencies, typically in the parietal or occipital regions of the brain. As the training session progresses, the alpha eventually decreases and becomes lower than theta wave activity. This process is referred to as a "crossover" and is associated with the resolution of traumatic memories and the integration of previously suppressed psychodynamic material [7]. The alpha/theta protocol was found to have very positive outcomes with substance dependence and PTSD [8,9].

11.2.4 Fehmi, Hardt, and Crane: Multichannel Alpha Synchrony Training

Les Fehmihad played a prominent role in the early days of biofeedback, helping to found what is now known as the Association for Applied Psychophysiology and Biofeedback (AAPB). Dr. Fehmi was initially interested in studying alpha brain wave synchrony as this state was associated with autonomic nervous system balance and increased efficiency in brain performance [10].

Fehmi discovered that synchronous alpha band activity increased dramatically when a person was able to be attentive and surrender control simultaneously [11]. After experimenting with this type of training, Fehmi and his students reportedly felt a deep state of relaxation and focus that was often described as "transcendent" or "in the zone." It was noted that the subjective quality of attention seemed to change after engaging in alpha synchrony training. Attention became more open and capable of taking in the gestalt of a situation. This was qualitatively different from the more typical narrow focus that is common during a stress response.

By utilizing multiple EEG channels, Fehmi helped establish a new kind of neurofeedback that focused on the impact of shaping a global alpha state [12]. Other early innovators, including Jim Hardt and Adam Crane, also utilized multichannel synchrony training, advancing the notion that such a protocol strengthened access to deeper states of consciousness and improved self-regulation when compared to single-channel training [3].

Clients trained with EEG synchrony and verbally guided "open focus" techniques report significant changes in perception and behaviors [13]. This work has led to the development of protocols and applications used for conditions including pain, depression, anxiety, and attention deficit hyperactivity disorder (ADHD) [14].

11.2.5 Lubar: Theta/Beta Training

Joel Lubar, a professor at the University of Tennessee, published the first article describing the use of neurofeedback to address hyperactivity [15]. Lubar expanded on Sterman's earlier work that reported on the successful use of SMR neurofeedback to reduce hyperactivity [16]. Lubar helped establish an arousal model of brain activity. The model recognized that reduced vigilance is associated with increases in theta activity, and increases in beta activity are associated with increases in cognitive processing [7,17]. Along with his wife Judy Lubar, Dr. Lubar began systematically using neurofeedback designed to decrease theta and increase beta as a treatment for ADHD. His protocols have been refined over the years and have become a standard [18], helping to establish validity and credibility in the field.

11.3 Description of the Neurofeedback Process

The process of neurofeedback consists of recording a brain-related signal, typically the EEG, and using electronics and computers to create a representation of that signal to teach the brain to change. Neurofeedback provides another channel of information for the brain to understand its own process (Figure 11.1).

Figure 11.2 shows a view of the various tissues, bone, scalp, and hair that comprise the human head. Because of the intervening tissue and bone, the signal on the scalp is thousands of times smaller than the potential if it were measured inside the head.

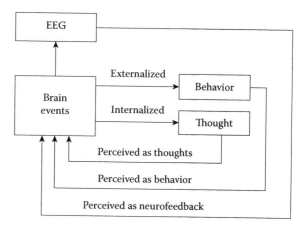

FIGURE 11.1
The process of neurofeedback.

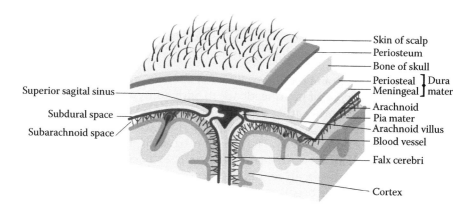

FIGURE 11.2
Anatomical view of the brain, skull, and scalp.

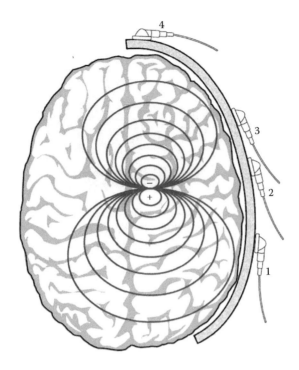

FIGURE 11.3
An EEG electrical source and the resulting electrical field in the brain.

In order to record the EEG, sensors must be attached firmly to the skin of the scalp, so that an accurate reading of the electrical potentials on the surface can be recorded (Figure 11.3).

The representation of brain electrical activity is known as feedback, hence the name neurofeedback. In practice, neurofeedback requires a practitioner to apply EEG sensors to the client's scalp, and to then use the equipment to acquire a live EEG signal. This signal must be inspected to ensure that it is free of artifacts and thus suitable for neurofeedback. Once a quality EEG signal has been achieved, the system is operated in such a manner as to provide feedback in the form of visual, auditory, or vibrotactile stimulation (Figure 11.4).

11.3.1 Signal Processing

In order to derive useful feedback information from the raw EEG signal, it is necessary to subject it to some signal processing. This consists of mathematical operations that are implemented either in hardware or in software, and which provide a suitable measure of the relevant EEG parameters. It is important that the signal processing be done in "real time," which

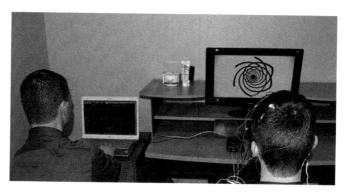

FIGURE 11.4
A subject being trained with z-score neurofeedback.

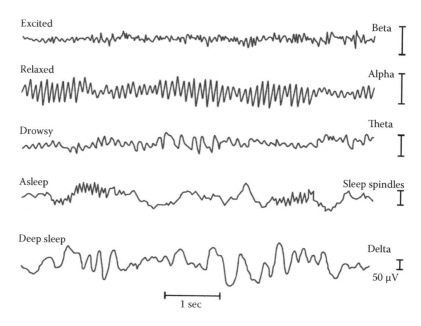

FIGURE 11.5
The basic types of EEG bands and their behavioral correlate.

means that the derived information must be computed and made available quickly (typically within less than 1/10th of a second). The most common transform of the signal in neurofeedback training is the Fourier transform. Others employ a joint time frequency analysis, and fewer still apply the Hilbert transform.

Signal processing for neurofeedback is most often based on a mathematical analysis of the EEG, producing estimates of the amount of energy

TABLE 11.1

Band-Pass Filters Used for Neurofeedback

Typical EEG Component Band Ranges

- Delta 1–3 Hz
- Theta 4–7 Hz
- Alpha 8–12 Hz
- Lo Beta 12–15 Hz (SMR)
- Beta 15–20 Hz
- High Beta 20–35 Hz
- Gamma 40 Hz (35–45 Hz)

(or connectivity) in the EEG for particular frequency ranges. For historical as well as physiological reasons, the EEG bands (Figure 11.5) are generally divided into certain standard bands shown in Table 11.1.

While neurologists and other EEG analysts view these waveforms directly, neurofeedback requires the reduction of these patterns to some type of number, so that the software can produce feedback signals related to quantitative patterns. It is beyond the scope of this chapter to detail the signal processing used for neurofeedback. However, the basic operations yield estimates of important parameters such as the amount of power in a certain frequency band. More elaborate computations provide information such as the speed of each EEG signal, the presence or absence of signals, and how the signals reflect the quality of connections between brain regions.

11.3.2 Operant Conditioning

There is a variety of mechanisms involved in the process of any learned behavioral change, and neurofeedback employs more than one (Table 11.2). Of these, the most often cited, and one of the most basic, is operant conditioning. This occurs whenever an organism is provided with feedback in the form of a reward (or punishment), and consequently learns to perform a desired behavior. In the case of neurofeedback, the "behavior" is the production of a particular type of EEG wave or waves.

It is important to view neurofeedback within the broader context of brain self-regulation mechanisms [19]. The brain undergoes normal cycles of activation, separated by periods of deactivation or relaxation [16]. Figure 11.6 shows the range that the brain may work within, as it moves from a state of concentration (activation) to a state of relaxation (deactivation).

While it might appear that the objective of neurofeedback is to increase brain waves that are deficient or decrease brain waves that are excessive, the goal is much broader and often involves attempts to encourage the brain into a more efficient and flexible state (Table 11.3).

TABLE 11.2

Neurofeedback Learning Mechanisms

- Classical conditioning
- Concurrent learning
- Habituation
- Self-efficacy
- Generalization
- Transference
- Nonlinear dynamic adaptation

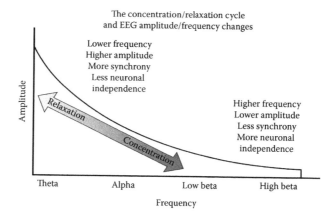

FIGURE 11.6
The synchronization/desynchronization cycle.

TABLE 11.3

Goals of Neurofeedback

Improve self-regulation
Achieve flexible and appropriate brain states
Normalize connectivity
Address functionality, not symptoms
Provide lasting change

11.4 Neurofeedback Modalities: qEEG-Guided Neurofeedback

The field of neurofeeback can be loosely divided into two categories: clinicians who employ a symptom-based approach and a treatment approach guided by a pretreatment qEEG. Clinicians who apply neurofeedback to present complaints without a quantitative analysis are symptom-based practitioners and those who utilize age-normed data in assessment and training are said to be qEEG based. Quantitative analysis is based on the use of a reference database.

One dominant theme in the analysis of EEG, and in the design of neuro-feedback protocols, is the application of reference databases. These consist of EEG data in various reduced forms that provide nominal or target values for key parameters. Databases employed for this purpose contain what are considered "normal" or "typical" EEGs, used to create statistical references or "norms" for comparison. In addition, a database may contain a number of abnormal or disordered subjects, used for comparison purposes. When abnormal as well as normal EEG is available, it is further possible to create "discriminant" functions that provide statistical estimates of the likelihood that a subject comes from one of the clinical populations.

There is a variety of ways to employ qEEG when planning and performing neurofeedback training. The traditional approach uses a qEEG analysis to determine which deviations from "normal" are relevant to presenting symptom expression, and designs a neurofeedback protocol that encourages the brain toward a more normative state. This approach seeks to make "large things small" and "small things larger." While this approach makes sense, training decisions should be made combining client complaint with functional neuroanatomy. Simple linear decisions may not be consistent with the complexity of brain function. qEEG deviations may be coping or compensatory mechanisms that facilitate normal function. In order to determine whether an observed deviation is pathological or compensatory, it is important that the neurofeedback practitioner have an understanding of functional neuroanatomy.

Despite the common observation that qEEG maps may become more normal or typical after treatment, it is not necessary for this to occur for the client to experience benefits. The brain is not a simple, linear mechanism that always responds in a predictable or consistent manner. Clients may have various ingrained tendencies or patterns, coping mechanisms, compensating strategies, or other behavioral or physiological mechanisms at play that can interact with the brain and the neurofeedback process in complex ways.

11.4.1 Description of qEEG Metrics

The particular computed values that are derived from the EEG are referred to as *metrics*. This simply means that something is being measured. As with any phenomenon or object, there are various ways to compute and use different metrics. When we ask "how large" an EEG signal is, we refer to its magnitude in microvolts. The metrics related to signal amplitude (or power) in a single channel include absolute power (value in microvolts), relative power (percentage of total band), and ratios (between bands). These values are descriptors for the activation processes in the cortex. These metrics reflect the quality of a response to a stimulus or the ease of shifting behavioral states. As absolute power increases or decreases, it reflects the activation dynamics of groups of cortical neurons to meet a behavioral demand. For instance, when one moves from actively reading to a drowsy state before

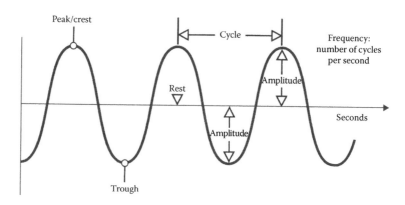

FIGURE 11.7
Basic EEG metrics.

sleep, groups of cortical neurons move from a low-amplitude desynchronized fast activity to a higher-amplitude slow-wave synchronous activity.

In addition to evaluating the activation processes, qEEG metrics allow the clinician to evaluate network dynamics. This is done through metrics that relate to the relationships between pairs of channels. These indices most often include amplitude asymmetry (a measure of the ratio of energy from one site to another), coherence (a measure of the amount of shared energy between network pairs), and phase (the metric used to determine the speed of information transfer). These metrics fall under the rubric of connectivity. Other connectivity metrics include comodulation and the spectral correlation coefficient, which reflect the magnitude consistency between two signals and the correlation between amplitude spectra, respectively. Connectivity metrics are described in detail by Collura [20]. These metrics (Figure 11.7) are generally computed for a given band, so that one generally refers to them, for example, as "the coherence between C3 and C4 in beta or the phase lag at T4 and P4in theta."

11.4.2 Databases

11.4.2.1 BrainDX

BrainDX is a database that is derived from the original NXLink database created and reported by E. Roy John and his laboratory at the Brain Research Laboratory of New York University [21–24]. Originally, it largely consisted of eyes-closed data from over 700 normal subjects, and also included a large number of clinical (abnormal) cases. It has recently been supplemented with adult eyes-open data, with plans to add child eyes-open data in 2014. There is also an associated software system that can conduct EEG analyses and generate reports. This system provides a complement of discriminant functions, sLORETA z-score images, and other report capabilities. The BrainDX database is incorporated into neurofeedback products by one vendor, Brainmaster Technologies, Inc.

11.4.2.2 *Applied Neuroscience/Neuroguide*

The Neuroguide database was created at the University of Maryland, as the Lifespan Database, and is currently provided by the Applied Neuroscience Institute (ANI) [25]. It contains approximately 625 subjects, and spans the age range from birth to 80, with both eyes-open and eyes-closed data. The Neuroguide application includes artifacting, as well as, optional discriminant functions, LORETA analysis, and other capabilities.

The ANI database is also used for z-score neurofeedback, by a number of providers, who incorporate the database into their own neurofeedback products. The consistency between the Neuroguide and the NXLink (now BrainDX) database has been established and documented, showing that essentially identical results can be expected regardless of which database is used [25].

11.4.2.3 *SKIL*

The SKIL database was created by David Kaiser and Barry Sterman [26] and is primarily an adult database. It contains eyes-open, eyes-closed, and task conditions. There is also a SKIL application that can read in EEG files from various providers and provide analysis functions. SKIL also provides neurofeedback software that operates on several types of EEG equipment.

11.4.2.4 *HBI/WinEEG*

The Human Brain Institute (HBI) database was developed by a team in Russia, led by Dr. Juri Kropotov [27]. The database comprises 885 healthy subjects of both sexes aged 7 to 89 years. The subjects were divided into 20 age groups. It is incorporated into the WinEEG system, and can be used to create maps and reports from EEGs read in from standard EEG files.

11.4.2.5 *BRID*

The Brain Resource International Database (BRID) is very extensive and includes a wide range of abnormal as well as normal EEGs. The database consists of more than 5,000 subjects that range in age from 6 to 100 years. BRID provides a report service for subscribers, and primarily uses equipment provided by the company to affiliated practitioners.

11.5 qEEG-Based Neurofeedback-Inhibit/Enhance Training

Inhibit or enhance training, also known as "traditional" neurofeedback, is based on providing rewards when particular frequency bands are either present in excess and so are "inhibited" or when they are insufficient and

TABLE 11.4

Standard Inhibit/Enhance Protocols

Protocol-Explanation of Protocol
• Alert C3—beta up; theta, high beta down
• Deep Pz—(Penniston) alpha up, theta up
• Focus C4—SMR up; theta, high beta down
• Peak2 C3-C4—alert and focus combined
• Relax Oz—alpha up; theta, high beta down
• Sharp Cz—broadband squash

so are "enhanced." This type of training is commonly done using the traditional bands for training, such as theta, alpha, low beta, beta, or high beta. For example, after consulting a brain map, a protocol for relaxation may include enhancing when alpha absolute power is insufficient compared to the age-normed population mean, but inhibiting when theta or high beta are large. The rewarding or reinforcing is commonly done when the chosen EEG band is above a specified threshold during enhance training and below a specified threshold during inhibit training. Table 11.4 shows some common protocols based on enhancing or inhibiting components. Enhance training is also sometimes called *uptraining*, while inhibit training is known as *downtraining*.

11.5.1 z-Score Training

z-Score training was created by the collaboration between two software and equipment vendors: Applied Neuroscience Inc and Brainmaster Technologies Inc with the contributions of several clinicians [28–30]. The training is based on the principle of using statistical z-scores in real time, rather than using raw signal amplitudes or other variables. In order to perform z-score training, it is necessary to select a suitable reference database (see Section 11.4.2), as well as to have software that can compute z-scores in real time. By using z-score parameters during the process of training, it is possible to inform the brain whenever any of one or more (often many more) variables are within a certain range of normal (e.g., +/−1.0 standard deviation). This type of training allows for the training of large numbers of variables across multiple brain sites simultaneously.

A typical set of pre- and post-treatment qEEG maps collected before and after z-score training is shown in Figure 11.8, as an example. It shows that there may be particular deviations from normal (excesses or deficits) in the initial map, and that these deviations may be partially or wholly resolved (normalized) in the post-treatment assessment [28]. In order to aid in understanding statistical deviance in the brain maps, an illustration of the relationship between color and standard deviations is provided in Figure 11.9. The qEEG and neurofeedback community accepts two standard deviations and greater as a representation of abnormal activity.

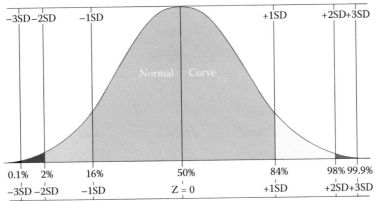

FIGURE 11.8
Bell curve: Relationship of color to standard deviation in Neuroguide.

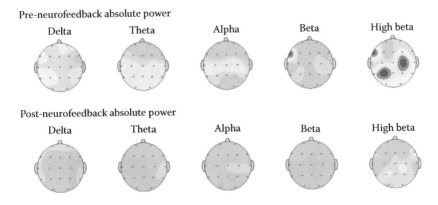

FIGURE 11.9
Pre-/post-neurofeedback training brain mapping.

The following coherence maps (Figure 11.10) taken before treatment and after treatment show a clear normalization of coherence, indicating that brain connectivity is more typical of normal following the treatment.

11.5.2 sLORETA/LORETA

LORETA and sLORETA are methods that can be used to extrapolate the activity of deeper brain regions from surface EEG data. This has been called the inverse solution as it applies an algorithm that utilizes the surface amplitudes to identify current sources below the outer layer of the cortex. While this generally requires a minimum of 19 sensors, it allows the client to receive

Pre-neurofeedback coherence

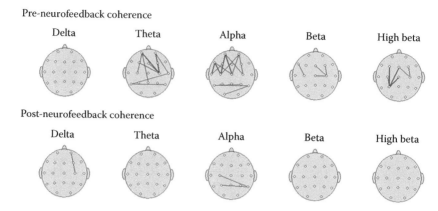

Post-neurofeedback coherence

FIGURE 11.10
Pre-/post-neurofeedback training demonstrating normalization of coherence.

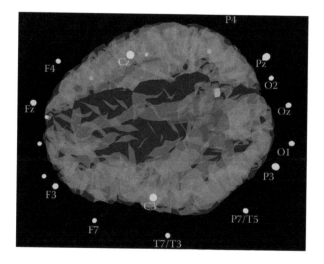

FIGURE 11.11
An sLORETA display of z-scored current source density in the cingulate gyrus.

feedback that is specific to deeper structures in the brain, such as the cingulate gyrus or the right insula. It measures the magnitude of current with current source density (CSD):nanoamperes per cubic millimeter. In addition, sLORETA/LORETA coherence and phase are currently being calculated and trained in some applications.

LORETA and sLORETA can also be combined with z-scores, so that each brain region is evaluated in comparison with a database of brain activity.

Figure 11.11 shows an example of an sLORETA display, showing brain activity in specific brain areas. In this image, excess z-scored CSD activity is visible in the cingulate gyrus.

11.6 Neurofeedback Modalities: Symptom-Based Approaches

Symptom-based approaches rely on applying the appropriate training protocols to the specific presenting problems. This paradigm was established at the commencement of the field and is applied differentially determined by client complaint, functional location in brain, and clinical experience. The symptom-based approach began with alpha band training, followed shortly after by SMR and alpha/theta training. More recently, symptom-based approaches have included slow cortical potential (SCP) training, alpha asymmetry and infra-slow fluctuation/infra-low frequency training. These neurofeedback training protocols are also utilized by qEEG-based clinicians.

11.6.1 Alpha Training

Alpha training is most often used for relaxation, peak performance, pain relief, and anxiety. It is performed in an eyes-closed condition in one or more channels [13,31,32]. The client attempts to increase electrical activity in the 8–12 Hz range while simultaneously inhibiting slower frequencies, usually 4–7 Hz and a higher band in the 20–30 Hz range. Alpha activity is trained in the posterior regions of the cortex, most frequently in the parietal regions. Jim Hardt, Les Fehmi, and others train alpha synchrony in two or more channels. In this form of alpha training, channels are summed together and microvolt increases rewarded under the assumption that synchronous neuronal assemblies firing simultaneously are responsible for amplitude increases.

11.6.2 SMR Training

SMR training was developed by Barry Sterman for seizure disorders [5]. It was later utilized by Joel Lubar and others for ADHD [15]. It has been used for a host of other clinical presentations and may be the most frequently studied and utilized neurofeedback training. The SMR band is commonly defined as 12–15 Hz. This range of frequencies is rewarded for amplitude increases while simultaneously inhibiting lower and higher frequency bands. The protocol is executed on the sensorimotor strip at one site or a combination of sites, usually C3, C4, and Cz. Ordinarily, it is performed in a single channel montage. Less frequently, two or more channels or a bipolar montage electrode array is implemented. The result of the training is often improved attention and reduction of restlessness or motoric stillness.

11.6.3 Alpha-Theta Training

Alpha/theta training, also known as deep states work, emerged from the work of Elmer Green, Steven Farhion, and Dale Walters at the Menninger

Clinic over 30 years ago [6,33]. Eugene Penniston and Paul Kulkowski later applied the intervention to alcoholics in recovery and Vietnam veterans with post-traumatic stress disorder (PTSD) [8,9]. It is executed with the eyes closed, usually with the client in a reclining chair. Alpha/theta training is performed with one or more electrodes that are most frequently placed in the posterior areas of the cortex. Both the alpha and theta bands are enhanced while inhibiting the delta band and a higher frequency band, typically 20–30 Hz. The goal of the training is to enter a "crossover" state wherein the theta band is greater in amplitude than the alpha band. Physiologically, crossover resembles a hypnagogic/hypnopompic state: the state of consciousness entered just before sleep or just before awakening. This crossover state is associated with spontaneous imagery and autobiographical memories and is said to enable individuals to process emotionally salient material. Alpha/theta training is frequently used with hypnotic induction or a scripted guided imagery that addresses the subject's presenting problem.

11.6.4 Alpha Asymmetry Training

Alpha asymmetry training was developed by Peter Rosenfeld and Elsa and Rufus Baehr in the mid-1990s [34]. They found a strong correlation between asymmetries in alpha absolute power in the frontal lobes and mood disorders. They proposed that right anterior activation is related to increases in withdrawal-related emotions, and left anterior activation is related to approach-related behaviors. Their technique was predicated on the assumption that the alpha rhythm reflected hypoactivation and that a greater alpha magnitude in the left frontal area indicated a deficit of approach-related behaviors mediated by that area. They did not use normative data but relied on an alpha asymmetry score obtained by calculating the ratio of the left and right prefrontal alpha. They used a two-channel montage at F3/F4 referenced to Cz.

11.6.5 Slow Cortical Potential

Slow cortical potential (SCP) training has been employed, largely in Europe, for more than 30 years [35]. It has been used to train epilepsy [36], brain–computer interface [37], and ADHD [38]. As originally applied, SCP training was a "one size fits all protocol" with the majority of the training done at the vertex of the brain. In this early manifestation, the client attempted to produce cortical excitation, a negative DC shift, in the case of ADHD; or cortical inhibition, a positive DC shift, in the case of epilepsy over the course of many short trials [36]. More recently, SCP training has evolved to focus on areas other than the vertex and trains both negative and positive gradient shifts in the same session with proportions of negative and positive trials dependent on client presentation [39].

11.6.6 Infra-Slow Fluctuation/Infra-Low Frequency

Infra-slow fluctuation (ISF) or infra-low frequency (ILF) training is a frequency-based training. The frequency bands addressed are different in the two techniques. However, both paradigms filter and train the very low frequency band known as the ultradian rhythm, which is less than 0.01 Hz [40]. These slow rhythms are influenced by changes in amplitude and polarization of the direct current (DC) signal [41]. Infra-slow training owes its lineage to the early beta/SMR training of Susan Othmer and EEG Spectrum [42]. That intervention was done with a single channel of EEG and an intra-hemispheric bipolar montage on the motor strip and temporal lobes. The technique was modified by replacing a set reward band with a 3 Hz window that was shifted up or down depending on the client response.

Infra-slow training produces immediate state shifts in the client. The trainer targets state regulation in real time by discovering an optimum frequency through trial and error for each client. Success is defined by the immediate improvement in affect and arousal regulation in session and ultimately by generalized improvements in behavioral and state regulation in life.

Over time, it became apparent that the vast majority of clients were finding an optimum clinical response at lower and lower frequencies. As the optimum frequencies trained went below 1 Hz, it became clear that the optimum response was more readily achieved when training was executed with a DC-coupled amplifier. This technique has begun to incorporate normative data and simultaneous bipolar and referential montages [41].

11.7 Patient Assessment

Neurofeedback clinicians utilize a range of strategies for assessing client symptoms including subjective symptom questionnaires, objective neuropsychological testing, and qEEG. The use of assessment instruments tends to divide along professional boundaries, with access to neuropsychological tests often determined by professional practice laws.

The vast majority of neurofeedback clinicians use some form of a subjective symptom questionnaire that is often set up in a Likert Scale format. The Continuous Performance Test is an instrument also in frequent use among practitioners. The use of pre/post-qEEGs is becoming a more frequent part of many practices. A smaller group of practitioners, largely psychologists, utilize neuropsychological testing.

11.7.1 Symptom Assessment

Although a formal assessment or psychological evaluation is not always necessary, it can be extremely helpful in the process of developing an efficient treatment plan and tracking symptom change. For example, if a client were seeking neurofeedback for "memory problems," objective testing may reveal differences between auditory and visual memory; or working memory versus long-term memory; or it may reveal that the problem stems more from problems with attention or anxiety. These distinctions may be too subtle for many clients to accurately perceive or articulate. This clarification may be important in relation to electrode placement, protocol selection, and any adjunct strategies that may be implemented.

Neurofeedback clinicians utilize a range of strategies for assessing client symptoms including objective neuropsychological testing, subjective symptom questionnaires, and qEEG integrated assessments. A sampling of these types of evaluations will now be discussed.

11.7.2 Neuropsychological Assessments

Neuropsychological assessments involve completion of a well-defined set of tasks known to be linked to specific brain regions. These assessments are based on standardized administration procedures and scored in relation to a normative database. Such testing can reveal specific deficits in neuropsychological functioning and help identify brain areas that may be involved in the deficits. These tests commonly examine constructs such as memory, intelligence, language skills, sensorimotor, visual/spatial processing, and executive function tasks. Traditionally, these assessments consist of a battery of tests that would be administered to the client by a trained psychologist or psychometrist. Examples of this type of testing might include the Wechsler Adult Intelligence Scale (WAIS), the Test of Memory and Learning (TOMAL), or the Tower of London Test. Once administered, these tests would require individual scoring and interpretation.

11.7.3 Computerized Neuropsychological Batteries

Programs such as MicroCog, IntegNeuro, and CNS Vital Signs are specifically designed for clinical use and allow the neurofeedback practitioner to administer the assessment in their office in 30–90 min. These programs often have both short and extended versions, include multiple domains of concern, require minimal computer literacy, and provide nearly instant scoring and report generation. Among the most commonly used assessments in this category is MicroCog.

MicroCog was the first computerized neuropsychological test battery developed for clinical use. It has measures in five domains, including attention/mental control, memory, reasoning/calculation, spatial processing, and reaction time. It is modestly priced and has been shown to be an effective screen for early dementia among elderly subjects [43]. It has also been shown to be effective in distinguishing between dementia and depression and measuring cognitive impairment related to drug abuse [44].

11.7.4 Targeted Neuropsychological Tests

An alternative to a battery of tests is to utilize tests specific to one area of concern. Potentially the most popular type of assessment utilized by neurofeedback practitioners is the computerized test of attention or continuous performance test (CPT). These tests present the client with a series of simple target and nontarget cues either through visual and auditory information. The client's performance is measured by examining variables such as reaction time, reaction time standard error, errors of omission, errors of commission, and signal detection to a normative sample. Performance on these tests has repeatedly been shown to differentiate ADHD from nonsymptomatic groups [45]. The most commonly used of these tests include the T.O.V.A. (Test of Variables of Attention), Conners', and the IVA (Integrated Visual and Auditory) continuous performance tests. This type of assessment is beneficial for symptom clarification, treatment planning, and progress tracking.

11.7.5 Subjective Assessment Questionnaires

Subjective assessment questionnaires utilized by neurofeedback practitioners are extremely varied in focus, length, and target population. Some are inventories that measure multiple dimensions such as the Minnesota Multiphasic Personality Inventory (MMPI) and The Millon Clinical Multiaxial Inventory (MCMI). Others are very specific to a concern, such as anxiety, depression, or autism spectrum disorders. For adult assessments, these questionnaires are typically self-report. As such, they must be interpreted cautiously, taking into consideration issues such as social desirability, malingering, secondary gain, acquiescence, and the capacity for a distressed individual to report accurately. Despite these concerns, self-report questionnaires have consistently been found to be a valuable tool for clinical assessment. Many of the scales designed to assess children, such as the Behavior Assessment System for Children, offer teacher and parent versions of the scale in addition to a self-report. This allows cross-comparison of ratings from different perspectives and from different environments.

11.7.6 qEEG Integrated Assessments

Several companies offer assessment tools that utilize qEEG data in conjunction with additional assessments to assist neurofeedback practitioners in protocol selection. Thatcher [46] has developed a symptom checklist with the goal of linking brain structure to neuropsychological function based on fMRI, PET, and EEG/MEG research studies. Other systems such as the MiniQ Analysis by New Mind or the CNC 1020 system by Brownback and Mason offer neurofeedback recommendations based on matching reported symptoms to implicated brain regions.

Effective and efficient treatment with neurofeedback is predicated on a clear understanding of the symptoms involved, related brain areas, and an analysis of the client's qEEG. As this brief section illustrates, this information can be obtained through many avenues and must be considered alongside issues such as accessibility and cost. For many practitioners, finding the right combination of tools is a trial-and-error process that evolves with the technology and scope of practice.

11.7.7 Event-Related Potential Assessment

Event-related potentials (ERPs) have been utilized as an assessment tool to measure the impact of neurofeedback training for almost two decades [47]. ERP data that has been combined with auditory oddball tasks and other testing paradigms has been largely limited to the treatment of ADHD [47–52]. Despite its early adoption by Joel Lubar [47], the assessment strategy is employed mainly outside of the United States. Professor Yuri Kropotov suggests that the difference between ERP and qEEG analysis is that ERPs reflect information flow within cortical networks, whereas qEEG reflects mechanisms of cortical self-regulation. According to Dr. Kroptov, it is quite possible to have a normal self-regulatory regime but an abnormal information flow or vice versa [50].

11.8 Addressing Conditions with Neurofeedback

From simple ADHD to complex disorders of children on the autism spectrum, to those with affective disorders, case reports of the positive effects of EEG biofeedback abound. Far fewer controlled studies exist. The dearth of controlled studies has begun to change recently with ADHD being the most thoroughly researched disorder.

11.8.1 ADHD

Lubar and Shouse published the first reports of neurofeedback training for ADHD in 1976. Employing an ABA research design, they demonstrated that hyperactive symptoms decreased when SMR was enhanced and increased when SMR was inhibited [15]. The outcomes in this study led to scores of studies employing SMR and several variations of this training to address ADHD. For a more detailed description of these protocols, also see Monastra et al. [53]. Since that time, attention problems have been the most thoroughly investigated issue to date. Many controlled studies have demonstrated a large effect size for neurofeedback [54].

Slow cortical potential and theta/beta ratio training have been studied in the successful treatment of ADHD [38,55]. Both neurofeedback protocols show comparable results on the three features of ADHD: inattention, impulsivity, and hyperactivity [55]. The theta/beta ratio training was actually an inhibit/enhance protocol that encouraged beta and reduced theta. A few of the studies included frontal sites for training, whereas in the majority of studies the theta/beta training addressed central strip locations (C3, Cz, C4) as did the SCP training.

In 2009, Arns et al. published a meta-analysis that concluded that neurofeedback can be considered an efficacious and specific intervention for ADHD with a large effect size for inattention and impulsivity and a medium effect size for hyperactivity [54]. The designation of Efficacious and Specific was determined by guidelines jointly accepted by the International Society for Neurofeedback and Research (ISNR) and the Association for Applied Psychophysiology and Biofeedback (AAPB), and is analogous to guidelines accepted by the American Psychological Association (APA).

The American Academy of Pediatrics has recently declared EEG biofeedback a "Level 1 Best Support" intervention for ADHD. Their October 2012 report on Evidence-Based Child and Adolescent Psychosocial Interventions concluded that neurofeedback had the highest level of support based on at least two randomized, placebo-controlled studies [56]. The academy's decision was based on three studies [55,57,58], two that utilized fMRI to measure the change in the neural substrates of selective attention after neurofeedback training.

Other protocols that have not been systematically investigated have demonstrated promise in the training of this disorder. Recently, several cases involving children with attention problems were treated with z-score training with positive results [28]. The enhancement of regional cerebral blood oxygenation (Hemoenphalography) in specific cerebral locations has been shown to increase sustained attention [59]. The training of the infralow frequency band has produced normalization on the T.O.V.A in a recent study [60].

11.8.2 Other Conditions

Other conditions have not yet reached a Best Support level of research. However, there are substantial studies for a variety of complaints that demonstrate the effectiveness of neurofeedback training. From traumatic brain injury to autism, the list of conditions addressed continues to grow. A variety of neurofeedback protocols have proved their effectiveness with a wide assortment of clinical presentations. These interventions have been applied by providers whose specialties cut across the human service delivery spectrum, including mental health practitioners, educators, physical therapists, and medical personnel [61].

Anxiety, the main staple of most psychotherapy practices, has been addressed effectively by neurofeeback with dozens of small studies [32,62,63], as have depression [64,65], obsessive compulsive disorder [66], and PTSD [8]. This effectiveness has led many mental health practitioners to include neurofeedback in their practice, making psychologists, social workers, and other mental health professionals the largest group of providers in the field. At the same time, EEG biofeedback has expanded the typical client base. Many in the mental health professions have utilized neurofeedback for substance abuse [9,67], autism spectrum disorders [60,68–70], and sleep disorders [71]. Promoting peak performance in athletes, artists, and other professions has become the primary focus of some practices [72–74]. Mental health practitioners have traditionally been involved with educators in eliminating impediments to learning. Neurofeedback has been central to addressing ADHD and more recently has been used to address other learning disabilities [75–77].

Doctors, nurses, occupational therapists, physical therapists, and other allied health practitioners utilize neurofeedback in their practices, and some have done so since the beginning of the field. Each brings a special set of skills that is complemented by neurofeedback. Seizure disorder, possibly the first disorder treated with neurofeedback [5,78], has recently become an intervention of interest to neurologists and other health professionals [29,60,79]. Migraine headaches have been the focus of neurofeedback for over 20 years [80]. Recently, a large study confirmed the early promise of this intervention guided by a neurologist who employed qEEG in the treatment paradigm [81]. Traumatic brain injury has been the focus of many in the mental health field who developed neurofeedback for brain injury to supplement the sparse treatment options of traditional medicine. Jonathan Walker, a neurologist and pioneer in the neurofeedback field, has been instrumental in developing neurofeedback strategies for the brain injured [82].

In order to understand the magnitude of the research, a viewing of the Comprehensive Neurofeedback Bibliography amassed by the International Society for Neurofeedback and Research is suggested [61].

11.9 Conclusion

A large body of research spanning several decades contends that neurofeedback has proved useful for a wide variety of psychological and medical conditions. An assortment of protocols has been applied to accomplish these successful outcomes. It is reasonable to propose that neurofeedback does not treat anything specific but rather optimizes the central nervous system, thereby improving general function. Recent research recognizes that EEG biofeedback accomplishes this task by conditioning brain electrical activity and so is more aptly referred to as training rather than treatment.

The idea that central nervous system regulation is the result of neurofeedback training may help to explain the reported positive impact of neurofeedback on a host of disorders. The growing body of research that supports this perception underscores neurofeedback's recent acceptance by mainstream health practitioners and the general public. From its beginnings at the University of Chicago in the latter half of the last century, neurofeedback has become a worldwide phenomenon. The growth of the industry has occurred largely without government or academic support. Kept alive by the efforts of a few pioneers who recognized the efficacy of this intervention, neurofeedback has only recently begun to garner the popular recognition, professional acceptance, and institutional support that it deserves.

Frank H. Duffy, M.D., Professor and Pediatric Neurologist at Harvard Medical School, stated in an editorial in the January 2000 issue of the journal *Clinical Electroencephalography* that the scientific literature advocates for the application of neurofeedback to address many disorders. "In my opinion, if any medication had demonstrated such a wide spectrum of efficacy it would be universally accepted and widely used ..... It is a field to be taken seriously by all" [83].

References

1. J. Kamiya, Conscious control of brain waves, *Psychology Today*, pp. 57–60, 1968.
2. J. Kamiya, Ed., *Operant Control of the EEG Alpha Rhythm and Some of Its Reported Effects on Consciousness (Altered States of Consciousness)*, New York: Wiley, 1969.
3. S. Othmer, Neuromodulation technologies, in *Introduction to Quantitative EEG and Neurofeedback: Advanced Theory and Applications*, T. Budzynski et al., Eds., 2nd ed., Amsterdam: Elsevier, pp. 3–14, 2009.
4. M. B. Sterman, et al., Behavioral and neurophysiological studies of the sensorimotor rhythm in the cat, *Electroencephalography and Clinical Neurophysiology*, vol. 27, pp. 678–679, 1969.
5. M. B. Sterman and L. Friar, Suppression of seizures in an epileptic following sensorimotor EEG feedback training, *Electroencephalography and Clinical Neurophysiology*, vol. 33, pp. 89–95, 1972.
6. E. Green and A. Green, *Beyond Biofeedback*, New York: Dellacorte Press, 1977.
7. R. Soutar and R. Longo, *Doing Neurofeedback*, San Rafael, CA: ISNR Research Foundation, 2011.
8. E. Penniston and P. Kulkosky, Alpha-theta brainwave neurofeedback for Vietnam veterans with combat related post-traumatic stress disorder, *Medical Psychotherapy*, vol. 4, pp. 47–60, 1991.

9. E. G. Penniston and P. J. Kulkosky, Alpha-theta brainwave training and beta-endorphin levels in alcoholics, *Alcoholism Clinical and Experimental Research*, vol. 13, pp. 271–279, 1989.
10. L. Fehmi and S. Shor, Open focus attention training, *Psychiatric Clinics of North America*, vol. 36, pp. 153–162, 2013.
11. L. Fehmi and J. Robbins, *The Open Focus Brain*, Boston: Trumpeter Books, 2007.
12. T. Budzynski, Ed., *From EEG to Neurofeedback (Introduction to Quantitative EEG and Neurofeedback)*, San Diego, CA: Academic Press, 1999.
13. L. Fehmi, Multichannel EEG phase synchrony training and verbally guided attention training for disorders of attention, in *Handbook of Neurofeedback*, J. Evans, Ed., New York: Informa Healthcare, pp. 301–319, 2009.
14. J. T. McKnight and L. Fehmi, Attention and neurofeedback synchrony training: Clinical results and their significance, *Journal of Neurotherapy*, vol. 5, pp. 4–62, 2001.
15. J. F. Lubar and M. N. Shouse, EEG and behavioral changes in a hyperkinetic child concurrent with training of the sensorimotor rhythm (SMR): A preliminary report, *Biofeedback and Self-Regulation*, vol. 1, pp. 293–306, 1976.
16. M. B. Sterman, Physiological origins and functional correlates of EEG rhythmic activities: Implications for self-regulation, *Biofeedback and Self-Regulation*, vol. 21, pp. 3–33, 1996.
17. L. Thompson and M. Thompson, QEEG and neurofeedback for assessment and effective intervention with attention deficit hyperactivity disorder (ADHD), in *Introduction to Quantitative EEG and Neurofeedback: Advanced Techniques and Applications*, T. Budzynski et al., Eds., 2nd ed., Burlington, AM: Academic Press, pp. 337–360, 2009.
18. J. Lubar and M. N. Shouse, Neurofeedback for the management of attention deficit disorders, in *Biofeedback: A Practitioners Guide*, M. S. Schwartz and F. Andrasik, Eds., New York: Guilford Press, pp. 409–437, 2003.
19. T. F. Collura, Neuronal dynamics in relation to normative electroencephalography assessment and training, *Biofeedback*, vol. 36, pp. 134–139, 2009.
20. T. F. Collura, *Technical Foundations of Neurofeedback*, New York: Routledge, 2013.
21. E. R. John, et al., Developmental equations for the electroencephalogram, *Science*, vol. 210, pp. 1255–1258, 1980.
22. E. R. John, et al., Neurometrics: Numerical taxonomy identifies different profiles of brain functions within groups of behaviorally similar people, *Science*, vol. 196, pp. 1393–1410, 1977.
23. E. R. John and L. S. Prichep, Eds., *Normative Data Banks and Neurometrics: Basic Concepts, Methods and Results of Norm Construction (Handbook of Electroencephalography and Clinical Neurophysiology: Volume III)*, Amsterdam: Elsevier, 1987.
24. E. R. John, et al., Neurometrics: Computer-assisted differential diagnosis of brain dysfunctions, *Science*, vol. 239, pp. 162–169, 1988.
25. R. Thatcher and J. Lubar, History of the scientific standards of QEEG normative databases, in *Introduction to Quantitative EEG and Neurofeedback: Advanced Theory and Applications*, T. Budzynski, Ed., 2nd ed., Amsterdam: Elsevier, pp. 29–62, 2009.
26. D. Kaiser and M. B. Sterman, *SKIL Analysis*, 2013.
27. E. P. Tereshchenko, et al., Normative EEG spectral characteristics in healthy subjects aged 7 to 89 years, *Human Physiology*, vol. 36, pp. 1–12, 2010.

28. T. F. Collura, et al., EEG biofeedback case studies using live Z-score training (LZT) and a normative database, *Journal of Neurotherapy*, vol. 14, pp. 22–46, 2010.

29. T. F. Collura, et al., Eds., *EEG Biofeedback Training Using Z-Scores and a Normative Database (Introduction to QEEG and Neurofeedback: Advanced Theory and Applications)*, New York: Elsevier, 2009.

30. M. L. Smith, A father finds a solution: Z-score training, *Neuroconnections*, pp. 22–25, 2008.

31. J. V. Hardt and J. Kamiya, Conflicting results in EEG alpha feedback studies: Why amplitude integration should replace percent time, *Biofeedback and Self-Regulation*, vol. 1, pp. 63–75, 1976.

32. J. V. Hardt and J. Kamiya, Anxiety change through electroencephalographic alpha feedback seen only in high anxiety subjects, *Science*, vol. 201, pp. 79–81, 1978.

33. S. L. Fahrion, et al., Alterations in EEG amplitude, personality factors, and brain electrical mapping after alpha-theta brainwave training: A controlled case study of an alcoholic in recovery, *Alcoholism: Clinical and Experimental Research*, vol. 16, pp. 547–552, 1992.

34. J. P. Rosenfeld, et al., Preliminary evidence that daily changes in frontal alpha asymmetry correlate with changes in affect in therapy sessions, *International Journal of Psychophysiology*, vol. 23, pp. 137–141, 1996.

35. T. Elbert, et al., Biofeedback of slow cortical potentials, *Electroencephalography and Clinical Neurophysiology*, vol. 48, pp. 293–301, 1980.

36. N. Birbaumer, et al., Biofeedback of slow cortical potentials in epilepsy, in *Clinical Applied Psychophysiology. Plenum Series in Behavioral Psychophysiology and Medicine*, A. R. Seifert, N. Birbaumer, J. G. Carlson, Ed., New York: Plenum Press, pp. 29–42, 1994.

37. N. Birbaumer and L. G. Cohen, Brain–computer interfaces: Communication and restoration of movement in paralysis, *The Journal of Physiology*, vol. 579, pp. 621–636, 2007.

38. H. Heinrich, et al., Training of slow cortical potentials in attention-deficit/hyperactivity disorder: Evidence for positive behavioral and neurophysiological effects, *Biological Psychiatry*, vol. 55, pp. 772–775, 2004.

39. U. Strehl, et al., Self-regulation of slow cortical potentials: A new treatment for children with attention-deficit/hyperactivity disorder, *Pediatrics*, vol. 118, pp. e1530–1540, 2006.

40. J. M. Palva and S. Palva, Infra-slow fluctuations in electrophysiological recordings, blood-oxygenation-level-dependent signals, and psychophysical time series, *NeuroImage*, vol. 62, pp. 2201–2211, 2012.

41. M. L. Smith, Infra-slow fluctuation training: On the down-low in neuromodulation, *Neuroconnections*, pp. 38–46, 2013.

42. D. A. Kaiser and S. Othmer, Effect of neurofeedback on variables of attention in a large multi-center trial, *Journal of Neurotherapy*, vol. 4, pp. 5–15, 2000.

43. R. W. Elwood, MicroCog: Assessment of cognitive functioning, *Neuropsychological Review*, vol. 11, pp. 89–100, 2001.

44. V. Di Sclafani, Neuropsychological performance of individuals dependent on crack-cocaine or crack-cocaine and alcohol at 6 weeks and 6 months of abstinence, *Drug Alcohol Dependence*, vol. 66, pp. 161–171, 2002.

45. J. Epstein, Relations between continuous performance test performance measures and ADHD behaviors, *Journal of Abnormal Child Psychology*, vol. 33, pp. 543–554, 2003.

46. R. W. Thatcher, *Handbook of Quantitative Electroencephalography and EEG Biofeedback*, Applied Neuroscience Inc, Seminole, FL, 2012.

47. J. F. Lubar, et al., Quantitive EEG and auditory event-related potentials in the evaluation of attention-deficit/hyper-activity disorder: Effects of methylphenidate and implications for neurofeedback training, *Journal of Psychoeducational Assessment*, pp. 143–160, 1995.

48. T. Egner and J. H. Gruzelier, Learned self-regulation of EEG frequency components affects attention and event-related brain potentials in humans, *Neuroreport*, vol. 12, pp. 4155–4159, 2001.

49. T. Egner and J. H. Gruzelier, EEG Biofeedback of low beta band components: Frequency-specific effects on variables of attention and event-related brain potentials, *Clinical Neurophysiology: Official Journal of the International Federation of Clinical Neurophysiology*, vol. 115, pp. 131–139, 2004.

50. J. D. Kropotov, *Quantitative EEG, Event Related Potentials and Neurotherapy*, London: Academic Press, 2009.

51. J. D. Kropotov, et al., Changes in EEG spectrograms, event-related potentials and event-related desynchronization induced by relative beta training in ADHD children, *Journal of Neurotherapy*, vol. 11, pp. 3–11, 2007.

52. S. Wangler, et al., Neurofeedback in children with ADHD: Specific event-related potential findings of a randomized controlled trial, *Clinical Neurophysiology: Official Journal of the International Federation of Clinical Neurophysiology*, vol. 122, pp. 942–950, 2011.

53. V. Monastra, et al., Electroencephalographic biofeedback in the treatment of attention-deficit/hyperactivity disorder, *Applied Psychophysiology and Biofeedback*, vol. 30, pp. 95–114, 2005.

54. M. Arns, et al., Efficacy of neurofeedback treatment in ADHD: The effects on inattention, impulsivity and hyperactivity: A meta-analysis, *Clinical EEG and Neuroscience*, vol. 40, pp. 180–189, 2009.

55. H. Gevensleben, et al., Is neurofeedback an efficacious treatment for ADHD? A randomised controlled clinical trial, *J Child Psychol Psychiatry*, vol. 50, pp. 780–789, 2009.

56. AAP, Appendix S2: Evidence-based child and adolescent psychosocial interventions, *Pediatrics*, vol. 125, p. S128, 2010.

57. M. Beauregard and J. Levesque, Functional magnetic resonance imaging investigation of the effects of neurofeedback training on the neural bases of selective attention and response inhibition in children with attention-deficit/hyperactivity disorder, *Applied Psychophysiology and Biofeedback*, vol. 31, pp. 3–20, 2006.

58. J. Levesque, et al., Effect of neurofeedback training on the neural substrates of selective attention in children with attention-deficit/hyperactivity disorder: A functional magnetic resonance imaging study, *Neuroscience Letters*, vol. 394, pp. 216–221, 2006.

59. H. Toomim, et al., Intentional increase of cerebral blood oxygenation using hemoencephalography (HEG): An efficient brain exercise therapy, *Journal of Neurotherapy*, vol. 8, pp. 5–21, 2004.

60. S. B. Legarda, et al., Clinical neurofeedback: Case studies, proposed mechanism, and implications for pediatric neurology practice, *Journal of Child Neurology*, vol. 26, pp. 1045–1051, 2011.

61. ISNR, *Comprehensive Bibliography*, D. C. Hammond and D. A. Novian, Eds., International Society for Neurofeedback and Research, 2013.

62. L. Huang-Storms, et al., QEEG-guided neurofeedback for children with histories of abuse and neglect: Neurodevelopmental rationale and pilot study, *Journal of Neurotherapy*, vol. 10, pp. 3–16, 2006.

63. D. C. Hammond, Neurofeedback with anxiety and affective disorders, *Child and Adolescent Psychiatric Clinics of North America*, vol. 14, pp. 105–123, 2005.

64. E. Baehr, et al., The clinical use of an alpha asymmetry protocol in the neurofeedback treatment of depression: Two case studies, *Journal of Neurotherapy*, vol. 2, pp. 10–23, 1997.

65. E. Baehr, et al., Clinical use of an alpha asymmetry neurofeedback protocol in the treatment of mood disorders: Follow-up study one to five years posttherapy, *Journal of Neurotherapy*, vol. 4, pp. 11–18, 2001.

66. D. C. Hammond, QEEG-Guided neurofeedback in the treatment of obsessive compulsive disorder, *Journal of Neurotherapy*, vol. 7, pp. 25–52, 2003.

67. W. C. Scott, et al., Effects of an EEG biofeedback protocol on a mixed substance abusing population, *The American Journal of Drug and Alcohol Abuse*, vol. 31, pp. 455–469, 2005.

68. M. E. J. Kouijzer, et al., Neurofeedback treatment in autism. Preliminary findings in behavioral, cognitive, and neurophysiological functioning, *Research in Autism Spectrum Disorders*, vol. 4, pp. 386–399, 2010.

69. E. Jarusiewicz, Efficacy of neurofeedback for children in the autistic spectrum: A pilot study, *Journal of Neurotherapy*, vol. 6, pp. 39–49, 2002.

70. R. Coben and I. Padolsky, Assessment-guided neurofeedback for autistic spectrum disorder, *Journal of Neurotherapy*, vol. 11, pp. 5–23, 2007.

71. B. Hammer, et al., Neurofeedback for insomnia: A pilot study of Z-score SMR and individualized protocols, *Applied Psychophysiology and Biofeedback*, vol. 36, pp. 251–264, 2011.

72. W. A. Edmonds and T. Tenenbaum, Eds., *Case Studies in Applied Psychophysiology*, Chichester, UK: Wiley-Blackwell, 2012.

73. J. Gruzelier, A theory of alpha/theta neurofeedback, creative performance enhancement, long distance functional connectivity and psychological integration, *Cognitive Processing*, vol. 10, Suppl 1, pp. S101–109, 2009.

74. J. H. Gruzelier, et al., Beneficial outcome from EEG-neurofeedback on creative music performance, attention and well-being in school children, *Biological Psychology*, vol. 95, pp. 86–95, 2013.

75. K. E. Thornton and D. P. Carmody, Electroencephalogram biofeedback for reading disability and traumatic brain injury, *Child and Adolescent Psychiatric Clinics of North America*, vol. 14, pp. 137–162, 2005.

76. M. A. Nazari, et al., The effectiveness of neurofeedback training on EEG coherence and neuropsychological functions in children with reading disability, *Clinical EEG and Neuroscience*, vol. 43, pp. 315–322, 2012.

77. T. Ros, et al., Facilitation of procedural learning directly following EEG neurofeedback, *Biological Psychology*, vol. 95, pp. 54–58, 2013.

78. A. R. Seifert and J. F. Lubar, Reduction of epileptic seizures through EEG biofeedback training, *Biological Psychology*, vol. 3, pp. 157–184, 1975.

79. J. E. Walker and G. P. Kozlowski, Neurofeedback treatment of epilepsy, *Child and Adolescent Psychiatric Clinics of North America*, vol. 14, pp. 163–176, 2005.

80. M. A. Tansey, A neurobiological treatment for migraine: The response of four cases of migraine to EEG biofeedback training, *Headache Quarterly: Current Treatment and Research*, vol. 2, pp. 90–96, 1991.

81. J. E. Walker, QEEG-guided neurofeedback for recurrent migraine headaches, *Clinical EEG and Neuroscience*, vol. 42, pp. 59–61, 2011.
82. J. E. Walker, et al., Impact of qEEG-guided coherence training for patients with a mild closed head injury, *Journal of Neurotherapy*, vol. 6, pp. 31–43, 2002.
83. F. H. Duffy, The state of EEG biofeedback therapy (EEG operant conditioning) in 2000: An editor's opinion, *Clinical Electroencephalography*, vol. 31, pp. V–VII, 2000.

12

Future Integration of EEG-fMRI in Psychiatry, Psychology, and Cultural Neuroscience

Mohd Nasir Che Mohd Yusoff, Mohd Ali Md Salim, and Mohamed Faiz Mohamed Mustafar
Universiti Sains Malaysia

Jafri Malin Abdullah
Universiti Sains Malaysia
Universiti Sains Malaysia Health Campus

Wan Nor Azlen Wan Mohamad
Universiti Sains Malaysia

CONTENTS

12.1 Overview of Cultural Neuroscience Framework 270
 12.1.1 Benefits and Challenges 270
 12.1.2 Suggestions for Future Work 271
12.2 Color and Psychology 274
 12.2.1 Depression: Color as Biological Marker 275
 12.2.2 Methodology for Future Predictors of Cognitive State 276
  12.2.2.1 Eye Tracking System 276
  12.2.2.2 Imaging Techniques 278
12.3 EEG/ERP in Psychopathy and Criminal Behavior 279
 12.3.1 P140 279
 12.3.2 P200 279
 12.3.3 Error-Related Negativity/Feedback-Related Negativity 280
 12.3.4 Error Positivity (Pe) 280
 12.3.5 P300 280
12.4 EEG/ERP and EEG-fMRI in Rehabilitation for Developmental and Psychological Disorders 281
 12.4.1 Specific Language Impairment 281
 12.4.2 Cerebral Palsy 281
 12.4.3 Autism Spectrum Disorders 281
 12.4.4 Obsessive Compulsive Disorder (OCD) 281
References 282

12.1 Overview of Cultural Neuroscience Framework

In general, cultural neuroscience (CN) could be understood as a paradigm that put forward the framework of culture to understand the influence of culture's elements such as values, practices, and beliefs regarding human thought. Uniquely, the construction of a CN framework relies on theories from various disciplines such as anthropology, psychology, and genetics. In a recent study, CN was framed in multiple time scales (situation, ontogeny, and phylogeny) in order to explain how the diversity of the cultural values and genetic factors contour the complexity of the human mind and behavior [1]. Similar to the previous contribution in CN Ref. [2] states that cultural capacities and their transmission that arose from complex human mental and neurobiological processes were critically determined from bidirectional indicators (culture and gene) across two timescales: macro and micro timescales. As an example, the macro timescale determines phylogeny and lifespan, whereas the micro timescale determines the situation. In this regard [3], "culture" is viewed as a set of traits that is inflexible and has specificities. In other words, the variation of population culture can manifest in neural activation patterns. In Ref. [4], the specific importance of culture for the brain is highlighted as a fundamental value in the development of racial identity and ideology.

12.1.1 Benefits and Challenges

Looking at culture not only from the behavioral point of view but also at the neural level will prove beneficial when answering questions such as, how do cultural traits affect neural processes and shape human behavior? [1,5]. A cultural approach, through interdisciplinary educational infrastructure and research [6], may provide a better understanding of health disparity across racial groups [7,8]. Besides of notifying interethnic ideology, the advantages of the cultural neuroscience approach include merging the health quality across diverse cultural populations [2]. In addition, CN could reflect crucial implication in health disparities and public policy development [1]. The CN framework should be expanded into integration of other neuroscience tools as well (such as a combination of event-related potential (ERP) and fMRI) and, thus, more useful data on human neural activities could be obtained. The existing fMRI data in culture studies that has been done mostly among Western, industrialized, rich, educated, and democratic populations can be used as a reference for further action such as exploring the CN of an underdeveloped community and integrating powerful tools of neuroscience (i.e., ERP and fMRI) to develop a more specific CN framework within this community. Within this CN framework, the trait of "individualism–collectivism" in cultural dimensions that modulate the psychological neural basis can be highlighted, together with identification of other cultural traits such as a preference for social hierarchy and racial identification [1]. However, bringing culture into the research mainstream for a national agenda in the future is not an easy task. The main challenge in the area of CN, indeed, is to empower it as an educational medium that is able to provide a comprehensive and culturally sensitive framework to educate the future generation. This objective could be achieved by improving the research infrastructure, increasing research capacity, and establishing appropriate ethical standards. It is important to note that CN is an interdisciplinary framework. Thus, in order to implement CN, interdisciplinary training is critically needed across the social and natural sciences to meet the nation's agenda.

12.1.2 Suggestions for Future Work

Neuro-culture interaction permits understanding of culturally patterned neural activities through neuroplasticity at the cellular level, which accelerates biological adaptation and social survival [9]. In order to understand the interaction of neuro-culture, cultural neuroscientists used various approaches (i.e., quantitatively and qualitatively), whether at the societal or individual cellular level [10]. An anthropologist, for example, implements a behavioral survey to detect cultural characteristics (i.e., values, practices, and beliefs) of a group of people from different societies [11]. Behavioral surveys such as using open-ended interviews and ethnography provides rich information on values, practices, and beliefs held by oneself and others [11,12]. From the cultural neuroscience toolbox [5], there are three major levels that can be used to study and explore culture. These are level 1, a behavioral paradigm (to observe the interaction between culture and mind); level 2, which consists of using fMRI, PET, ERP, and TMS (to observe the interaction between brain and mind); and level 3, which involves using candidate functional polymorphisms (to observe the interaction between brain and genes). All these classification are seen through ontogeny and phylogeny. In a recent critical review, the approach to the CN framework was implemented by using fMRI to detect the "universalism and differentialism" in culture practice [3]. It was reported that the neural response is more sensitive to cultural value such as "individualism–collectivism" than cultural origin, especially during one's self-appraisal [2]. Another culture construct is "tightness–looseness," which refers to one's obedience to the social norms prevailing in society. The tightness–looseness may reflect the differentiation of mental patterns across cultures [13]. For example, people who live under a tight culture may have a great tendency to adhere to and value their cultural norms. However, those who are not under this kind of society (i.e., loose culture) are more likely to be not sensitive to their societal norms. To take another example, it is indicated that people's morality is a product, at least in part, of the coevolution between the "tightness–looseness" value of culture with the serotonin transporter gene [14]. Meanwhile, culture–gene coevolutionary models of human behavior established in the literature help us to understand the culture–biology interactions in adaptive behavior [15–17]. As a result, in human reaction to adaptive behavior, it was observed that the serotonin transporter gene (*SLC6A4*) indicates reciprocal evolutionary change and interaction with the cultural values of "individualism–collectivism" [15]. In addition, the S allele of 5HTTLPR, the polymorphism that occurs in the promoter region of the gene, is likely to be more prevalent in "collectivistic" nations, compared to the long (L) allele [15]. Implementation of the CN framework in underdeveloped countries is urgently needed, in part because nearly all the neuroculture evidence from the neuroimaging method was gathered from the Western industrialized population, which is educated and has high income levels [5,18]. Thus, the CN framework should be tested and established in Eastern underdeveloped populations in order to depict a clear picture of the neuroculture patterns and constructs. From the above observations, authors are hoping that some neuro-based society action could be undertaken in future.

First, the integration of the combination of quantitative EEG, qualitative EEG, and fMRI to detect values such as "individualism–collectivism" and "tightness–looseness" that construct culture is done in order to observe the interaction between culture and social problems in society. Thus, authors suggest using a comprehensive CN tool (that encompasses behavioral surveys, genetic studies, qualitative and quantitative EEG, as well as fMRI) to gain a comprehensive understanding of "culture-social problem" interaction. To cater to social problems in society, it is suggested to pay attention to social problems such as loitering in public places by adolescents and abandonment of babies by mothers, as these two issues are still understudied, and more evidence is needed to prove hypotheses such as this one: "culture differences in a multi-ethnic population may construct the human mind and behavior, and thus, some races may be predisposed to anti-social activities such as loitering and abandonment." In a multi-ethnic population such as Malaysia, the occurrence of loitering and abandonment is likely to be popular among certain races. This raises question of whether a neural substrate is playing a role in these differences. The authors are hoping to explore some implications from this suggestion, in particular, the development of a pre-CN framework prior to implementing the full framework of CN. The main aim of this pre-framework of CN is to detect any abnormality of the brain pattern that may exist in social misfits (i.e., loitering and abandonment) before the interaction between culture and the social problem is observed (i.e., the CN framework), in order to detect "individualism–collectivism" and "tightness–loitering in public places by adolescents and abandonment of babies by mothers' looseness" in culture practice. Implementation of neuroscience tools (qualitative and quantitative EEG, as well as fMRI) to discover this social phenomenon can be done in two settings: the laboratory or a real setting. In order to observe loitering activity among adolescents, the authors suggest using wireless EEG-MINDO in a real setting (such as on the stairs of a shopping complex, other public places, etc.) to identify the brain pattern during the real event while adolescents are loitering. This would help to discover whether loitering activity has a positive or negative effect to the EEG pattern of the brain, than pre-occupied occasion. Meanwhile, for the pre-CN framework in the cases of abandonment, the authors suggest using neuroscience tools (qualitative and quantitative EEG, as well as fMRI) to detect a woman's "intention" to abandon her baby. This means that all pregnant women should be screened to detect any abnormality of the EEG pattern in the brain as a biomarker of their intention to abandon their babies. Of course, ethical issues will be raised and debated. In addition, a study should be implemented in a very large population to develop normative data as a reference and determine the cutoff value in order to differentiate between positive and negative EEGs.

Second, NGOs have a role to play. It is hoped that NGOs (together with the government) around the world will be using wireless EEG-MINDO effectively as a screening tool at the community level, especially to develop education packages among adolescents who like loitering and wasting time outside their home. Similarly, it is hoped that NGOs (together with government) will actively take action to prevent the cases of baby abandonments by implementing a screening program for pregnant women, by using wireless or non-wireless EEG (or in combination with fMRI).

At this stage, incorporating the element of neuromarketing in the CN framework could prove very timely. Neuromarketing is a subfield in the neuroscience discipline that offers an understanding of the human neural mechanisms underlying the cognitive response to marketing stimuli. NGOs and neuromarketers may work together, going beyond the scope of the research or community service, looking at the potential market of EEG as an effective tool to detect the abnormality of brain activities among social misfits. In this case, the social problem is manipulated as a "product." This means that wireless EEG should be able to reflect any changes in brain activities while subjects are performing relevant tasks related to social problems. For example, in the cases of loitering, subjects are being tested to choose between two products (in this case, "product" refers to "social problem"), the first product being "During my free time, I like loitering at the shopping complex with my friends" and the second product "During my free time, I like sitting at home and watching TV." Similarly, in the abandonment issue, some tricky task can be tested to detect if the EEG can detect any abnormalities in brain activities to reflect "unacceptance value" in society. This task could be directed to pregnant women, and their response to items such as "I love my baby so much" (as Product 1), and "I love my baby but this is not the right time to have a baby" (as Product 2) could be observed.

By incorporating neuromarketing in the CN framework, the EEG marketer could narrow the target through gaining a comprehensive understanding of the underlying neurobiology from the interaction between psychological and marketing phenomena [19], within the specific market target: social problem or social violence market. Although there are debates on the effect of a product's brand (I would like to refer to the product's brand here as a value and a belief held in society) on consumer decision making

(I would like to refer "consumer" here as a unit in society who has to make a choice between accepted behavior and non-accepted behavior) [20]. The neurobiological basis of brand preference such as the dorsolateral prefrontal cortex (DLPFC) might instigate reward-motivated behavior through connections to the mesolimbic system [21]. The ventromedial prefrontal cortex (VMPFC) was suggested to be involved in making choices from among the brands offered, even though this suggestion has been challenged [22].

The success of a CN-neuromarketing framework indeed relies on the element of trustworthiness in making any choices in the social market. Using the combination of EEG and fMRI to detect trustworthiness in brand preference is a great approach and could be implemented as there is evidence indicating the activation of amygdala and the insula cortex from the fMRI scans [23]. Neurobiology modulation of trusting also involves several neurotransmitters and hormones such as oxytocin [23,24]. However, in order to market EEG in this social problem market, the gender perspective should be taken into consideration as women might activate the brain region differently, especially related to this "trustworthiness" issue [25]. The challenges of CN-neuromarketing may also arise from the appropriate ethical guidelines and potential ethical dilemmas to maximize the profit [26]. After all, the growing interest among neurologist in neuromarketing (and implementation of CN-neuromarketing in the future market target) should ensure that the marketing environment will prevent the market from being a manipulative way of selling unnecessary goods and services [27].

In summary, marketing research questions from neuromarketing (and neuroconsumer as well) can be a great agenda in the world of neuroscience and economics [28]. To achieve any of the objectives of neuromarketing, developing the model to understand the biological and behavioral reactions in marketing should be the primary focus. Owing to this, exhibiting special consumer vulnerability seems to be important to the success of neuromarketing competition [28]. By taking one step ahead, implementing the CN-neuromarketing (or CN-neuroconsumer as an alternative) offers EEG marketers the opportunity to provide specialty in their service to the nation. Collaboration among neurologists, psychiatrists, and the general medical neuroscientific community may work to achieve the target [29].

12.2 Color and Psychology

The visual system has very complex neural circuitry compared to other sensory systems. The complexity of the visual system has attracted much attention from researchers across multidisciplinary backgrounds including neuroscience, biomedicine, physiology, physics, psychology, and economics. One aspect of the visual system that gains attention is color vision. We experience color every day in our daily routine and even in our dreams. The color experience that we encounter is dependent on the amount or spectral composition of the light that reaches the eye and the contextual influence. The visibility of the light to human eye can be measured by wavelengths ranging from 400 nm to 700 nm. Researches in color and psychology have revealed new discoveries that shed light on how color affects our life and the way we use color, especially in the health and clinical setting. There are many studies that report color has a significant effect on the psychological functioning of an individual. Even though much work has been put in to understand the fundamental issues of the color visual system, the development of the research remains at an early stage [30].

Studies on the color visual system and psychology research are developing. However, certain challenges are encountered in past studies, which can be improved to understand the color visual system holistically. This was discussed in detail in Ref. [30], which listed a few important points including the effect of color to a specific cultural background orientation, the limited usage of sampling (limited to only university students), lacking in the aspect of variability of color (chroma) used in the experiment [30].

12.2.1 Depression: Color as Biological Marker

Depression is a common and severe mental disorder. From the artistic point of view, this disorder is frequently associated and illustrated with certain colors like gray and black that evoke the feeling of sadness. Even though this association is just an expression of the state of depression in the form of color, physiologically there is some evidence of a possible biological link between vision and depression.

The current practice of diagnosing depression is based on the symptoms listed in the *Diagnostic and Statistical Manual of Mental Disorder* (DSM). However, the lists of symptoms are rather subjective to some extent, even though they have been validated and recognized. Among those symptoms are significant weight loss, changes in sleep pattern, and fatigue. Recent research on visual science and major depressive disorder (MDD) has provided an alternative diagnosis method that can support the current practice and that is claimed to be more objective [31].

The study was conducted by a group of researchers at the University Hospital of Freiburg, Germany and focused on the possibility of a linkage between depression and visual contrast [32]. They were measuring cells at the retina level, such as photoreceptors, in patients with MDD by using an electroretionogram, which enables the electrical response in the retina cell to be recorded and is usually used by the ophthalmologist to detect and diagnose retinal disease. This study revealed that, among patients with depression, whether they were on medication or not, there was significant reduction in their contrast gain [31]. This indicates a significant link between the visual system and depression.

EEG and ERP studies on color and emotion also revealed consistent findings. Different colors were reported to cause significant changes in brain waves that resemble the changes of the mood especially in terms of arousal (8–13 Hz relaxation state, 15–30 Hz excitation state) [33]. The participants of the research were exposed to different colors: green, blue, and red. The color stimuli were presented for 1 s at 2 s intervals between the colors [33]. Another finding was that different color exposures stimulate different response areas and different frequency levels [33]. These recordings were made at the V1, V2, and V3 visual areas of the brain. This shows that mood changes caused by the color visual experience can be detected by certain patterns of brain waves as the biological marker. Therefore, prediction of the emotional state can be made based on the color that was seen, which is characterized by certain brain waves.

An animal study on color and depression also came up with interesting findings. Red light was found to be less depressive compared to the other light colors at night [34]. This study was conducted on hamsters, which were exposed to four different types of light including blue, red, white, and no light. Their behavior and the density of the dendritic spines in the hippocampus were analyzed. They reported that red light and no light caused less depressive behavior and higher density of the dendritic spines than the blue and white light [34].

The foregoing studies are some examples of evidence that major depressive disorders could be detected through the color visual system, which serves as a biological marker. The future challenge for this field is to provide more physiological evidence that contributes to the detection of clinical disorder.

12.2.2 Methodology for Future Predictors of Cognitive State

12.2.2.1 Eye Tracking System

The eye tracking system is one of the well-known methods in the study of the cognitive visual system. This method uses the properties of eye movement behavior to understand the underlying cognitive process involved in respect of certain visual tasks including color visual tasks. The primary properties of eye movement behavior are the number of fixations, the duration of fixation time, and the rapid eye movements (saccade) [35]. One of the advantages of using eye movement data is that it is a more natural and spontaneous type of interaction with the environment and does not require training [35]. The development of eye movement studies has helped us to understand the fundamental issues related to eye movement in our daily behavior, the role of internal reward, and the development of theoretical advancements in reinforcement learning and graphic simulation. The latest research has also revealed that by understanding eye movement behavior, we are able to predict the individual cognitive state [35].

A number of previous studies have reported that eye movement behaviors are highly influenced by the experimental visual task engaged in by the participant. For instance, reading and viewing a scene have different saccade amplitudes and fixation durations. The same situation also happens in a visual searching task and a memorizing task [32]. In other words, specific visual tasks can determine and influence the behavior of eye movements. However, the changes in eye movement behavior during a specific task overlaps greatly across different tasks. Therefore, it is hard to really identify the task that an individual is engaged in based on the eye movement data alone. A proper classification of eye movement data is crucial to make sense of the cognitive state involved. Henderson and colleagues [35] have proved to some extent that this problem can be overcome by proper classification of the eye movement data. In their research [35], used multivariate pattern analyses (MVPA) as the method of classification in order to understand and predict the task engaged by the participants and their cognitive state through their eye movements. The participants were asked to complete four different tasks including memorizing, reading, visual searching, and pseudo-reading. Before the process of classification is made, some of the data acquired were removed such as fixation longer than 1500 ms, and some other eye signal noises such as due to the eye blinking. Furthermore, a naïve Bayes classifier was used for each participant to capture the eye movement pattern. The result demonstrated that eye movement behaviors are significantly dependent on the task engaged and the cognitive state.

The classification accuracies reported were highly reliable above chance. The same reliability was also obtained by using four eye movement properties, the means and standard deviation of the saccade amplitude, and the duration of fixation [35]. This indicated that different viewing tasks have their own features of eye movement based on the four-way classification of eye movement [35]. Even though the cognitive state defined in this study refers to the viewing task (e.g., skimming, reading, and memorizing), other classifications of cognitive states might also be possible, such as confusion and inattention [35], which can resemble a more clinical aspect of the cognitive state (e.g., anxiety, depression, phobia, arousal). However, the challenge is that a broader classification is needed that is not limited to only the viewing task. In order to achieve this goal, comprehensive and continuous eye movement recording is needed. The development of the technology of smart glass (e.g., Google glass) seems to be promising in this regard. The prototype of Google glass recently introduced is a more lightweight product and more comfortable for daily wear and this technology fundamentally based on the eye tracking system which is dependent on the eye movement behavior [36].

This development may provide a solution for continuous recording and classification to cover a more general cognitive state. Through a comprehensive classification of the cognitive state through eye movement behavior, early detection of potential psychological illnesses can be taken into account, precautions can be taken, and a preventive treatment plan can be implemented. This classification should be extended to color-specific stimuli because color always has a greater sensitivity to the center of fixation.

12.2.2.2 Imaging Techniques

There is an increased trend of using imaging techniques in color and psychology research to provide physiological evidence of the effect of color on psychological functioning. The technique used enables us to understand the underlying process in the brain that has been the center of attention of recent studies.

Color-related information was found to be processed in the ventral temporal lobe in humans and the ventromedial region of the occipital lobe and the posterior temporal cortex in non-human primates [37]. Recently, research on the color of Chinese characters with ERP imaging tools detected a strong association with late posterior negativity (LPN) in searching, retrieving, and evaluating color objects [38]. This finding was in agreement with the earlier finding that the LPN component was associated with picture color source retrieval. These examples indicate that different techniques or tools used in research might result in different or new knowledge on the same subject. It is not a question of right or wrong, but just a matter of retrieving the information on the same object from different perspectives.

Color has its own property to influence the level of attention. Some colors do have a higher degree of attention level, whereas others have a low level of attention. This variance of attention intensity in color has an impact on our physiological and cognitive system [39]. Chang and Huang [39] used two types of visual task (low level and high level of attention) that they defined by different colors, and they used EEG to record the brain activity at different band waves. They found that theta synchronization in Fz, Cz, and Pz was stronger in a high-attention-level task than in a low-attention-level task [39]. The high-attention-level task also was found to decrease the LF component and LF/HF ratio compared to the low-attention-level task [39]. This indicated that variation in the attention level produced by color can induce different changes in the brain wave activity.

The combination of these two types of methodologies has huge potential in scientific color visual research in providing a better understanding of human cognitive and psychological states and functioning. For instance, research by Nikolaev, Jurica, Nakatani, Plomp, and Leeuwen (2013) on visual encoding and target selection fixation used a combination of eye movement and EEG recording data and reported fascinating results where they were trying to understand or predict the next fixation target in the natural scene by observing presaccadic electrical brain activity from the EEG [40]. In the near future, it is expected that multidisciplinary research and methodologies will gain greater attention in color and psychology research.

12.3 EEG/ERP in Psychopathy and Criminal Behavior

An approach to psychopathology that employs neurocognitive endophenotypes posits that behavioral and cognitive process abnormalities found in most psychiatric disorders are associated with distinct disturbances in the brain and neural systems [41]. For instance, the aberrations in affective functions and behavioral problems found in individuals with psychopathic traits are believed to be correlated with focal brain lesions as postulated by the paralimbic system dysfunction model [42]. Prior work has demonstrated that ERP provides an excellent window on the brain with a very good temporal resolution and is very useful in elucidating cognitive mechanisms and neural substrates of a particular event [43]. Indeed, ERP components such as P140, P200, error-related negativity (ERN), error-positivity (Pe), and P300 are increasingly becoming important neurophysiological markers in assessing several aspects (e.g., pathological reward sensitivity [44]) of criminal behaviors and other related psychiatric disorders.

12.3.1 P140

P140 is an early ERP component of index attentional processing under threat-focus conditions. Prior studies highlighted that greater P140 indicates superior selective attention. Psychopathy was found to be hypersensitive to threat-related stimuli, and there was difficulty in reallocating attention in goal-directed behavior. Therefore, this ERP component could be a useful marker in assessing abnormal threat processing reported in individuals with psychopathic traits [45].

12.3.2 P200

P200 is an orbitofrontal ERP marker to reflect attention capture modulation and salience detection based on a task relevance item [46–49]. The P200, which peaks at about 200 ms after feedback onset, has been found to be associated with reward sensitivity [50], and is known as a salience pathway [49] signaling reward and absence of reward [51,52]. Psychopathy has been associated with impulsivity and novelty-seeking indices of heightened sensitivity to reward. This ERP component could be a useful marker indexing reward sensitivity found in individuals with addictive behaviors [53] and criminal behaviors.

12.3.3 Error-Related Negativity/Feedback-Related Negativity

Error-related negativity (ERN) and/or feedback-related negativity (FRN) are ERP components reflecting performance monitoring and feedback processing. The ERN is a negative deflection that peaks shortly after an erroneous response in speeded performance task, whereas the FRN peaks following the presentation of feedback information. These components were found to reflect individual learning abilities; they are believed to be generated in the anterior cingulate cortex (ACC) of the brain. Previous research on ERN/FRN indicated that psychopathy is associated with impaired learning abilities and error monitoring [54,55]. Current evidence suggests that the FRN could possibly assess the pathological reward sensitivity associated with reward-related striatal activation [56]. Thus, the FRN could be a significant electrophysiological marker for pathological reward sensitivity reported in individuals with psychopathic traits.

12.3.4 Error Positivity (Pe)

Post-error positivity (Pe) is a positive deflection that occurs between 100 and 400 ms after the ERN and is associated with recognition of error. A reduced Pe amplitude would indicate that participants are unaware that they have committed errors or are unable to identify errors [57–60]. Because previous studies highlighted that psychopathy is associated with impaired performance monitoring and difficulty in conforming to social norms, the Pe could be a useful component for assessing deficits in awareness among psychopaths while they commit errors.

12.3.5 P300

The P300 amplitude is a positive ERP deflection that peaks between 300 and 600 ms post-onset feedback. This ERP component is elicited by novel and salient stimuli associated with attention and orienting processes [61] and to high-level motivational evaluation [62]. Mixed findings concerning the P300 amplitude and psychopathy using different paradigms (e.g., oddball task) are reported in the field [42]. Despite the inconsistent results with regard to smaller or larger P300 amplitudes, most studies highlighted a different scale in the assessment of psychopathic traits (e.g., PCL-R, TriPM) and yielded different P300 results depending on the target stimuli. Therefore, future research with regard to the P300 amplitude is needed because this component is useful to identify deficiencies or abnormalities in attention and orienting processes found in psychopathy.

12.4 EEG/ERP and EEG-fMRI in Rehabilitation for Developmental and Psychological Disorders

12.4.1 Specific Language Impairment

ERPs may guide researchers in describing phonological processing's role in language comprehension and speech production. The influence of lexical processing on sentences would be very important in setting the baseline and tracking the progress during rehabilitation for children with a specific language impairment [63].

12.4.2 Cerebral Palsy

The future use of EEG would move from the traditional concept to portable EEG, which is suitable for measuring brain waves during active movement such as walking, driving, and speaking. During physical therapy for children with cerebral palsy, portable EEG would be able to capture real-time data with better time resolution. Clinicians will be able to localize cortical electrical activation areas in probing neuroplastic changes associated with motor recovery during and after rehabilitation training. Real-time cortical activation neurofeedback could be the basis for facilitating motor imagery training in the course of neurorehabilitation for children with cerebral palsy [64].

12.4.3 Autism Spectrum Disorders

In children with autism, neurofeedback training could be targeted to improve executive functions in connectivity-guided neurofeedback by remediating connectivity disturbances. Researchers have found that significant improvements were noted in attentional control and cognitive flexibility in children with autism by increasing slow wave brain activity. Changes in qEEG could be seen as improvements in set-shifting, reciprocal social interactions, and communication skills in children with autism [65].

12.4.4 Obsessive Compulsive Disorder (OCD)

Neurofeedback training using qEEG could be an alternative advancement in reducing obsession thoughts and compulsive behaviors among patients with OCD. The mechanism modifies brain function and improves clinical symptoms by decreasing obsession and compulsion behaviors. Different EEG waves could be discriminated by the patients by identifying alpha or beta waves in various triggers in the treatment program. Learning to associate the desired EEG patterns and learning to dissociate unwanted EEG patterns could be part of the treatment modules in rehabilitation processes [66].

References

1. Chiao, J. Y., Cheon, B. K., Pornpattananangkul, N., Mrazek, A. J., and Blizinsky, K. D., Cultural neuroscience: Progress and promise, *Psychological Inquiry: An International Journal for the Advancement of Psychological Theory*, 24(1), 1–19, 2013.
2. Chiao, J. Y., Harada, T., Komeda, H., Li, Z., Mano, Y., Saito, D. N., et al., Neural basis of individualistic and collectivistic views of self, *Human Brain Mapping*, 30, 2813–2820, 2009.
3. Mateo, M. M., Cabanis, M., Loebell, N. C. E., and Krach, S., Concerns about cultural neurosciences: A critical analysis, *Biobehavioral Reviews*, 36, 152–161, 2012.
4. Choudhury, S. and Kirmayer, L. J., Cultural neuroscience and psychopathology: Prospects for cultural psychiatry, In: Joan, Y. C. (Ed.), *Progress in Brain Research*, Elsevier, New York, pp. 263–283, 2009.
5. Chiao, J. Y., Cultural neuroscience: A once and future discipline, *Progress in Brain Research*, 178, 287–304, 2009.
6. Chiao, J. Y. and Cheon, B. K., The weirdest brains in the world, *Behavioral and Brain Sciences*, 33, 28–30, 2010.
7. Dovidio, J. F. and Fiske, S. T., Under the radar: How unexamined biases in decision-making processes in interactions can contribute to health care disparities, *American Journal of Public Health*, 102, 945–952, 2012.
8. Williams, D. R., John, D. A., Oyserman, D., Sonnega, J., Mohammed, S. A., and Jackson, J. S., Research on discrimination and health: An exploration study of unresolved conceptual and measurement issues, *American Journal of Public Health*, 102, 975–978, 2012.
9. Kitayama, S. and Uskul, A. K., Culture, mind, and the brain: Current evidence and future directions, *Annual Review of Psychology*, 62, 419–449, 2011.
10. Chiao, J. Y., Hariri, A. R., Harada, T., Mano, Y., Sadato, N., Parrish, T. B., et al., Theory and methods in cultural neuroscience, *Social Cognitive and Affective Neuroscience*, 5, 356–361, 2010.
11. Chiao, J. Y., Harada, T., Komeda, H., Li, Z., Mano, Y., Saito, D. N., et al., Dynamic cultural influences on neural representations of the self, *Journal of Cognitive Neuroscience*, 22, 1–11, 2010.
12. Gelfand, M. J., Shteynberg, G., Lee, T., Lun, J., Lyons, S., Bell, C., et al., The cultural contagion of conflict, *Philosophical Transactions of the Royal Society B: Biological Sciences*, 367, 692–703, 2012.
13. Gelfand, M. J., Raver, J. L., Nishii, L., Leslie, L. M., Lun, J., Lim, B. C., et al., Differences between tight and loose cultures: A 33-nation study, *Science*, 332, 1100–1104, 2011.
14. Mrazek, A. J., Chiao, J. Y., Blizinsky, K. D., Lun, J., and Gelfand, M. J., Culture-gene coevolution of tightness-looseness and the serotonin transporter gene, Manuscript submitted for publication, 2013.
15. Chiao, J. Y. and Blizinsky, K. D., Culture-gene coevolution of individualism-collectivism and the serotonin transporter gene (5-HTTLPR), *Proceedings of the Royal Society B: Biological Sciences*, 277, 529–537, 2010.
16. Nikolaidas, A. and Gray, J. R., ADHD and the DRD4 exon III 7-repeat polymorphism: An international meta-analysis, *Social Cognitive and Affective Neuroscience*, 5(2–3), 188–193, 2010.

17. Way, B. and Lieberman, M. D., Is there a genetic contribution to cultural differences? Collectivism, individualism, and genetic markers of social sensitivity, *Social Cognitive and Affective Neuroscience*, 5, 203–211, 2010.
18. Henrich, J., Heine, S., and Norenzayan, A., The weirdest people in the world? *Behavioral and Brain Sciences*, 33, 61–83, 2010.
19. Reimann, M., Schilke, O., Weber, B., Neuhaus, C., and Zaichkowsky, J., Functional magnetic resonance imaging in consumer research: A review and application, *Psychology and Marketing*, 13(6), 608–637, 2011.
20. Park, C. W., MacInnis, D. J., Priester, J., Eisingerich, A. B., and Iacobucci, D., Brand attachment and brand attitude strength: Conceptual and empirical differentiation of two critical brand equity drivers, *Journal of Marketing*, 13, 1–17, 2010.
21. Ballard, I. C., Murty, V. P., Carter, R. M., MacInnes, J. J., Huettel, S. A., and Adcock, R. A., Dorsolateral prefrontal cortex drives mesolimbic dopaminergic regions to initiate motivated behavior, *Journal of Neuroscience*, 13, 10340–10346, 2011.
22. Santos, J. P., Seixas, D., Brandão, S., and Moutinho, L., Investigating the role of the ventromedial prefrontal cortex in the assessment of brands, *Front Neuroscience*, 13, 77, 2011.
23. Riedl, R. and Javor, A., The biology of trust: Integrating evidence from genetics, endocrinology, and functional brain imaging, *Journal of Neuroscience, Psychology, and Economics*, 13, 63–91, 2012.
24. Shahrokh, D. K., Zhang, T. Y., Diorio, J., Gratton, A., and Meaney, M. J., Oxytocin-dopamine interactions mediate variations in maternal behavior in the rat, *Endocrinology*, 13(5), 2276–2286, 2010.
25. Riedl, R., Hubert, M., and Kenning, P., Are there neural gender differences in online trust? An fMRI study on the perceived trustworthiness of ebay offers, *MIS Quarterly*, 13, 397–428, 2010.
26. Madan, C. R., Neuromarketing: The next step in market research? *Eureka*, 13(1), 34–42, 2010.
27. Eser, Z., BaharIsin, F., and Tolon, M., Perceptions of marketing academics, neurologists, and marketing professionals about neuromarketing, *Journal of Marketing Management*, 13, 854–868, 2011.
28. Javor, A., Koller, M., Lee, N., Chamberlain, L., and Ransmayr, G., Neuromarketing and consumer neuroscience: Contributions to neurology, *BMC Neurology*, 13, 1471–2377, 2013.
29. Sharp, C., Monterosso, J., and Montague, P. R., Neuroeconomics: A bridge for translational research, *Biological Psychiatry*, 13(2), 87–92, 2012.
30. Elliot, A. J. and Maier, M. A., Color psychology: Effects of perceiving color on psychological functioning in humans, *Annual Review of Psychology*, 65, 95–120, 2014.
31. Bubl, E., Kern, E., Ebert, D., Bach, M., and van Elst, L. T., Seeing gray when feeling blue? Depression can be measured in the eye of the diseased, *Journal of Biopsychology*, 68, 205–208, 2010.
32. Castelhano, M. S., Mack, M., and Henderson, J. M., Viewing task influence eye movement control during active scene perception, *Journal of Vision*, 9(3), 1–15, 2009.
33. Zhang, H. and Tang, Z., To judge what color the subject watched by color effect on brain activity, *International Journal of Computer Science and Network Security*, 11(2), 80–83, 2011.

34. Ellis, M., Depressed mood could be lifted by color of nightlight, p. 284, August 8, 2013. Retrieved from http://www.medicalnewstoday.com/articles/264507.php

35. Henderson, J. M., Shinkareva, S. V., Wang, J., Luke, S. G., and Olejarczyk, J., Predicting cognitive state from eye movements, *PLoS One*, 8(5), e64937, 2013.

36. Miller, C. C., Google searches for style, *The New York Times*, February 20, 2013. Retrieved 10 August 2013 from http://www.nytimes.com/2013/02/21/technology/google-looks-to-make-its-computer-glasses-stylish.html?pagewanted = 1&_r = 0

37. Liebe, S., Fischer, E., Logothetis, N. K., and Rainer, G., Color and shape interactions in the recognition of natural scenes by human and monkey observers, *Journal of Vision*, 9(5), 14.1–16, 2009.

38. Nie, A., Guo, C., Liang, J., and Shen, M., The effect of lateral posterior negativityin retrieving the color of Chinese characters, *Neuroscience Letters*, 534(8), 223–227, 2013.

39. Chang, Y. and Huang, S., The influence of attention levels on psychophysiological responses, *International Journal of Psychophysiology*, 86, 39–47, 2011.

40. Nikolaev, A. R., Jurica, P., Nakatani, C., Plomp, G., and Leeuwen, C. V., Visual encoding and fixation target selection in free viewing: Presaccadic brain potentials, *Frontier in System Neuroscience*, 7(26), 1–12, 2013.

41. Robbins, T. W., Gillan, C. M., Smith, D. G., de Wit, S., and Ersche, K. D., Neurocognitive endophenotypes of impulsivity and compulsivity towards dimensional psychiatry, *Trends in Cognitive Sciences*, 16(1), 81–91, 2012.

42. Kiehl, K. A., A cognitive neurosciences perspectives on psychopathy: Evidence for paralimbic system dysfunction, *Psychiatry Research*, 142(2–3), 107–112, 2006.

43. Luck, S. J., *An Introduction to the Event-Related Potential Technique*, The MIT Press, London, England, 2005.

44. Foti, D., Weinberg, A., Dien, J., and Hajcak, G., Event-related potential activity in the basal ganglia differentiates rewards from non-rewards: Temporospatial principal component analysis and source localization of the feedback negativity, *Human Brain Mapping*, 32, 2207–2216, 2011.

45. Baskin-Sommers, A., Curtin, J. J., Li, W., and Newman, J. P., Psychopathy-related differences in selective attention are captured by an early event-related potential, *Personality Disorder*, 3(4), 370–378, 2012.

46. Potts, G. F., Liotti, M., Tucker, D. M., and Posner, M. I., Frontal and inferior temporal cortical activity in visual target detection: Evidence from high spatially sampled event-related potentials, *Brain Topography*, 9, 197–209, 1996.

47. Potts, G. F. and Tucker, D. M., Frontal evaluation and posterior representation in target detection, *Cognitive Brain Research*, 11, 147–156, 2001.

48. Riis, J. L., Chong, H., McGinnnis, S., Tarbi, E., Sun, X., Holcomb, P. J., et al., Age-related changes in early novelty processing as measured by ERPs, *Biological Psychology*, 82, 33–44, 2009.

49. San Martin, R., Manes, F., Hurtado, E., Isla, P., and Ibanez, A., Size and probability of rewards modulate the feedback error-related negativity associated with wins but not losses in a monetarily rewarded gambling task, *Neuroimage*, 51, 1194–1204, 2010.

50. Martin, L. E. and Potts, G. F., Reward sensitivity in impulsivity, *Neuroreport*, 15(9), 1519–1522, 2004.

51. Rolls, E. T., *The Brain and Emotion*, Oxford University Press, Oxford, 1999.

52. Schultz, W., Dayan, P., and Montague, P. R., A neural substrate of prediction and reward, *Science*, 275, 1593–1599, 1997.
53. Franken, I. H. A., van den Berg, I., and van Strien, J. W., Individual differences in alcohol drinking frequency are associated with electrophysiological responses to unexpected nonrewards, *Alcoholism: Clinical and Experimental Research*, 34(4), 702–707, 2010.
54. Brazil, I. A., de Bruijn, E. R., Bulten, B. H., van Borries, A. K., van Lankveld, J. J., Buitelaar, J. K., and Verkes, R. J., Early and late component of error monitoring in violent offenders with psychopathy, *Biological Psychiatry*, 65(2), 137–143, 2009.
55. Brazil, I. A., Jan Verkes, R., Brouns, B. H. J., Buitelaar, J. K., Bulten, B. H., and de Bruijn, E. R. A., Differentiating psychopathy from general antisociality using the P3 as a psychophysiological correlate of attentional allocation, *PLoS One*, 7(11), 1–8, 2012.
56. Carlson, J. M., Foti, D., Mujica-Parodi, L. R., Harmon-Jones, E., and Hajcak, G., Ventral striatal and medial prefrontal BOLD activation is correlated with reward-related electrocortical activity: A combined ERP and fMRI study, *Neuroimage*, 57, 1608–1616, 2011.
57. Endrass, T., Reuter, B., and Kathmann, N., ERP correlates of conscious error recognition: Aware and unaware errors in an antisaccade task, *The European Journal of Neuroscience*, 26(6), 1714–1720, 2007.
58. Niewenhuis, S., Ridderinkhof, K. R., Blom, J., Band, G. P., and Kok, A., Error-related brain potentials are differentially related to awareness of response errors: Evidence from an antisaccade task, *Psychophysiology*, 38, 752–760, 2001.
59. O'Connell, R. G., Bellgrove, M. A., Dockree, P. M., Lau, A., Hester, R., Garavan, H., et al., The neural correlates of deficient error awareness in attention-deficit hyperactivity disorder (ADHD), *Neuropsychologia*, 47(4), 1149–1159, 2009.
60. Shalgi, S., Barkan, I., and Deouell, L. Y., On the positive side of error processing: Error-awareness positivity revisited, *European Journal of Neurosciences*, 29, 1522–1532, 2009.
61. Polich, J., Updating p300: An integrative theory of P3a and P3b, *Clinical Neurophysiology*, 118, 2128–2148, 2007.
62. Polich, J. and Criado, J. R., Neuropsychology and neuropharmacology of P3a and P3b, *International Journal of Psychophysiology*, 60(2), 172–185, 2006.
63. Shaheen, E. A., Shohdy, S. S., Raouf, M. A. A., Abd, S. M. E., and Elhamid, A. A., Relation between language, audio-vocal psycholinguistic abilities and P300 in children having specific language impairment, *International Journal of Pediatric Otorhinolaryngology*, 75(9), 1117–1122, 2011.
64. Lee, N. G., Kang, S. K., Lee, D. R., Hwang, H. J., Jung, J. H., You, J. H., et al., Feasibility and test-retest reliability of an electroencephalography-based brain mapping system in children with cerebral palsy: A preliminary investigation, *Archives of Physical Medicine and Rehabilitation*, 93(5), 882–888, 2012.
65. Karimi, M., Haghshenas, S., and Rostami, R., Neurofeedback and autism spectrum: A case study, *Procedia Social and Behavioral Sciences*, 30, 1472–1475, 2011.
66. Barzegary, L., Yaghubi, H., and Rostami, R., The effect of qEEG-guided neurofeedback treatment in decreasing of OCD symptoms, *Procedia Social and Behavioral Sciences*, 30, 2659–2662, 2011.

13

Future Use of EEG, ERP, EEG/MEG, and EEG-fMRI in Treatment, Prognostication, and Rehabilitation of Neurological Ailments

Jafri Malin Abullah and Zamzuri Idris
Universiti Sains Malaysia Health Campus

Nor Safira Elaina Mohd Noor, Tahamina Begum, Faruque Reza, and Wan Ilma Dewiputri
Universiti Sains Malaysia

CONTENTS

13.1 Introduction 288
13.2 Two Main Aspects of Future EEG: Brain Waves and
Their Maps, and EEG Technology 289
   13.2.1 Brain Waves and Their Maps 289
   13.2.2 EEG Technology 292
      13.2.2.1 EEG Technology without Surgery (Noninvasive
         or Scalp EEG) 292
      13.2.2.2 EEG Technology with Surgery (Intracranial EEG)..... 293
      13.2.2.3 Other Technologies 293
13.3 The Future Aspects of Clinical Applications Using EEG 293
   13.3.1 Noninvasive Mapping and Creation of Ideal Networks
      for the Brain 293
   13.3.2 Brain–Computer Interface and Robotic Technology 294
   13.3.3 Diagnosis and Monitoring 295
   13.3.4 Therapy 295
   13.3.5 Neurofeedback and Neurocognition 296
13.4 Future of fMRI-EEG and Its Clinical Applications 296
13.5 Future of EEG-MEG and Its Clinical Applications 297
13.6 Clinical Use of EEG/EEG Monitoring, ERP, and EEG-fMRI in
Epilepsy 298
13.7 Clinical Use of EEG/qEEG, ERP, and EEG-fMRI in Traumatic
Brain Injury (TBI) 300
13.8 Future Clinical Application of EEG/ERP in Stroke 300
13.9 Neurofeedback as a Future Aspect of fMRI-EEG 303
References 304

13.1 Introduction

The applications of electroencephalography (EEG) in combination with event-related potential (ERP) and functional magnetic resonance imaging (fMRI) are gaining importance in the field of both social neurosciences as well as medical neurosciences. The current research on resting-state brain networks looks into the science of activation of different brain and the rise and fall in synchronicity while the brain is assumed to be at rest. Many neuroscientists used fMRI to locate, characterize, and analyze data received from these networks. Unfortunately, a disadvantage is the relative slowness of fMRI, which is limited to the data collection of activities that change every 10 s [1]. Another disadvantage is that the data from fMRI shows that the spatial and topographical pattern of activity is similar at rest and while performing tasks [1].

A more cost-effective and robust method is magnetoencephalography (MEG), which detects activities at millisecond levels, recording waves of activity at frequencies from slow (0.1–4 Hz) to fast (greater than 50 Hz). The value of the EEG in medical neurosciences after one has studied its effects in fMRI is the robustness, time saving, and patient friendliness, especially for the younger age group.

This technology is used for prognostication, treatment, rehabilitation, and most importantly as an electrical biomarker for physicians both in the acute setting in intensive care units, sub-acute settings for use in high dependency wards, and later for rehabilitation in acute neurological emergencies. It is used in developmental behavior disorders such as autism and dyslexia as well as in psychological and psychiatric conditions such as stress, anxiety neurosis, and psychosis as an objective method of following up a patient in a clinical setting.

Stroke and traumatic brain injury management both in the acute setting and in the rehabilitation phase makes this an important tool for the future. The classical use of this technology coupled with MEG and fMRI for both applied therapy in the field of epilepsy treatment and follow-up will make it useful both at tertiary- and secondary-level hospitals.

EEG holds great promise for the study of brain activity. Obvious advantages compared to MEG, which is also capable in studying brain waves, are: (1) hardware costs are lower, (2) the scalp EEG is noninvasive and cheap, while intracranial EEG (iEEG or ECoG) offers higher temporal and spatial resolutions than EEG. These two advantages enable EEG technology to be widely used to study brain activities in two major circumstances: (1) without craniotomy (healthy subjects and patients) and (2) with craniotomy or burr holes (patients). Therefore, the future aspects of clinical applications using EEG can be classified based on these two considerations.

Regarding EEG and its combination with other techniques, the two most widely used combinations nowadays are fMRI-EEG [2–5] and MEG-EEG [6,7]. The current clinical applications using these combinations are mainly focused on epilepsy [8–10]. In our opinion, wider clinical application is likely in the near future if we gain better knowledge in understanding and mapping brain waves. Similar to genetic coding (a frontier in the 20th century), which gives rise to heterogeneous and complex phenotypical manifestations, brain waves and rhythms are similarly viewed as being able to contribute to a myriad of diverging brain functions (a frontier for the 21st century). Therefore, one of the areas that is in need of vigorous exploration in future is understanding brain waves and their significance (or codes).

13.2 Two Main Aspects of Future EEG: Brain Waves and Their Maps, and EEG Technology

The future aspects of clinical applications using EEG may evolve together with advances in two important areas of EEG: (1) a better understanding of brain waves recorded by EEG and (2) advances in EEG technology itself.

13.2.1 Brain Waves and Their Maps

There are five major brain waves distinguished by their different frequencies, ranging from high to low: gamma (above 30 Hz), beta (13–30 Hz), alpha (8–13 Hz), theta (4–8 Hz), and delta (0.1–4 Hz). Some basic and salient points regarding the brain waves are as follows [11,12]:

1. Hertz is a unit of frequency that equals cycles per second.
2. Slow waves are waves with frequencies less than 8 Hz, which include theta and delta waves.
3. Focal slow wave activity indicates focal neuronal damage.
4. Delta waves are primarily associated with deep sleep and may also be present in the waking state.
5. Theta waves are presumed to originate from the thalamus. They appear as consciousness slips toward drowsiness. Theta waves have been associated with deep meditation, creative inspiration, and unconscious material. They are often accompanied by other frequencies and seemed to be related to the level of arousal. They also play an important role in infancy and childhood.
6. Larger amount of theta waves in a waking adult are abnormal and are caused by various pathological problems.
7. Alpha waves appear in the posterior half of the head and are usually found over the parieto-occipital region of the brain. They can be detected in all parts of the posterior lobe of the brain. They may appear as round, sinusoidal, or sometimes as sharp waves. Alpha waves could

indicate a relaxed awareness without any attention or concentration. It is the most prominent rhythm in brain activity.

8. Alpha waves have higher amplitudes in the occipital region.

9. Beta waves are the usual waking rhythm of the brain associated with active thinking, active attention, focusing on the outside world, or solving concrete problems and are found in normal adults mainly in the frontal region. A panic state may also induce a high level of beta waves.

10. Rhythmical beta activity is encountered chiefly over the frontal and central regions.

11. Fast beta waves or gamma waves have low amplitude, and their occurrence is rare. Detection of these waves can be used for confirmation of certain brain diseases.

12. Gamma waves are also a good indicator of event-related synchronization of the brain and can be used to demonstrate the locus for right and left index finger movement, right toes, and tongue movement (Figure 13.1).

13. Waves with frequency higher than 300 Hz have been found in cerebellar structures.

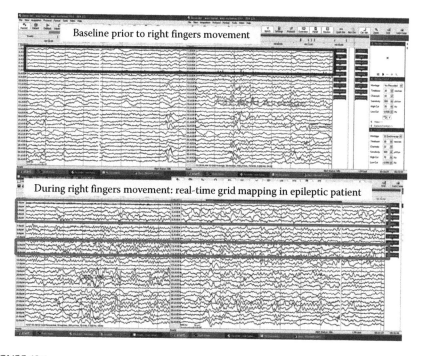

FIGURE 13.1

Gamma wave activity (electrode 1 inside light gray boxed) during initiation of right finger movement identified on electrocorticography (ECoG).

14. Bilateral persistent EEG (burst of seizure activity/ictal wave pattern/ spikes/polyspikes) is usually associated with impaired conscious cerebral reaction.

15. Focal persistent EEG (burst of seizure activity/ictal wave pattern/ spikes/polyspikes) or epileptiform discharges are either interictal (seen between seizures episodes) or ictal (seen during acute seizure episode) and manifest in either spikes or sharp wave form, which is typically followed by a slow wave. It is usually associated with focal cerebral disturbance such as a seizure.

The future clinical applications of EEG obviously depend on our understanding of brain waves. Besides the morphological aspects of brain waves briefly mentioned already, future mapping of brain waves with certain human behaviors and activities undoubtedly seem crucial to gain an insight into how the brain works (Figure 13.2). The information gained from this

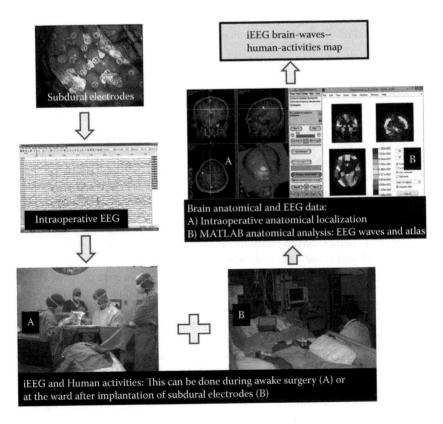

FIGURE 13.2
Brain-waves and human-activities map. Current EEG technology allows only iEEG as a reliable mapping technique. Future advances in EEG technology may also incorporate scalp EEG as a reliable technique in brain mapping.

"brain-waves–human-activities map" can be used in various sectors such as clinical, military, robotic, or communications technology. Mapping of the brain waves should ideally be first obtained via iEEG because of its current high signal resolution. Hopefully, in the near future, with advances in EEG technology, high-resolution EEG signals will be directly obtained via the scalp EEG and thus the three major drawbacks of the iEEG technique will be overcome: (1) invasiveness, (2) restricted sampling area, and (3) mostly applied to the diseased brain. Therefore, the prudent conclusion is that collateral development in EEG technology as well as better understanding of brain waves are required to completely and reliably map the brain waves with human activities.

13.2.2 EEG Technology

We may expect to see big changes in the field of EEG in the coming years. EEG technology is currently widely used in two main situations: (1) with surgery that normally requires a craniotomy or burr hole and (2) without surgery. Therefore, improvements in EEG technology can be separately discussed based on these two main classifications.

13.2.2.1 EEG Technology without Surgery (Noninvasive or Scalp EEG)

In order to gain a better understanding of brain activities, EEG technology should be able to explore the short-lived spatiotemporal dynamics of many of the underlying brain processes. This is currently possible only with iEEG (ECoG). Intraoperative EEG provides brain signals that have an exceptionally high signal-to-noise ratio, less susceptibility to artifacts than current scalp EEG, and have a high spatial and temporal resolution (i.e., <1 cm and <1 ms, respectively). Therefore, the current scalp EEG should be improved in terms of (1) higher signal resolution and (2) greater reliability and convenience.

Because there is bone and dura covering the brain, higher signal resolution for scalp EEG can be achieved by modifying the number of electrodes, the electrode contact and sensor, recording system such as the amplifier and filter, and wireless system. The future EEG should make the numbers higher in order to sample nearly all areas of the brain. Currently, EEG channels are commonly available in 16, 32, and 64 channels, but lately EEG with 256 channels is available (EEG-1200 Neurofax, Nihon Kohden, Japan, and dense array EEG, Geodesic Sensor Net, EGI Inc, Eugene, Oregon, USA). Dry electrode contact and wireless system are already in the market (Mindo, Contec, etc.) [13,14]. Certainly, improvements in EEG sensor and recording technology would improve the prospects for the technology. Nanotechnology may change many aspects of EEG technology, for example, if "each" electrode sensor is fitted with a nano-amplifier system [15]. The greater the use of miniaturized and more sensitive materials for sensors, the higher the number of electrode sensors and signal detection that can be made. With advances in signal processing, the short-lived brain signals that are normally obtained via iEEG can be obtained via scalp EEG.

13.2.2.2 EEG Technology with Surgery (Intracranial EEG)

The brain wave signals obtained via iEEG are regarded as a gold standard for EEG analysis, mainly because of their high signal-to-noise ratio, reduced susceptibility to artifacts, and high spatial and temporal resolution. Further advances are likely to be similar to those for scalp EEG: improvement in sensors and their numbers, recordings, as well as signal analysis systems. The current frontier in this technology is a high-density microelectrode array iEEG system (hdMEA) [16], which has been manufactured and utilized only at the research level. Through this technology, researchers are able to identify high-frequency pathological oscillations localized to sub-millimeter scale tissue volumes and hence can diagnose subclinical seizures and localize the microdischarges.

13.2.2.3 Other Technologies

Besides the aforementioned advances and developments in scalp and iEEG, future EEG systems should also be able to not only record brain waves but also be capable of generating them via neural stimulation. The improvement should be made to enable scalp EEG to also be used as a brain waves generator without causing significant side effects to the person. Another advance is the EEG-cap system, which can be modified to tailor a site for craniotomy, especially during awake surgery with subdural EEG electrodes in situ and sterile scalp EEG electrodes at a non-craniotomy site. By having these technologies in place, brain waves and their relationship with human activities (in normal and diseased person) can easily be studied, and neural signals can be mapped.

13.3 The Future Aspects of Clinical Applications Using EEG

With better understanding of, and future advances in, EEG technology as described earlier, the clinical applications of EEG will expand, and the existing clinical applications can be made more refined to increase the expected yield. The future aspects of clinical applications using EEG can be classified into the following categories.

13.3.1 Noninvasive Mapping and Creation of Ideal Networks for the Brain

Anatomical mapping using MRI-DTI technology and brain atlas has been started and continues to evolve [17]. EEG technology may enhance those mapping processes and possibly create a human anatomical-brain-waves-activities map. This map incorporates brain anatomy and brain waves that correlate to human activities. Together with future developments in molecular imaging and genetics, additional brain-molecular/genetic mappings can be incorporated to the anatomical-brain-waves–activity map and create an ideal brain map inclusive of anatomy (gray and white matter, atlas), molecular profiles (dopamine, acethycholine, noradrenaline, serotonin), genetic profiles, brain waves and rhythms, and human activity.

13.3.2 Brain–Computer Interface and Robotic Technology

Brain–computer interface (BCI) is a promising medical–engineering–computer technology advancement. It is currently being widely investigated for various applications [18–21]. By having a normal and complete brain map for human brain activities, one can use these data to enhance the brain–computer interface or brain–prosthetic interface technology. With advances in EEG technology and the availability of normal network templates for certain brain activities, a shorter training time in obtaining the right brain waves for certain human activities can be achieved. A human-like robot is another possibility. A complete understanding of brain networks may allow scientists to create an "electronic brain," which could be used to create a human-like skillful robot. With this achievement, the robots can be used to perform surgery and to rehabilitate patients at hospital or at home (Figure 13.3).

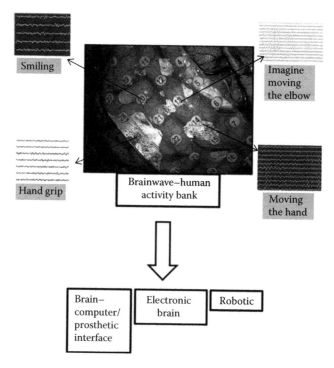

FIGURE 13.3
Knowledge gained from EEG–human activity map can be utilized in brain–computer or prosthetic interface, and the creation of electronic brains and robots.

13.3.3 Diagnosis and Monitoring

Epilepsy is one of the diseases that can be diagnosed based on the EEG. With developments in EEG technology (noninvasive techniques with high signal resolution, etc.) and a better understanding of brain signals and their analysis, other clinical diagnoses can be reliably and accurately made. These include stroke, brain tumors, and viral infections. The use of EEG to continuously monitor patients in the intensive care unit (ICU) has already been implemented [22,23], and further refinements in EEG technology and brain waves may give the intensivists and scientists the opportunity to study consciousness and sleep in wider contexts and correlate them with diseases. Besides studying consciousness, intensivists can achieve a better understanding of brain functions when they have a universal monitoring system that incorporates all important brain parameters. The ICU-EEG machine should have wireless capability, a high signal-to-noise ratio with good spatial and temporal resolution, scalp EEG system together with intracranial pressure waves, and blood flow and oxygenation parameters, all of which can be analyzed at the bedside. The availability of such technology will certainly improve the care of critical patients (Figure 13.4).

13.3.4 Therapy

Implanted EEG to detect abnormal electrical discharges prior to seizures has been developed [24]. This therapy can possibly be improved to avoid the implantation surgery, by the wearing of an EEG cap. The detected EEG waves can also be transferred telemetrically to the epilepsy treating center, and hence further advice can be given. Neuromodulation plays a significant

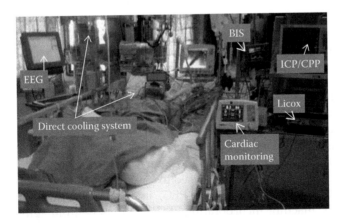

FIGURE 13.4
Multimodality monitoring for brain injury. The EEG machine is separated from other monitoring machines (ICP, CBF, brain temperature). The future should see one EEG machine incorporating all important parameters so that meaningful analysis at the right time scale can be made.

role in treating some neurological diseases such as Parkinson's, dystonia, tremors, pain, and depression. With the availability of a high signal-to-noise ratio for scalp EEG, the brain waves can be studied during deep-seated brain anatomical stimulation. The data obtained can be used to find ways to modulate their identified networks superficially and treat accordingly with neural stimulation (implanted superficial EEG or cap EEG) without requiring deep brain stimulation surgery. EEG and pharmacotherapy for brain diseases is obviously interesting, and clinicians and pharmacologists may learn enough about the brain mechanism to discriminate effective or ineffective drug therapy and identify brain wave biomarkers for certain disorders.

13.3.5 Neurofeedback and Neurocognition

Neurofeedback or EEG biofeedback is an EEG operant-conditioning training technique that helps individuals learn to control or modify their brain activity. It is used to treat a variety of conditions such as attention deficit hyperactivity disorder, anxiety, addictions, and epilepsy [25–29]. The person has to be trained to produce certain features of brain waves that fundamentally has been shown to correspond to certain beneficial cognitive or clinical outcomes. Wider application in fields of learning disabilities, schizophrenia, dementia, depression, and brain injury is likely with the advances in EEG technology and a better understanding of brain waves and rhythms.

13.4 Future of fMRI-EEG and Its Clinical Applications

EEG technology has evolved from its single acquisition method to complementary methods of measuring brain activity. The simultaneous acquisition of EEG and other functional neuroimaging such as fMRI and MEG are now increasingly available after crucial technical challenges have been met in terms of amplifier systems, biocompatible EEG systems, and correction of scanner-related artifacts in the EEG (Figures 13.5a and 13.5b). By combining EEG with fMRI, the clinician can confidently interpret the EEG findings and spatially map the area of interest. Currently, such advantages have been widely used in managing epilepsy [30–32]. Obviously, with the advances in scalp EEG, signal detection, and analysis, the information provided by simultaneous fMRI-EEG can be used to study brain networks and cognition.

With the great advantages of EEG, in the near future we may see a "universal" machine incorporating EEG and other neuroimaging techniques (Figure 13.5c). Ideally, the EEG sensors should have already been incorporated inside those neuroimaging techniques without the need for the technician to attach them to the head (just like the gamma knife head system used in radiosurgery).

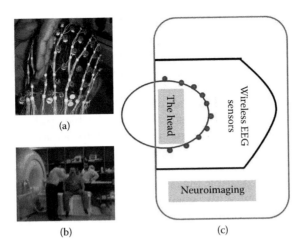

FIGURE 13.5
(a) Typical MRI-compatible scalp EEG. (b) fMRI-EEG session. (c) A sketched diagram of a universal machine incorporating the EEG and neuroimaging.

13.5 Future of EEG-MEG and Its Clinical Applications

The usage of interpretation of brain signals in neurorehabilitation applications has gained attention over the recent years. This is a rapidly emerging field of brain research which involves multidisciplinary research on the brain with computer interfaces (BCI) [33]. The current advance researches in brain signals analysis, noninvasive EEG with brain–computer interface (BCI) technologies, have been used to manage a computer cursor or a limb orthosis, forward processing and accessing the Internet, and for other functions including environmental control or entertainment. BCI technologies have improved the lifestyle of people suffering from advanced amyotrophic lateral sclerosis (ALS) [34].

The research and development of BCI is a multidisciplinary effort involving researchers from different backgrounds, and they are still looking forward to creating and inventing new technologies allowing for precise, relevant diagnosis and effective treatment. MEG has recently been proposed as a potential new source of brain-derived signals to operate with BCI. MEG is a noninvasive method that is able to detect frequency ranges above those available in EEG recordings with higher spatial resolution [35]. The combination of EEG-MEG and brain interfaces is regarded as another breakthrough within the relatively novel but rapid developing BCI technology. The signals from the computer interfaces of EEG-MEG generate reliable self-control of sensorimotor rhythm amplitude and satisfactory two-dimensional BCI control in the rehabilitation process of stroke patients [35,36].

The combination of EEG-MEG with BCI currently requires huge and expensive equipment housed in a safe environment. Future research and development of cryogen-free solutions such as the spin exchange relaxation-free (SERF) magnetometer, which is operated in a closed nonmagnetic field, can make this combination of modalities practical for widespread clinical use [37]. Reconstruction of models of EEG, MEG, and fMRI signal generation based on computational neuroscience is a possible application for patients in neurorehabilitation. These models may help achieve a deeper understanding of complex brain processes associated with particular functions. Further understanding of individually shaped features of brain processes and signals may be possible [37].

However, a patient with sensory and motor cortex impairment due to cerebral palsy might not be able to use BCIs with EEG or single-neuron activity from this cortical area. The application of ERP P300 component with the BCI system in other brain areas of interest might be excellent alternatives [35].

13.6 Clinical Use of EEG/EEG Monitoring, ERP, and EEG-fMRI in Epilepsy

The sensitivity of EEG is valuable for detecting spikes in epilepsy. EEG is an important tool for accurately identifying the ictal onset zone that generates interictal sharp waves or spikes [38]. By using scalp EEG tracings in a 10-20 system, clinicians/neurologists can recognize the type of epilepsy with visual analysis, which is important in an emergency. Additional T1/T2 electrodes in a 10-20 system are helpful for recognizing the ictal area accurately to avoid misdiagnosis or mistreatment. Therefore, EEG has diagnostic value for the clinicians/neurologists if additional T1 and T2 electrodes are used in a 10-20 system. As EEG is sensitive to activities from both the fissure and the gyrus, it may reveal multiple neuronal generators of electrical activity [39]. Some data appear to provide curious clinical evidence that spikes tangential to the scalp might be less apparent or undetected by scalp EEG [40]. As there have been both failures [41] and successes [42] in recording insular spikes, it is highly recommended to use simultaneous EEG-MEG to detect spikes in deeply set regions. The simultaneous EEG-MEG technique can detect insular spikes as MEG is exclusively sensitive to activities from the fissure cortex [39]. It is important to compare the sensitivity between MEG and intracranial recording using electrocorticography (ECoG). It was reported in Ref. [43] that all MEG spikes were observed with ECoG, but 56% of interictal ECoG spikes had a MEG counterpart [43]. They found that spike detectability varied according to the anatomical region; detectability was greatest in the orbitofrontal and interhemispheric regions (>90%), followed by the superior frontal, central, and lateral temporal regions (<75%), with mesial temporal spikes being the most difficult to detect (<25%). In Ref. [44], spike detectability was compared between stereostetic-EEG (SEEG) and MEG in one patient with occipital epilepsy and three patients with mesial temporal epilepsy. They reported that 95% of all interictal SEEG spikes had a MEG counterpart in the occipital case, while only 25 to 60% of all SEEG spikes had a MEG correlate in the mesial temporal cases [44].

Conventional EEG analysis is not adequate to recognize the irritative region and the ictal onset zone in the case of epilepsy surgery evaluation. Therefore, EEG source imaging (ESI) has been developed with computational techniques that can depict the presumed source of EEG activity. Clinical use of ESI in the past two decades has been highly successful. The ESI process will enhance the utility of ESI in the presurgical investigation of patients with refractory focal epilepsy [38]. At the time of localization of the ictal zone by ESI, it is necessary to define seizures accurately. Video-EEG monitoring is a good biomarker for the clinician in such cases. On the other hand, for the proper management of seizures in infancy, especially in some clinically generalized seizures, video-EEG monitoring is still popular among clinicians [45], and its utility and safety cannot be ignored in pediatric patients [46].

Over the last decade, epilepsy surgery has improved, though non-seizure-free outcomes in patients do occur. The major causes of non-seizure-free outcomes have been identified, and incomplete or nonresection of unrecognized seizure foci usually results in unfavorable seizure control [47]. Accurate localization of the epileptic foci is an important factor for good surgical outcomes, but it remains a challenge, especially for extra temporal lobe epilepsy (ETLE) and nonlesional epilepsy, which employ normal magnetic resonance imaging (MRI). EEG-fMRI makes it possible to discover the possible generators of interictal epileptiform discharges (IEDs) [4] and identify the accurate ictal location for diagnosis and further prognosis of epileptic patients. There is clinical value in detecting the irritating zone (IZ) with the EEG-fMRI technique in the presurgical evaluation of epilepsy, and EEG-fMRI often identifies scattered interictal spiking in epileptic patients who have frequent IEDs [48–50]. Zijlmans et al. [51] have convincingly demonstrated that EEG-fMRI could improve source localization and corroborate a negative conclusion concerning surgical candidacy in complex clinical cases (with either unclear focus or multifocality), and EEG-fMRI is a valuable tool in presurgical evaluation [51].

Though ERP is not frequently used in epilepsy patients for diagnosis, it can be helpful for cognitive function assessment as epilepsy can decrease cognitive function [52].

Therefore, scalp EEG of a 10-20 system with additional T1 and T2 electrodes, video-EEG monitoring, ERP, MEG, and EEG-fMRI are the most valuable biomarkers for epilepsy patients.

13.7 Clinical Use of EEG/qEEG, ERP, and EEG-fMRI in Traumatic Brain Injury (TBI)

While EEG is more suitable and much more cost effective than the gold standard imaging modalities, EEG data play a vital role in studying the severity of injury of TBI patients, level of awareness, and unconsciousness, and to predict patients' outcome. Differentiating cognitive deficits due to TBI is more challenging than developing EEG methods. Quantitative EEG (qEEG) is more useful for that purpose because it employs quantitative techniques to analyze EEG characteristics over time independently or in combination [53]. For diagnosis and for assessing the prognosis of mild TBI (mTBI), including the likelihood of its progressing to the post-concussive syndrome (PCS) phase, qEEG is a unique method [54]. Similar to qEEG, ERP is also important to assess the prognosis of mTBI. It is true that TBI is a significant risk factor for post-injury neuropsychiatric disorders. ERP component properties (e.g., timing, amplitude, latencies, and scalp distribution) are good indicators for detecting the severity of injury. Moreover, the complex relationships between brain injury and psychiatric disorders are receiving increased research attention, and ERP technologies are aiding this effort [55]. Diagnosis and prognosis in moderate and severe TBI using conventional imaging techniques is informative [56], but these techniques fail to detect the majority of mild brain injuries and provide little or no pathophysiological information related to the injury mechanism [57]. The prognostic importance of ERP is high as revealed previously that high education group had higher cognitive functions (examined by the P300 component) in compare to low education group [58, Figure 13.6], Therefore, ERP should be an important biomarker for prognosis of TBI for the clinicians.

EEG-fMRI coregistration and high-density EEG (hdEEG) can be combined to map noninvasively abnormal brain activation elicited by epileptic processes. Recently, these two different methods have become very popular to focus on the same epileptic area to map abnormal brain activation in patients with post-TBI [59]. For effective diagnosis of TBI, Tsallis entropy-based analysis of scalp EEG data has an important role as a noninvasive method. For indexing cognitive deficits due to neurological disorders, this method plays the role of a biomarker in other clinical applications [60].

13.8 Future Clinical Application of EEG/ERP in Stroke

A stroke is a cerebrovascular disease occurring when a focal deficit of brain function occurs owing to the interruption of blood supply to the brain tissues. Strokes can be categorized into two main forms: one is ischemic, which is related to the obstruction of a blood vessel supplying the brain,

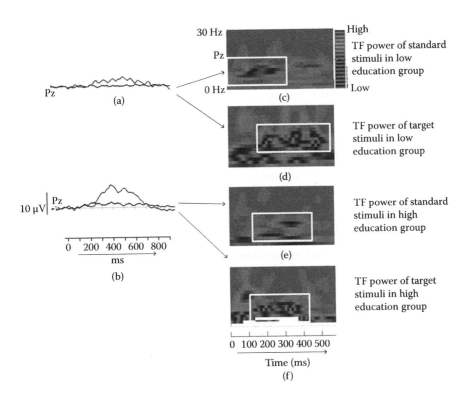

FIGURE 13.6
Grand average ERP waveforms comparing the responses of the standard (lower trace, (a) and (b)) and target (upper trace, (a) and (b)) stimuli in low (a) and high (b) educational groups at Pz location (left side). At the same time, time-frequency (TF) representation was exposed. Rectangle box indicates power of delta frequency at 300 ms after stimuli. Time is presented on the x-axis, stimulus onset being indicated by 0 ms. Frequencies between 1 and 30 Hz are represented on the y-axis. (From Begum, T. et al. Delta signal in high educational level in auditory oddball paradigm—A wavelet study. *Proceeding of the 4th International Congress on Image and Signal Processing*, 2756–2759, 2011.)

and the other form is hemorrhagic, which is connected with bleeding into or around the brain tissue. Hemiplegia is a common disability, a complete paralysis on one side of the body due to a stroke. A one-sided weakness or hemiparesis, which is not as severe as hemiplegia, may occur due to a stroke. There are several stroke risk factors that are not fixed, such as high blood pressure, heart disease, diabetes mellitus, hyperlipidemia, obesity, smoking, excessive alcohol consumption, etc.

In order to identify the underlying vascular disease and the nature of the lesion, an electrocardiogram (ECG) and brain imaging are used. In order to identify the common vascular risk factors and biomarkers, simple blood tests are done initially after a stroke. Genetic testing or counseling can also identify the stroke risk factors. However, the use of EEG, which monitors brain electrical activity through the skull, for stroke risk factor analysis is limited. Future clinical applications of EEG for predicting stroke risk factors on the basis of current research will be discussed here. One of the stroke risk factors, alcoholism, has been linked with a candidate gene SGIP1, which has an association with the EEG theta band power [61] as an end phenotype of alcoholism in the Plains Indians.

Although this study was not replicated using a genome-wide association study in a large but different group of European and/or African ancestry [62]. Another study showed the association between genetic variants in SGIP1 and obesity-related traits in humans [63]. On the other hand Ref. [64], has documented that abnormal cortical neural synchronization and desynchronization of alpha rhythms are related to healthy obese subjects [65]. We may expect that the qualitative and quantitative EEG findings together with the genome-wide association study in stroke patients may provide some valuable information in future.

EEG is hardly utilized in the diagnosis of stroke, but the prognostic value of EEG remains in an early stage. However, ambulatory EEG (AEEG) is valuable for early diagnosis of progressive stroke. The detective rate of progressive stroke with AEEG is comparatively high and is especially higher than CT in the early period of a stroke [66]. qEEG parameters such as pairwise derived brain symmetry index, (delta plus theta) or (alpha plus beta) ratio [67], and EEG phase synchronization [68] were associated with the functional outcome score of stroke patients, which is pertinent to proficient management. In order to predict cognitive and motor performance in patients with ischemic stroke, one study assesses resting-state functional connectivity (FC) with EEG and relates the observed topography of FC to cognitive and motor deficits in stroke patients [69].

ERP is not utilized in the diagnosis of stroke, but some comorbidities such as cognitive impairment and depression are common in stroke survivors. One study suggests that abnormal serum apolipoprotein E (ApoE), which is associated with some diseases with cognitive function defect, can be a useful biomarker together with abnormalities in some ERP components to assess post-stroke depression risk and clinical diagnosis [70].

The contribution of EEG and ERP in stroke rehabilitation is becoming significant. Translating thoughts into functional movements, which is the result of the integration of neural and muscular components of motor control through BCI, would play a potential role to allow people with brain and spinal cord injuries to live more independent lives in future. In this brain–machine or brain–computer interface, researchers need to record the brain electrical activities related to specific tasks from scalp or implanted sensors near the motor cortex of subjects paralyzed from a stroke, as an example. The neuronal firing patterns are recorded by the computer during imagery movements of the patients and later translated to the movement of a device, cursor, or prosthetic limb when the patients think of moving their hands to grasp some objects.

Clinical application of this type of EEG or ECoG based brain–machine interface for the neuro-rehabilitation in disabilities from neurological disorders, most commonly in strokes, has great value in future studies. An electrophysiologic brain–computer interface system can be based on the following three systems: scalp recorded EEG, ECoG, and intracortical recording [71]. The most studied signal is the scalp EEG, which is noninvasive and easy to record, but the spatial resolution is poor. Sensory motor rhythms and evoked-potential-based P300 speller have been used. ECoG activity is recorded from the cortical surface, which is less invasive and uses higher-frequency brain activity. Implanted intracortical microelectrodes, which are invasive in nature, are used to control neuroprosthetic limb in patients with tetraplegia [72]. One interesting future direction of BCI technology is mentioned in C.L. Eric's [73] editorial about the sites of ECoG recording, ipsilateral cortex which is also encode complex motor kinematics may contribute to control neuroprosthetic through transcallosal inhibition in stroke-induced hemiparesis. Besides BCI guided by the brain's electrical and magnetic field activity, there is limited use of fMRI and near-infrared imaging techniques for controlling BCI systems. However, simultaneous fMRI-EEG recording techniques to control BCI systems are not used, which may be a potential procedure to guide prosthetic limbs in the future.

13.9 Neurofeedback as a Future Aspect of fMRI-EEG

Neurofeedback (NFB) is a BCI technique that allows subjects to learn how to volitionally influence the neuronal activation in the brain. The principle behind NFB training, in general, is that brain activity is self-modifiable through operant conditioning where the subjects are provided with feedback about ongoing neuronal activation with the goal of regulating it, and a "reward" is given should a certain level of activity be achieved. NFB can be implemented using various neuroimaging modalities such as EEG, fMRI, and functional near infrared spectroscopy (fNIRS).

This volitional control of defined aspects of the central nervous system was successfully implemented at first using EEG, whereby healthy subjects learn to control their electric brain activity. The temporal resolution of EEG (in milliseconds) allows for monitoring of the brain electrical activity in real time.

Various EEG-NFB training protocols (as reviewed in Ref. [74]) that have been used to target various clinical conditions: beta/sensorimotor rhythm (SMR) in epileptic disorders [75], alpha-theta EEG in substance use disorder [76], slow cortical potential (SCP) and theta/beta in attention deficit hyperactivity disorder [77], and alpha asymmetry in depressive disorders [78]. EEG-NFB has a long history and has been effectively established as an alternative therapy to treat certain clinical conditions. However, EEG-NFB has the drawback of not being able to access the deeper subcortical regions such as the basal ganglia and the amygdala, which are often implicated in neuropsychiatric diseases.

This is where fMRI-NFB shows potential. In comparison to EEG, fMRI-NFB can offer a higher spatial resolution; therefore, subcortical brain areas could be targeted. fMRI measures the blood oxygenation level dependent (BOLD) response, that is, signal differences due to local changes in the concentration of deoxygenated hemoglobin in the brain tissue that depends on neuronal metabolism and activity. Advances in real-time fMRI, which was first envisioned two decades ago [79], have now led to the development of fMRI-NFB, which relies on providing feedback of the brain activity in real time. In fMRI-NFB, subjects try to achieve voluntary control of BOLD activity from a defined brain region, from a network of interest. Although starting to gain a foothold as an alternative treatment, fMRI-NFB is still mainly a research tool, whereas EEG-NFB makes up the majority of NFB therapy in the clinics due to EEG's lower cost and better availability [80].

EEG-NFB and fMRI-NFB can be viewed as complementary to each other. In fact, these two techniques have been combined in a recent study as a proof of concept of real-time EEG-fMRI NFB in which the subjects wear an EEG cap inside the MRI scanner, and EEG recordings are performed in parallel with fMRI data acquisition [81]. The integration of simultaneous EEG-fMRI NFB provides the best of both worlds by measuring the electric and metabolic neuronal activity concurrently, yielding information about when and where neural activity occurs. Although this novel system is still in its infancy, it is certainly promising in the field of NFB.

References

1. Betti, V., Della Penna, S., de Pasquale, F., Mantini, D., Marzetti, L., Romani, G. L., et al., Natural scenes viewing alters the dynamics of functional connectivity in the human brain, *Neuron*, 79, 782–797, 2013.
2. Jorge, J., van der Zwaag, W., and Figueiredo, P., EEG-fMRI integration for the study of human brain function, *Neuroimage*, 2013.
3. Neuner, I., Arrubla, J., Felder, J., and Shah, N. J., Simultaneous EEG-fMRI acquisition at low, high and ultra-high magnetic fields up to 9.4T: Perspectives and challenges, *Neuroimage*, June 2013.
4. Rosenkranz, K. and Lemieux, L., Present and future of simultaneous EEG-fMRI, *Magma*, 23(5–6), 309–316, 2010.
5. Vulliemoz, S., Carmichael, D. W., Rosenkranz, K., Diehl, B., Rodionov, R., Walker, M. C., et al., Simultaneous intracranial EEG and fMRI of interictal epileptic discharges in humans, *Neuroimage*, 54(1), 182–190, 2011.
6. Claus, S., Velis, D., Lopes da Silva, F. H., Viergever, M. A., and Kalitzin, S., High frequency spectral components after secobarbital: The contribution of muscular origin—A study with MEG/EEG, *Epilepsy Research*, 100(1–2), 132–141, 2012.
7. Lee, A. K., Larson, E., and Maddox, R. K., Mapping cortical dynamics using simultaneous MEG/EEG and anatomically-constrained minimum-norm estimates: An auditory attention example, *Journal of Visualized Experiments*, (68), e4262, 2012.

8. Stefan, H., Paulini-Ruf, A., Hopfengartner, R., and Rampp, S., Network characteristics of idiopathic generalized epilepsies in combined MEG/EEG, *Epilepsy Research*, 85(2–3), 187–198, 2009.

9. Hamandi, K., Salek-Haddadi, A., Laufs, H., Liston, A., Friston, K., Fish, D. R., et al., EEG-fMRI of idiopathic and secondarily generalized epilepsies, *Neuroimage*, 31(4), 1700–1710, 2006.

10. Tyvaert, L., Chassagnon, S., Sadikot, A., LeVan, P., Dubeau, F., and Gotman, J., Thalamic nuclei activity in idiopathic generalized epilepsy: An EEG-fMRI study, *Neurology*, 73(23), 2018–2022, 2009.

11. Bruce, J. F., *Fisch and Spehlmann's EEG Primer: Basic Principles of Digital and Analog EEG*, 3rd ed., Amsterdam: Elsevier, 1999.

12. Bernard, S. C., Donald, L. S., and Ernst, N., Normal EEG and sleep: Adults and elderly, In *Niedermeyer's Electroencephalography: Basic Principles, Clinical Applications and Related Fields*, L. S. Donald and H. L. Fernando, Eds., 6th ed., Philadelphia, PA: Lippincott Williams & Wilkins, 183–214, 2011.

13. Dias, N. S., Carmo, J. P., Mendes, P. M., and Correia, J. H., Wireless instrumentation system based on dry electrodes for acquiring EEG signals, *Medical Engineering and Physics*, 34(7), 972–981, 2012.

14. Filipe, S., Charvet, G., Foerster, M., Porcherot, J., Beche, J. F., Bonnet, S., et al., A wireless multichannel EEG recording platform, *Conference Proceedings of the IEEE Engineering in Medicine and Biology Society*, 2011, 6319–6322, 2011.

15. Hou, K. C., Chang, C. W., Chiou, J. C., Huang, Y. H., and Shaw, F. Z., Wireless and batteryless biomedical microsystem for neural recording and epilepsy suppression based on brain focal cooling, *IET Nanobiotechnology*, 5(4), 143–147, 2011.

16. Mathieson, K., Kachiguine, S., Adams, C., Cunningham, W., Gunning, D., O'Shea, V., et al., Large-area microelectrode arrays for recording of neural signals, *IEEE Transactions on Nuclear Science*, 51(5), 2027–2031, 2004.

17. Hasan, K. M., Ali, H., and Shad, M. U., Atlas-based and DTI-guided quantification of human brain cerebral blood flow: Feasibility, quality assurance, spatial heterogeneity and age effects, *Magnetic Resonance Imaging*, 31, 1445–1452, 2013.

18. Oishi, K., Faria, A. V., Yoshida, S., Chang, L., and Mori, S., Quantitative evaluation of brain development using anatomical MRI and diffusion tensor imaging, *International Journal of Developmental Neuroscience*, 31(7), 512–524, 2013.

19. Jarosiewicz, B., Masse, N. Y., Bacher, D., Cash, S. S., Eskandar, E., Friehs, G., et al., Advantages of closed-loop calibration in intracortical brain-computer interfaces for people with tetraplegia, *Journal of Neural Engineering*, 10(4), 046012, 2013.

20. Rodriguez-Bermudez, G., Garcia-Laencina, P. J., and Roca-Dorda, J., Efficient automatic selection and combination of EEG features in least squares classifiers for motor imagery brain-computer interfaces, *International Journal of Neural Systems*, 23(4), 1350015, 2013.

21. Wolpaw, J. R., Brain-computer interfaces as new brain output pathways, *Journal of Physiology*, 579(Pt 3), 613–619, 2007.

22. Lin, C. T., Chen, Y. C., Huang, T. Y., Chiu, T. T., Ko, L. W., Liang, S. F., et al., Development of wireless brain computer interface with embedded multitask scheduling and its application on real-time driver's drowsiness detection and warning, *IEEE Transactions on Biomedical Engineering*, 55(5), 1582–1591, 2008.

23. Freye, E. and Levy, J. V., Cerebral monitoring in the operating room and the intensive care unit: An introductory for the clinician and a guide for the novice wanting to open a window to the brain. Part I: The electroencephalogram, *Journal of Clinical Monitoring and Computing*, 19(1–2), 1–76, 2005.

24. Friedman, D., Claassen, J., and Hirsch, L. J., Continuous electroencephalogram monitoring in the intensive care unit, *Anesthesia and Analgesia*, 109(2), 506–523, 2009.

25. Stacey, W. C. and Litt, B., Technology insight: Neuroengineering and epilepsy-designing devices for seizure control, *Nature Clinical Practice Neurology*, 4(4), 190–201, 2008.

26. Angelakis, E., Stathopoulou, S., Frymiare, J. L., Green, D. L., Lubar, J. F., and Kounios, J., EEG neurofeedback: A brief overview and an example of peak alpha frequency training for cognitive enhancement in the elderly, *Clinical Neuropsychology*, 21(1), 110–129, 2007.

27. Wang, J. R. and Hsieh, S., Neurofeedback training improves attention and working memory performance, *Clinical Neurophysiology*, 124, 2406–2420, 2013.

28. Hillard, B., El-Baz, A. S., Sears, L., Tasman, A., and Sokhadze, E. M., Neurofeedback training aimed to improve focused attention and alertness in children with ADHD: A study of relative power of EEG rhythms using custom-made software application, *Clinical EEG and Neuroscience*, 44(3), 193–202, 2013.

29. Dehghani-Arani, F., Rostami, R., and Nadali, H., Neurofeedback training for opiate addiction: Improvement of mental health and craving, *Applied Psychophysiology and Biofeedback*, 38(2), 133–141, 2013.

30. Nagai, Y., Biofeedback and epilepsy, *Current Neurology and Neuroscience Reports*, 11(4), 443–450, 2011.

31. Vulliemoz, S., Rodionov, R., Carmichael, D. W., Thornton, R., Guye, M., Lhatoo, S. D., et al., Continuous EEG source imaging enhances analysis of EEG-fMRI in focal epilepsy, *Neuroimage*, 49(4), 3219–3229, 2010.

32. Carney, P. W., Masterton, R. A., Flanagan, D., Berkovic, S. F., and Jackson, G. D., The frontal lobe in absence epilepsy: EEG-fMRI findings, *Neurology*, 78(15), 1157–1165, 2012.

33. Daly, J. J., Brain–Computer Interface Applied to Motor Recovery after Brain Injury. *Introduction to Neural Engineering for Motor Rehabilitation*. John Wiley & Sons, Inc., 463–476, 2013.

34. Janis, J. D. and Jonathan, R. W., Brain computer interfaces in neurological rehabilitation, *Lancet Neurology*, 7, 1032–1043, 2008.

35. Mak, J. N. and Wolpow, J. R., Clinical applications of brain computer interfaces: Current state and future prospects, *IEEE Reviews in Biomedical Engineering*, 2, 187–199, 2009.

36. Buch, E., Weber, C., Cohen, L. G., Braun, C., Dimyan, M. A., Ard, T., et al., Think to move: A neuromagnetic brain-computer interface (BCI) system for chronic stroke, *Stroke*, 39(3), 910–917, 2008.

37. Mikolajewska, E. and Mikolajewski, D., Magnetoencephalography in brain-computer interfaces—Current and future solutions, *Journal of Health Sciences*, 3(7), 15–20, 2013.

38. Kaiboriboon, K., Luders, H. O., Hamaneh, M., Turnbull, J., and Lhatoo, S. D., EEG source imaging in epilepsy—Practicalities and pitfalls, *Nature Reviews Neurology*, 8(9), 498–507, 2012.

39. Ebersole, J. S. and Ebersole, S. M., Combining MEG and EEG source modeling in epilepsy evaluations, *Journal of Clinical Neurophysiology*, 27, 360–371, 2010.
40. Kakisaka, Y., Iwasaki, M., Alexopoulos, A. V., Enatsu, R., Jin, K., Wang, Z. I., et al., Magnetoencephalography in fronto-parietal opercular epilepsy, *Epilepsy Research*, 102, 71–77, 2012.
41. Goldenholz, D. M., Ahlfors, S. P., Hämäläinen, M. S., Sharon, D., Ishitobi, M., Vaina, L. M., et al., Mapping the signal-to-noise-ratios of cortical sources in magnetoencephalography and electroencephalography, *Human Brain Mapping*, 30, 1077–1086, 2009.
42. Park, H. M., Nakasato, N., and Tominaga, T., Localization of abnormal discharges causing insular epilepsy by magnetoencephalography, *Tohoku Journal of Experimental Medicine*, 226, 207–211, 2012.
43. Agirre-Arrizubieta, Z., Huiskamp, G. J., Ferrier, C. H., van Huffelen, A. C., and Leijten, F. S., Interictal magnetoencephalography and the irritative zone in the electrocorticogram, *Brain*, 132, 3060–3071, 2009.
44. Santiuste, M., Nowak, R., Russi, A., Tarancon, T., Oliver, B., Ayats, E., et al., Simultaneous magnetoencephalography and intracranial EEG registration: Technical and clinical aspects, *Journal of Clinical Neurophysiology*, 25, 331–339, 2008.
45. Yu, H. J., Lee, C. G., Nam, S. H., Lee, J., and Lee, M., Clinical and ictal characteristics of infantile seizures: EEG correlation via long-term video EEG monitoring, *Brain and Development*, 35(8), 771–777, 2013.
46. Arrington, D. K., Ng, Y. T., Troester, M. M., Kerrigan, J. F., and Chapman, K. E., Utility and safety of prolonged video-EEG monitoring in a tertiary pediatric epilepsy monitoring unit, *Epilepsy & Behavior*, 27(2), 346–350, 2013.
47. Harroud, A., Bouthillier, A., Weil, A. G., and Nguyen, D. K., Temporal lobe epilepsy surgery failures: A review, *Epilepsy Research and Treatment*, 2012, 201651, 2012.
48. Gotman, J., Epileptic networks studied with EEG-fMRI, *Epilepsia*, 49(Suppl. 3), S42–S51, 2008.
49. Krakow, K., Imaging epileptic activity using functional MRI, *Neurodegenerative Diseases*, 5(5), 286–295, 2008.
50. Flanagan, D., Abbott, D. F., and Jackson, G. D., How wrong can we be? The effect of inaccurate mark-up of EEG/fMRI studies in epilepsy, *Clinical Neurophysiology*, 120(9), 1637–1647, 2009.
51. Zijlmans, M., Huiskamp, G., Hersevoort, M., Seppenwoolde, J. H., van Huffelen, A. C., and Leijten, F. S., EEG-fMRI in the preoperative work-up for epilepsy surgery, *Brain*, 130(Pt 9), 2343–2353, 2007.
52. Helmstaedter, C., The impact of epilepsy on cognitive function, *Journal of Neurology, Neurosurgery and Psychiatry*, 84(9), e1, 2013.
53. O'Neil, B., Prichep, L. S., Naunheim, R., and Chabot, R., Quantitative brain electrical activity in the initial screening of mild traumatic brain injuries, *Western Journal of Emergency Medicine*, 13(5), 394–400, 2012.
54. Haneef, Z., Levin, H. S., Frost, J. D., Jr., and Mizrahi, E. M., Electroencephalography and quantitative electroencephalography in mild traumatic brain injury, *Journal of Neurotrauma*, 30(8), 653–656, 2013.
55. Rapp, P. E., Rosenberg, B. M., Keyser, D. O., Nathan, D., Toruno, K. M., Cellucci, C. J., et al., Patient characterization protocols for psychophysiological studies of traumatic brain injury and post-TBI psychiatric disorders, *Frontiers in Neurology*, 4, 91, 2013.

56. Shenton, M. E., Hamoda, H. M., Schneiderman, J. S., Bouix, S., Pasternak, O., Rathi, Y., et al., A review of magnetic resonance imaging and diffusion tensor imaging findings in mild traumatic brain injury, *Brain Imaging and Behavior*, 6, 137–192, 2012.

57. Bettermann, K. and Slocomb, J. E., Clinical relevance of biomarkers for traumatic brain injury, In *Biomarkers for Traumatic Brain Injury*, S. Dambinova, R. L. Hayes, and K. K. W. Wang, Eds., Cambridge: Royal Society of Chemistry, pp. 1–18, 2012.

58. Begum, T., Reza, F., Ahmed, A. L., Elaina, S., and Abdullah, J. M., Delta signal in high educational level in auditory oddball paradigm—A wavelet study, *Proceeding of the 4th International Congress on Image and Signal Processing*, pp. 2756–2759, October 15–17, 2011.

59. Storti, S. F., Formaggio, E., Franchini, E., Bongiovanni, L. G., Cerini, R., Fiaschi, A., et al., A multimodal imaging approach to the evaluation of post-traumatic epilepsy, *Magma*, 25(5), 345–360, 2012.

60. McBride, J., Zhao, X., Nichols, T., Vagnini, V., Munro, N., Berry, D., et al., Scalp EEG-based discrimination of cognitive deficits after traumatic brain injury using event-related Tsallis entropy analysis, *IEEE Transactions on Biomedical Engineering*, 60(1), 90–96, 2013.

61. Hodgkinson, C. A., Enoch, M. A., Srivastava, V., Cummins-Oman, J. S., Ferrier, C., Iarikova, P., et al., Genome-wide association identifies candidate genes that influence the human electroencephalogram, *Proceedings of the National Academy of Sciences of the United States of America*, 107(19), 8695–8700, 2010.

62. Derringer, J., Krueger, R. F., Manz, N., Porjesz, B., Almasy, L., Bookman, E., et al., Nonreplication of an association of SGIP1 SNPs with alcohol dependence and resting theta EEG power, *Psychiatric Genetics*, 21(5), 265–266, 2011.

63. Cummings, N., Shields, K. A., Curran, J. E., Bozaoglu, K., Trevaskis, J., Gluschenko, K., et al., Genetic variation in SH3-domain GRB2-like (endophilin)-interacting protein 1 has a major impact on fat mass, *International Journal of Obesity (London)*, 36(2), 201–206, 2012.

64. Babiloni, C., Marzano, N., Lizio, R., Valenzano, A., Triggiani, A. I., Petito, A., et al., Resting state cortical electroencephalographic rhythms in subjects with normal and abnormal body weight, *NeuroImage*, 58(2), 698–707, 2011.

65. Del Percio, C., Triggiani, A. I., Marzano, N., Valenzano, A., De Rosas, M., Petito, A., et al., Poor desynchronisation of resting-state eyes-open cortical alpha rhythms in obese subjects without eating disorders, *Clinical Neurophysiology*, 124(6), 1095–1105, 2013.

66. Liqin, M. A., Xueqing, Z., Deshu, W. E. N., and Xiaoyan, L. A. N., Analysis of clinical application of ambulatory electroencephalogram on early diagnosis of progressive stroke, *Chinese Journal of Contemporary Neurology and Neurosurgery*, 11(4), 4, 2011.

67. Sheorajpanday, R. V. A., Nagels, G., Weeren, A. J. T. M., van Putten, M. J. A. M., and De Deyn, P. P., Quantitative EEG in ischemic stroke: Correlation with functional status after 6 months, *Clinical Neurophysiology*, 122(5), 874–883, 2011.

68. Wu, W., Sun, J., Jin, Z., Guo, X., Qiu, Y., Zhu, Y., et al., Impaired neuronal synchrony after focal ischemic stroke in elderly patients, *Clinical Neurophysiology*, 122(1), 21–26, 2011.

69. Dubovik, S., Ptak, R., Aboulafia, T., Magnin, C., Gillabert, N., Allet, L., et al., EEG alpha band synchrony predicts cognitive and motor performance in patients with ischemic stroke, *Behavioural Neurology*, 26(3), 187–189, 2013.

70. Zhang, Z., Mu, J., Li, J., Li, W., and Song, J., Aberrant apolipoprotein E expression and cognitive dysfunction in patients with poststroke depression, *Genetic Testing and Molecular Biomarkers*, 17(1), 47–51, 2013.
71. Shih, J. J., Krusienski, D. J., and Wolpaw, J. R., Brain-computer interfaces in medicine, *Mayo Clinic Proceedings*, 87(3), 268–279, 2012.
72. Collinger, J. L., Wodlinger, B., Downey, J. E., Wang, W., Tyler-Kabara, E. C., Weber, D. J., et al., High-performance neuroprosthetic control by an individual with tetraplegia, *The Lancet*, 381(9866), 557–564, 2013.
73. Leuthardt, E. C., Engineering new treatments for stroke with brain–computer interfaces, *Future Neurology*, 4(2), 133–136, 2009.
74. Niv, S., Clinical efficacy and potential mechanisms of neurofeedback, *Personality and Individual Differences*, 54(6), 676–686, 2013.
75. Sterman, M. B. and Egner, T., Foundation and practice of neurofeedback for the treatment of epilepsy, *Applied Psychophysiology and Biofeedback*, 31(1), 21–35, 2006.
76. Sokhadze, T. M., Cannon, R. L., and Trudeau, D. L., EEG biofeedback as a treatment for substance use disorders: Review, rating of efficacy, and recommendations for further research, *Applied Psychophysiology and Biofeedback*, 33(1), 1–28, 2008.
77. Gevensleben, H., Holl, B., Albrecht, B., Schlamp, D., Kratz, O., Studer, P., et al., Distinct EEG effects related to neurofeedback training in children with ADHD: A randomized controlled trial, *International Journal of Psychophysiology: Official Journal of the International Organization of Psychophysiology*, 74(2), 149–157, 2009.
78. Choi, S. W., Chi, S. E., Chung, S. Y., Kim, J. W., Ahn, C. Y., and Kim, H. T., Is alpha wave neurofeedback effective with randomized clinical trials in depression? A pilot study, *Neuropsychobiology*, 63(1), 43–51, 2011.
79. Cox, R. W., Jesmanowicz, A., and Hyde, J. S., Real-time functional magnetic resonance imaging, *Magnetic Resonance in Medicine: Official Journal of the Society of Magnetic Resonance in Medicine/Society of Magnetic Resonance in Medicine*, 33(2), 230–236, 1995.
80. Dewiputri, W. I. and Auer, T., Functional magnetic resonance imaging (fMRI)-based neurofeedback: Implementations and applications, *Malaysian Journal of Medical Sciences*, 20(5), 5–15, 2013.
81. Zotev, V., Phillips, R., Yuan, H., Misaki, M., and Bodurka, J., Self-regulation of human brain activity using simultaneous real-time fMRI and EEG neurofeedback, *Neuroimage*, 85, 985–995, 2014.

Index

A

Absence seizures, *see* Petit mal seizures
AC artifacts, 14
Action potential, 5–6
Adaptive mean amplitude, 76
Affective computing, 162–163
Akaike information criteria (AIC), 45, 82
Alpha asymmetry, 164, 167, 180
Alpha asymmetry training, 256
Alpha band, 37
Alpha rhythm, 151
Alpha synchrony training,
    multichannel, 243
Alpha/theta training, 242–243, 255–256
Alpha training, 241, 255
Alpha waves, 7, 8
Alzheimer's disease, 17
Ambulatory EEG (AEEG), 302
Amygdala and memory, 226–227
Analysis equation, 28
Analyzing waveforms, 51
Applied Neuroscience Institute (ANI)
    database, 251
Approximate entropy, 153–154
Arousal-valence detection, 163–165
Artifacts, 11–14
    biological, 11–13
    defined, 11
    technical, 14
Artificial neural network (ANN), 180
Asymmetric activation, 164
Asymmetric approaches, 164, 198
Asymmetric studies, 202
Asymmetry, 184
    gender differences in, 151
    interhemispheric, 183
Atkinson-Shiffrin memory model,
    221–222
Atonic seizures, 125
Attention
    measurement of, P300 as a tool for,
      158–161
    and memory, 228

Attention-deficit/hyperactivity disorder
    (ADHD), 85
    ADHD feature monitoring by
      qEEG, 170
    neurofeedback training for, 168,
      260–261
    subtypes, 85
Attractors
    defined, 61
    types of, 61
Autism spectrum disorders, 281
Autocorrelation, 31, 34
Autocorrelation ergodic processes, 33
Autocorrelation function of a random
    process, 31, 34
    defined, 30
Autocorrelation sequence (ACS), 32
Autocovariance function of a stochastic
    process, defined, 30
Autoregressive (AR), 43–44
    time and frequency domains
      representation of EEG signals
      using, 44–45
Autoregressive (AR) model, selecting
    the order of the, 45–46
Autoregressive modeling, full-brain, 204
Autoregressive (AR) parameters,
    MATLAB functions for, 49–50
Autoregressive (AR) process, and linear
    prediction, 46–48
    modified covariance method, 48–49
Average, moving, 43, 44
Average between subjects (ERP
    extraction), 73, 76–77
Average reference montage, 10
Average within subject (ERP extraction),
    73–77
Averaged file, 73

B

Backward linear prediction (BLP), 46
Backward linear prediction error
    (BLPE), 46

Beta/alpha ratio, 167
Beta band, 37; *see also* Theta/beta
training
Beta waves, 7, 8
Bioelectric signals, characteristic
properties of, 14
Bipolar montage, 10
Bispectrum, 56–59
Bivariate analysis of EEG signals, 65–68
Brain activity, monitoring abnormal, 149
monitoring mechanism, 150
Brain anatomy, 92, 244
areas associated with memory
functions, 225–227
Brain-computer interface (BCI), 303
EEG and, 293
ERPs for, 158–162
and the future of EEG-MEG, 297–298
novel BCIs, 162–163
therapeutic usage, 168
Brain dysfunctions, human, 143–144;
see also specific dysfunctions
Brain injury, 144; *see also* Traumatic
brain injury (TBI)
multimodality monitoring for, 295
Brain lobes, coherency functions
calculated between the
different, 67
Brain lobes, functional diagram of, 3
Brain mapping, 233
pre-/post-neurofeedback
training, 253
Brain Resource International Database
(BRID), 251
Brain rhythms, 6–8
Brain wave patterns, 6, 8
Brain waves and their maps, 289–292
Brain-waves-human-activities map,
291–292
BrainDX database, 250
BRID (Brain Resource International
Database), 251
Burg algorithm, 47–48

C

Causal (discrete-time) systems, 27
Cerebral cortex and memory, 225–226
Cerebral palsy (CP), 281

Chaos theory and dynamical analysis,
59–60
correlation dimension, 62–63
definitions, 61
Lyapunov exponents, 63–64
reconstruction of state space, 61–62
Cognitive load theory, 223
Coherence function, 66–67; *see also*
Normalization of coherence
Color
as biological marker for depression,
275–276
and psychology, 274–276
methodology for future predictors
of cognitive state, 276–279
Concussion, 147; *see also* Post-concussion
syndrome; Traumatic brain
injury
Conditioning, *see* Operant conditioning;
Reward and memory
Continuous-time signals, 23–24
Convolution property, 28–29
Correction dimension, 62
Correlation coefficient, 31
Correlogram for power spectrum
estimation, 42
Criminal behavior, EEG/ERP in,
279–280
Cross-correlation, 31
Cross-variance, 65
Crossover, 242
Cultural neuroscience (CN) framework,
overview of, 270
benefits and challenges, 270–271
suggestions for future work, 271–274
Current source density (CSD), 254

D

DC noise, 14
DCM-based generative models, 203–204
Default mode network (DMN), 145
Delirium, 144
Delta band, 36
Delta waves, 6, 7
Depression, 177
causes, 178
color as biological marker, 275–276
diagnosis, 178–179

differential diagnostic classification, 185–186
  EEG and, 183–186
    discriminant functions, 183–184
  subtypes, 177–178
  treatment, 179–180
    prediction of treatment response, 186
Depression detection, 184–185
Detection rates, *see* False detection rate; Good detection rate
Developmental disorders, EEG/ERP and EEG-fMRI in rehabilitation for, 281
Diagnosis; *see also specific topics*
  EEG technology and, 295
Digital EEG (DEEG), 176
Dilation parameter, 51
Discrete-time Fourier transform (DTFT), 27–28
Discrete-time sequences, typical and their representations, 24–25
Discrete-time signals, 23–25
  classification, 25
Discrete-time signals in requirements domain, representation of, 27–29
Discrete-time systems, 25–26
  classification, 26–27
Downtraining, 252
Drug abuse, 86
Dual coding theory of memory, 222
Dynamic ARX (dARX) models, 208
Dynamic causal models (DCMs), 203–204, 208, 211
  development, 203
Dynamical analysis; *see also* Chaos theory and dynamical analysis of EEG signals using entropy, 64

E

EEG (electroencephalography); *see also specific topics*
  defined, 93–94
  future aspects of clinical applications using, 293–296
  main aspects of the future of, 289–293
EEG analysis techniques, 152–154

EEG and ERP recording, techniques of, 8–10
EEG and ERP signals, 15–16
  history, 2–3
  source of neural activities, 3
    human brain, 3–4
EEG band, types of, 245–246
  behavioral correlates of various, 246
EEG data analysis, classical methods in, 37–42
EEG-fMRI; *see also* fMRI-EEG
  clinical use of
    in rehabilitation, 281, 302–303
    in traumatic brain injury, 300
EEG inverse problem, *see* Inverse problem
EEG-MEG and its clinical application, future of, 297–298
EEG modeling, parametric models in, 43–46
EEG patterns, 169
EEG signals, 36–37
  characteristics, 36–50
EEG software, commonly used
  windows in, 39–41
  properties, 40
EEG source imaging (ESI), 299
EEG technology, 292, 293
  with surgery, 293
  without surgery, 292
Electrical source and resulting electrical field in brain, EEG, 245
Electrocardiogram (ECG) artifacts, 13
Electrocorticography (ECoG), 212, 290, 298, 302–303
Electrodes, 8–9
  artifacts due to improper placement of, 14
  locations, 93–95
Electroencephalography, *see* EEG
Electromyogram (EMG) artifacts, 11–13
Electrooculographic (EOG) artifacts, 11
eLORETA (exact low resolution electromagnetic tomography), 106–107
Embedding dimension, 62
Embedding vectors, 61
Emotion, types/categories of, 181–182
Emotional features, qEEG platform for visualizing, 165–167

Encephalopathy, 144
Ensemble of a stochastic process,
      defined, 29
Entropy, 153
Epilepsy, 18, 123–124; *see also* Seizure
      detection; Seizures
   clinical use of EEG/EEG monitoring,
      ERP, and EEG-fMRI in, 298–299
Epileptic EEG data
   2D bispectrum of, 56, 57, 59
   3D bicoherence of, 60
   3D bispectrum of, 56, 58
   third-order cumulants of, 54, 55
Ergodic in the mean, 33
Ergodicity of random processes, 31–33
Error positivity (Pe), 280
Error-related negativity (ERN), 202, 280
Event-related potential (ERP); *see also*
      specific topics
   brain activity assessment using,
      83–87
Event-related potential (ERP) extraction
   averaging technique, 73–76
   single-trial subspace-based
      technique, 77–83
Exact low resolution electromagnetic
      tomography (eLORETA),
      106–107
Exercise and memory, 229–230
Expected value of a stochastic process,
      defined, 30
Eye movements and blinks, effect on
      EEG recordings, 11, 12
Eye tracking system, 276–278

F

False detection rate (FDR), 128–131
FDM (finite difference method), 117, 118
FDM-LORETA, activation map for,
      117, 118
FDM-sLORETA, 118–119
   activation map for, 119
Feedback-related negativity (FRN), 280
Filters, frequency responses of ideal
      types of, 28, 29
Finite difference method, *see* FDM
First-order distribution, 31
First-order stationary, 32

fMRI and M-EEG, *see* M-EEG and
      fMRI data fusion
fMRI-EEG; *see also* EEG-fMRI
   and its clinical application,
      future of, 296
   neurofeedback as a future aspect of,
      303–304
Focal underdetermined system solution,
      see FOCUSS
FOCUSS (focal underdetermined
      system solution), 101–102,
      108–110
Forward and inverse problem, 96–98;
      see also Inverse problem
Forward linear prediction (FLP), 46
Forward linear prediction error
      (FLPE), 46
Fourier transform
   discrete-time, 27–28
   inverse discrete-time, 27
   short-time, 50, 51
Full-brain autoregressive modeling
      (FARM), 204
Functional connectivity (FC), 302

G

Gamma band, 37
Gamma waves, 7, 8
Gaussian processes, 33–34
   defined, 33
Generalized seizures, 124–125
Generalized subspace approach (GSA),
      77, 82–83
   implementation of GSA algorithm, 83
Genetics and genetic testing, 301–302
Good detection rate (GDR), 128, 129, 131
Grand average ERP waveform, 76
Grand average file, 73
Granger-causality-based models,
      204–205

H

Heart rate (HR), *see* Stress: case study of
      mental stress
Hemodynamic response function
      (HRF), 201, 204
Hippocampus and memory, 226

Human Brain Institute (HBI)
database, 251
Hyperpolarization, 5

I

Imaging techniques; *see also*
Neuroimaging techniques
in color, 278–279
Impedance, artifacts due to high, 14
Independent component analysis (ICA),
200–201
Information entropy, 153
Inhibit/enhance protocols; *see also*
qEEG-based neurofeedback-
inhibit/enhance training
standard, 252
Interictal epileptiform discharges
(IEDs), 299
Interpolation technique, 152
Intra-slow fluctuation (ISF)/intra-low
frequency (ILF), 257
Intracranial EEG (iEEG), 133, 291–293
Inverse discrete-time Fourier transform
(IDTFT), 27
Inverse Karhunen-Loeve transform
(IKLT), 80
Inverse problem, 90, 92, 94, 96–99,
103, 116; *see also* FOCUSS;
Forward and inverse problem;
Localization using EEG
signals; *specific algorithms*
hybrid algorithm as efficient
solution of, 108
M-EEG, 202, 207, 212
techniques for solution of, 111–113,
see also Localization
techniques/algorithms
Inversion for a physical model, 97

J

Joint ICA (jICA), 200
Joint time-frequency representation of
EEG signals, 50–52

K

Karhunen-Loeve transform (KLT), 80
Kuhn-Tucker conditions, 79

L

Lag window, 42
Lagrange multiplier,
optimum value of, 83
Language impairment, specific, 281
Laplacian montage, 10
Levinson recursion, 47, 48
Linear analysis, 152
Linear (discrete-time) systems, 26
Linear time-invariant (LTI) systems
(LTIs), 26–28, 35–36
Localization techniques/algorithms
comparison of, 110–115
eLORETA, 106–107
FOCUSS, 101–102
hybrid WMN, 103–104
LORETA, 100–101
minimum norm, 99–100
MUSIC, 102
recursive sLORETA-FOCUSS,
108–109
shrinking LORETA-FOCUSS, 110
sLORETA, 104–105
WMN-LORETA, 107–108
Localization using EEG signals, brain
source, 91–96
recommendations for the future,
115–119
of seizure onset, 93–96
Long-term potentiation (LTP) and
memory retention, 227–228
LORETA (low resolution
electromagnetic tomography),
99–101, 111, 114, 117–119,
253–254
defined, 100
exact, 106–107
standardized, 104–105
Low resolution
electromagnetic tomography,
see LORETA
Lyapunov exponents, 63–64

M

M-EEG and fMRI data fusion, 195–198,
212; *see also* M-EEG/fMRI data
integration

methods for, 198–199
   correlation between BOLD and
      M-EEG signals in particular
      bands, 199–200
   DCM-based generative models,
      203–204
   extract common spatial signatures
      for M-EEG and fMRI, 200–201
   Granger-causality-based models,
      204–205
   M-EEG- and fMRI-informed
      analyses, 201–203
   neural-mass-based generative
      models, 205–206
   significance of fusion in various
      modalities, 206–207
M-EEG/fMRI data integration, practical
      and realistic, 208–212; *see also*
      M-EEG and fMRI data fusion
Magnetic resonance imaging,
      see fMRI; MRI
Magnetoencephalography (MEG), 125;
      see also EEG-MEG and its
      clinical application
Maximal Lyapunov exponent, 63
MEG (magnetoencephalography), 125;
      see also EEG-MEG and its
      clinical application
Melancholic and nonmelancholic
      depression, 177
Memory consolidation, amygdala
      activation and, 226–227
Memory experimental design issues,
      233–234
Memory functions, 219–221, 234–235
   brain areas associated with, 225–227
Memory processes, 219–221, 234–235
   traditional cognitive and memory
      theories of, 221–224
Memory retention and recall, 224–227
   factors affecting, 228–232
   long-term potentiation and memory
      retention, 227–228
Mental stress, *see* Stress
MicroCog, 258–259
Mild cognitive impairment (MCI), 144
Minimum norm solution, 49
Mismatch negativity (MMN), 85
Mnemonics and memory, 230

Monitoring, medical
   EEG technology and, 295
Montages, 10
Mother wavelet, 51
Moving average (MA), 43, 44
MRI, functional, *see* fMRI
MRI-compatible scalp EEG, 296, 297
Mu waves, 7, 8
Multilayer perceptron artificial neural
      networks (MPL-NN), 128
Multilayer perceptrons (MLPs), 181–182
Multimedia learning,
      cognitive theory of, 223–224
Multiple signal classification (MUSIC)
      algorithm, recursive, 102
Multiple signal classification (MUSIC)
      method, 128
Multiple sparse priors (MSP) algorithm,
      202–203
Multivariate measures, 137
Mutual entropy, 153
Myasthenia gravis, 17
Myoclonic seizures, 125

N

N100, applications of, 84
N200 (or N2), applications of, 84–85
Neural mass models (NMMs), 205–206
Neurocognition, 296
Neurodegenerative disorder, 144
Neurofeedback (NF/NFB), 158, 240–241,
      262–263, 296
   addressing conditions with, 260–262
   applications, 168–171
   description of neurofeedback
      process, 244–245
      operant conditioning, 247, 248
      signal processing, 245–247
   early history of the field
      Barry Sterman, 241
      Eugene Penniston and Elmer
         Green, 242–243
      Fehmi, Hardt, and Crane, 243
      Joe Kamiya, 241
      John Lubar, 243
   as a future aspect of fMRI-EEG,
      303–304
   goals, 247, 248

learning mechanisms, 247, 248
modalities, *see also* qEEG-guided
        neurofeedback
    symptom-based approaches,
            255–257
    patient assessment, 257
        computerized neuropsychological
                batteries, 258–259
        ERP assessment, 260
        neuropsychological
                assessments, 258
        qEEG integrated assessments,
                259–260
        subjective assessment
                questionnaires, 259
        symptom assessment, 258
        targeted neuropsychological
                tests, 259
Neuroguide database, 251, 253
Neuroimaging techniques, 145;
        see also Imaging techniques
    temporal and spatial resolution of,
            92, 93
Neurological disorders, application of
        EEG in, 16–18, 302–303
Neuromodulation, 295–296
Neuronal dysfunction
    clinical schemes of severity
            assessment of brain's, 145, 146
    regional deficits, 145, 147
Neuronal injury, assessment
        criteria of, 145
    clinical evaluation tests, 145, 146
Neurons, structure and function of, 4–6
Neuropsychological assessments, 258
Neuropsychological batteries,
        computerized, 258–259
NeuroSurfer, 169–170
Neurotherapy, 148
Neurotransmission in brain, 220
Nongovernmental organizations
        (NGOs), 273
Noninvasive EEG, 292
Noninvasive mapping and creation
        of ideal networks for brain,
        293–294
Nonlinear analysis, 152
Nonlinear descriptors of EEG signals,
        52–53

higher-order statistics for EEG data
        analysis, 53
    definitions and properties, 53–60
Nontraumatic brain injury (nTBI), 144
Normalization of coherence, pre-/
        post-neurofeedback training
        demonstrating, 253, 254
Normalized cross-covariance function,
        65–66
Nutrition and memory, 229–230

O

Obsessive compulsive disorder (OCD), 281
Ocular artifacts, 11
Operant conditioning; *see also* Reward
        and memory
    in neurofeedback, 247, 248
Orthogonal random processes, 31

P

P100, applications of, 84
P140, 279
P200, 279–280
P300, 280
    applications, 85–87
    as a tool for attention measurement,
            158–161
Parametric methods for EEG modeling,
        43–46
Parkinson's disease (PD), 17–18
Partial seizures, 124
    simple and complex, 125
Pathological substance use (PSU), 86
Perceptrons, multilayer, 128, 181–182
Periodic attractor, 61
Periodogram for power spectrum
        estimation, 38–39
Petit mal seizures, 124
Phase locking, 67–68
Phase synchronization, 67–68
Point attractor, 61
Post-concussion syndrome (PCS), 147–148,
        300; *see also* Traumatic brain
        injury
Post-error positivity (Pe), 280
Posttraumatic stress disorder (PTSD), 256
Power line artifacts, 14

Power spectrum, 55–56
Power spectrum density/power spectral density (PSD), 35, 42, 128
Power spectrum estimation, 41–42
correlogram for, 42
Principal components analysis (PCA), 138
Pseudouniverse, 49
Psychiatric disorders, 176–179; *see also specific disorders*
clinical treatment methods, 179–180
Psychiatry, EEG in, 180–186
Psychological disorders, EEG/ERP and EEG-fMRI in rehabilitation for, 281
Psychomotor disturbance in depression, 178
Psychopathy, EEG/ERP in, 279–280
Psychotic depression, 178

Q

qEEG (quantitative EEG), 144
in human behavior studies after neuronal injury, 150–151
therapeutic usage, 168
qEEG-based neurofeedback-inhibit/enhance training, 251–254
qEEG-guided neurofeedback, 248–251; *see also* qEEG-based neurofeedback-inhibit/enhance training
databases, 250–251
qEEG integrated assessments, 259–260
qEEG metrics
basic, 250
description of, 249–250
qEEG parameters, 149
qEEG platform for visualizing emotional features, 165–167
qEEG signals, human diversity and the nature of, 151
Quantitative EEG, *see* qEEG

R

Random process(es); *see also* Stochastic processes
conceptual representation of a, 29, 30
linear filtering of, 35–36
power spectrum, 34–35

Rapid serial visual presentation (RSVP) paradigm, 158–159
Reconstruction of state space, 61–62
Recording and artifacts, 11–14
Referential montage, 10
Reflection coefficient, 46
Rehabilitation, EEG/ERP and EEG-fMRI in, 281, 302–303
Rehearsal and memory, 228–229
Repetitive visual stimulation (RVS), 161–162
Response-locked average, 76
Retrieval practice, *see* Testing effect
Reward and memory, 232
Robotic technology, 293
Rostral cingulate zone (RCZ), 202

S

Sample function, 33
defined, 29
Scalp EEG, 292
Second-impact syndrome (SIS), 148
Second-order stationary, 32
Seizure detection, epileptic EEG and, 125
seizure detection algorithms, 126–134
EEG features for, 137
Seizure detection algorithms, 126–134
types of, 126
Seizure detection system using singular values of EEG signals, 134
database, 134–135
machine-learning-based seizure detection system using SVD, 135–136
Seizure detector, evaluation parameters for singular-values-based, 136
Seizure event detection, 126–131
Seizure onset, source localization of; *see* Localization techniques/algorithms; Localization using EEG signals, brain source
Seizure onset detection, 131–135
Seizure prevention system, closed-loop, 138

Seizures, epileptic, 40–41
    EEG showing the start of an, 125
    types of, 124–125
Sensorimotor rhythm (SMR) training,
        242, 243, 255, 260
Short-time Fourier transform (STFT),
        50, 51
Signal subspace dimension,
        estimation of, 82
Singular value decomposition (SVD), 134
    machine-learning-based seizure
            detection system using, 135–136
SKIL database, 251
Sleep and memory, 229
Sleep spindle, 151
sLORETA, 104–105, 111, 114, 118–119,
        253–254
Slow cortical potential (SCP), 256–257
SMR (sensorimotor rhythm) training,
        242, 243, 255, 260
Source estimation, process of,
        see Inverse problem
Spectral analysis, 150
Spectrum estimation, MATLAB
        functions for, 49–50
Sports concussion, 148; *see also* Post-
        concussion syndrome;
        Traumatic brain injury
Stable (discrete-time) systems, 27
Standardized low resolution
        electromagnetic tomography,
        see sLORETA
State space
    defined, 61
    reconstruction of, 61–62
Stationarity of random processes, 31–32
Stationary processes, 31–32
Steady-state visual evoked potentials
        (SSVEPs), 161–163
Stimuli-locked average, 76
Stochastic processes, 29–33
    defined, 29
    mean and autocorrelation functions,
        30–31
Strange attractor, 61
Stress, 177, 179
    case study of mental stress, 187–191
Stress characterization, 180–183
Stress detection, 167–168

Stroke, future clinical application of
        EEG/ERP in, 300–303
Substance use, pathological, 86
Sun-Stok-LORETA, activation map for, 117
Sun-Stok-sLORETA, 117–118
    activation map for, 118
Support vector machines (SVMs), 130,
        132, 135–136
Symmetric vs. asymmetric
        approaches, 198
Synapse, types of, 220
Synaptic transmission in brain, 220
Synthesis equation, 28

T

Testing effect, 230–231
Theta band, 37; *see also* Alpha/theta
        training
Theta/beta training, 243
Theta waves, 7, 8
Time and frequency domains methods,
        37–42
Time and frequency domains
        representation of EEG signals
        using AR, 44–45
Time-frequency analysis, 76
Time-frequency representation of EEG
        signals, joint, 50–52
Time-invariant (discrete-time) systems, 26
Time-series analysis, EEG, 136–137
Tonic-clonic seizures, 124
Translation parameter, 51
Traumatic brain injury (TBI), 144, 147–148,
        300; *see also* Brain injury; Post-
        concussion syndrome
    clinical use of EEG/qEEG, ERP, and
        EEG-fMRI in, 300

U

Unbiased estimator, 66
Uptraining, 252

V

Visual evoked potentials (VEPs), 11
    steady-state, 161–163
Visually evoked potential (VEP)
        technique, 2

W

Wavelet analysis, 76
Wavelet coefficients of epileptic data, 52
Wavelet transform (WT), 51–52
Weighted minimum norm, *see* WMN
Welch spectrum estimation, 41
    MATLAB functions for, 42
White Gaussian noise (WGN), 34
White noise, power spectral
        density of, 35
White noise random process, 34
    autocorrelation, 34
Wide-sense stationary (WSS), 32

Wide-sense stationary (WSS) zero-mean
        random process, 34
WinEEG system, 251
WMN (weighted minimum norm), 99
    hybrid, 103–104
WMN-LORETA, 107–108
Working memory (WM), Baddeley's
        model of, 222–223

Z

z-score (neurofeedback) training,
        252–253
Zero-mean random process, 34

9781138077089